Progress in Phytochemistry

Volume 3

ADVISORY BOARD

Progress in Phytochemistry
Volume 3

Edited by

L. REINHOLD

Department of Botany
The Hebrew University of Jerusalem
Israel

Y. LIWSCHITZ

Department of Organic Chemistry
The Hebrew University of Jerusalem
Israel

1972

INTERSCIENCE PUBLISHERS
a Division of JOHN WILEY & SONS
London   New York   Sydney   Toronto

Library of Congress Catalog Card Number 68-24347
ISBN 0 471 71556 5

Printed in Great Britain by William Clowes & Sons Limited
London, Colchester and Beccles

The untimely death of **Yeheskel Liwschitz** prevented his seeing the completion of this volume. It is dedicated to the memory of this distinguished researcher, scrupulous editor and highly-valued friend.

L.R.

Foreword

In phytochemistry, as in other sectors of the rapidly advancing front of our knowledge, formidable problems face the worker who wishes to stay abreast of developments in his own field or keep himself informed about progress in others. One of the aims of this series is to help both the initiated and the uninitiated to find their way through the forests of rapidly accumulating literature in the various areas. The intention is that each chapter in these volumes, as opposed to reviews dealing only with recent advances in a particular field, will include sufficient background information to enable a non-specialist to obtain a clear picture of the problems involved and the present state of our knowledge in the field. For the specialist, we hope to provide critical and integrated evaluation of the results achieved.

Phytochemistry has a static and a dynamic side. It is our concern that both sides should be adequately represented in these volumes. We hope that the books will interest not only chemists and biochemists but workers in several of the plant sciences, notably taxonomy and physiology. We believe that all readers will profit from the knowledge and judgment of the authors of the various chapters, experts in their field from many different countries. It is perhaps hardly necessary to mention that we have not judged it desirable to attempt to impose any uniformity on the individuality of their various styles. We wish to thank these distinguished scientists for their readiness to cooperate with us in this project.

LEONORA REINHOLD
YEHESKEL LIWSCHITZ

Jerusalem, 1970

Contributing Authors

BASSETT, R. A.,   *Imperial College of Science and Technology, London, England*

BOULTER, D.,   *University of Durham, England*

BROWN, S. A.,   *Trent University, Peterborough, Ontario, Canada*

GOAD, L. J.,   *University of Liverpool, England*

GOODWIN, T. W.,   *University of Liverpool, England*

HANSON, J. R.,   *University of Sussex, Brighton, England*

TAKEDA, K.,   *Shionogi and Co., Ltd., Osaka, Japan*

THOMAS, R.,   *University of Surrey, Guildford, England*

WETTER, L. R.,   *National Research Council of Canada, Saskatoon, Saskatchewan, Canada*

Contents

Methods for Investigation of Biosynthesis in Higher Plants

STEWART A. BROWN   DEPARTMENT OF CHEMISTRY,
TRENT UNIVERSITY,
PETERBOROUGH, ONTARIO, CANADA

and

L. R. WETTER   PRAIRIE REGIONAL LABORATORY,
NATIONAL RESEARCH COUNCIL OF CANADA,
SASKATOON, SASKATCHEWAN, CANADA

1

I. INTRODUCTION

A notable development in biochemistry since the Second World War has been the great surge of experimentation on the metabolism of green plants. Of the three factors mainly responsible, two were directly related to methodology: the ready availability of reactor-produced radioisotopes and, concurrently, the revolutionary advances in separation techniques through new developments in chromatography. While these cut across the whole field of biochemistry, the fresh opportunity presented was seized with particular success by Melvin Calvin's group[9] in their elegant research on the path of carbon in photosynthesis. It is probably correct to state that this work was the foundation on which rests most of the plant tracer work done since that time. Indeed, we consider it to be the third factor mentioned above—the one which provided the chief impetus for extension of the new methodology to biosynthetic studies of virtually every class of plant constituent.

Following or sometimes coincident with tracer studies has come an increasing application of enzyme methodology to higher plants. Inherent difficulties of detection and isolation have frequently limited this approach, but in many instances it has helped to confirm or clarify reaction sequences indicated by tracer investigations. This is probably the only accurate way of verifying a biosynthetic pathway.

In writing this chapter we have considered especially the reader who, while well grounded in chemistry and biochemistry, has had little experience in the field of biosynthesis, particularly in higher plants; but we hope the discussions and references will not be devoid of interest to the specialist as well. Space limitations have imposed restrictions on what we can consider here, especially since the literature on methods of separation and purification of the components in plant-derived mixtures is vast and could occupy volumes. Our main purpose is to deal with other facets of methodology as they relate to higher plants, with emphasis on the theory and practical problems of tracer and enzymological approaches to biosynthetic investigation at the molecular level. Even here our treatment has been, for most aspects, selective rather than comprehensive. We have attempted to illustrate procedures by including references to work which we believe to be representative of the methods discussed. Further examples will doubtless occur to the experienced reader, and mention of some of these will often be found in the references quoted.

II. SCOPE AND LIMITATIONS OF BIOSYNTHETIC INVESTIGATION

It would be well at the outset to define biosynthesis for the purpose of this review, and on this question we concur in the restriction adopted by Swain[160]. He points out that usual practice has been to confine the term to the *in vivo* production of more complex molecules from simpler ones, although many substances formed in living organisms arise through degradative reactions. He distinguishes between the overall exothermic transformations which can lead, for example, to the 'synthesis' of malonic acid from carbohydrate, and biosynthesis proper, a term describing the production of molecules by an overall endothermic process. In discussions of methodology the distinction often becomes blurred, since many of the techniques we shall discuss are equally applicable to both types of process, but for the most part we shall draw from the literature dealing with biosynthesis in the more restricted sense.

Nearly 20 years ago Adelberg[1] recognized three broad experimental approaches for elucidation of biosynthetic pathways: the use of isotopic tracers, the use *in vitro* of enzyme systems, and the use *in vivo* of organisms with blocked biosynthetic pathways. To this list Swain[160] has more recently added the use of non-radioactive precursors and the study of sequential synthesis of apparently related substances. We may note that this last method can be combined with the first in cases where the sequence begins with an exogenously supplied precursor. Except in cell-free systems methods involving non-radioactive precursors are now seldom employed, having been almost entirely supplanted by the use of labelled compounds, and although of considerable historical importance, they will not be discussed further except in the context of enzyme studies.

In the ensuing discussion of these methods we shall conform to a terminology introduced by Davis[44] and now widely used. According to these definitions:

(1) a *precursor* is any compound, whether endogenous or exogenous, that can be converted by an organism into some product;

(2) an *intermediate* is a compound that is both formed and further converted by the organism under identical conditions; and

(3) an *obligatory intermediate* is a member of a path that is the only one by which an organism can synthesize a given product from given source materials.

Adelberg[1] has drawn attention to the great difficulties involved in deciding whether a given compound X occupies the position C or F in the pathway:

$$A \rightarrow B \rightarrow C \rightarrow D \rightarrow \text{End product}$$
$$\Updownarrow$$
$$F$$

The only known way to resolve this problem is to establish that a single enzyme catalyses the conversion of B to X and another single enzyme the conversion of X to D, in which case X must occupy the position C and not F in the scheme. While current criteria of enzyme purity have allowed this to be established with a high probability in some cases, the great difficulties of defining enzyme homogeneity have meant that no compound has really been identified with absolute finality as a metabolic intermediate. The problem of whether citric acid participates in the tricarboxylic acid cycle or is formed in a side-reaction was a notable example of the uncertainties which can arise. We can thus validly maintain only that intermediates in a metabolic pathway have been identified with varying degrees of probability, and the problems peculiar to plant enzymology, to be discussed later, have made it especially difficult to arrive at very high degrees of probability in plants.

Although it would seem unlikely that reversible interconversions of the $C \rightleftharpoons F$ type are common in metabolism, the proviso must be borne in mind that the results from techniques to be considered are subject to misinterpretation in the sense just discussed, particularly since relatively little work has been done with purified enzymes from higher plants.

Other limitations characteristic of individual methods will be mentioned from time to time.

III. APPLICATIONS OF MOLECULAR GENETICS

A. Organisms with Blocked Biosynthetic Pathways

The use of microbial mutants has been of enormous value in the study of metabolic pathways. Beadle and Tatum[14] first used this technique in studies with the mould *Neurospora sitophila*, and it has since been extensively utilized with numerous organisms and compounds[171]. The success of the method depends to a great extent on the feasibility of treating large microbial populations with mutagenic agents and selectively isolating a few of the desired mutants.

Such an approach is clearly much more limited with higher plants and few examples of it are to be found in the literature, but the work that has been done points up its value if a suitable mutant is available. Langridge and Brock[109] observed a spontaneous single-gene mutant of tomato (*Lycopersicon esculentum*) unable to synthesize thiamine, but able to grow almost normally when supplemented with this cofactor. By testing other supplements they were able to localize the block in the biosynthetic pathway to thiamine, and to gain information on the site of thiamine action in the elaboration of chlorophyll. Langridge[108] had earlier studied a radiation-induced mutant of *Arabidopsis thaliana* blocked at a different reaction in the pathway to thiamine. His theoretical discussion is recommended to the reader interested in this field. Miller, Garber, and Voth[120], working with the liverwort *Marchantia polymorpha*, were able to produce auxotrophic mutants by radiation and used them to study methionine formation and the ornithine cycle. Gemmae of this species, after X-irradiation, were grown on a defined medium until 2 weeks after thalli began to develop from them. Growth of thalli at this stage could be scored as either normal or impaired, and auxotrophic mutants could subsequently be selected from among the impaired plants with the aid of a defined medium supplemented in various ways.

B.   DNA Transformations

The above glimpses have indicated that studies at the level of molecular genetics are a potentially powerful tool for the elucidation of biosynthetic pathways in plants. Most of the work done thus far on higher plants has been concentrated on flower pigmentation, and has been reviewed fairly recently by Alston[2]. While the focus of these studies has been on molecular components the methodology has been almost entirely that of classical genetics, and since we have confined our coverage to investigations at the molecular level it is not our intention to discuss it in any detail.

Quite recently, however, some studies on the molecular genetics of flower pigmentation have been reported by Hess[89], who succeeded for the first time in accomplishing in plants the type of DNA transformation that had been previously demonstrated in other organisms. DNA was extracted from young leaves of flowering cyanidin-type *Petunia hybrida* according to a standard procedure[115]. Eight-day-old seedlings of a white petunia mutant were floated for 15 minutes on a solution of this DNA and then grown

48 hours in quartz sand impregnated with the same solution. About one-quarter of the flowers later produced by these seedlings were red, and the principal anthocyanin was the same as in crosses between the cyanidin-type petunia and the white mutant. The transforming principle has been partly characterized[90]; it is insensitive to RNase, but its activity is lowered by DNase and it has a molecular weight probably in excess of 8×10^5.

The methods used here are peripheral to biosynthetic investigation at present. Methodologically the application of molecular genetics to the investigation of biosynthesis in plants is not highly developed, but this is an area where significant advances can be expected, and it is with an eye to the future that the preceding discussion is included here.

IV.   STUDIES OF SEQUENTIAL SYNTHESIS

In many instances one can observe accumulation in a plant of several compounds with related structures. This has often led to speculation, largely on the basis of known chemical reactions, about the sequence of their formation, most notably by Robinson[140], and Geissman and Hinreiner[72, 73]. In theory it should be possible, by analysis of the components at intervals as the plant grows, to detect the order in which they first appear, which may correspond with the sequence in a biosynthetic pathway. In practice this approach is often frustrated by (1) the existence of branches in the pathway leading to simultaneous synthesis of two or more compounds, and (2) the great rapidity with which successive steps can occur during active synthesis, making it impossible, or at least impractical, to distinguish between sequential and simultaneous synthesis. Thus Steck and Bailey[157], analysing the furanocoumarins in the leaves of *Angelica archangelica* over the period 30–144 days after germination, found that they are apparently synthesized simultaneously, showing marked increases after 60 days, and in more or less fixed ratios one to another. In spite of these limitations studies of seasonal variations have yielded information of value in some cases. Instances from the fields of alkaloids and triterpenoids will serve as examples.

Investigations of the relationships among the alkaloids of hemlock (*Conium maculatum*) were carried out by Fairbairn and others[55, 56]. Analyses were made with the use of paper chromatography, first at weekly intervals, and later at intervals of 1 day, 4 hours, and 2 hours. The peak in the total content per fruit of γ-coniceine, now known to be **1**, occurred 1 week earlier than that of coniine (**2**), an observation consistent with derivation of the latter from the former. However, the very rapid changes in the levels of these two alkaloids revealed by the shorter-term experiments showed that the picture is much more complex than previously believed, and raised the

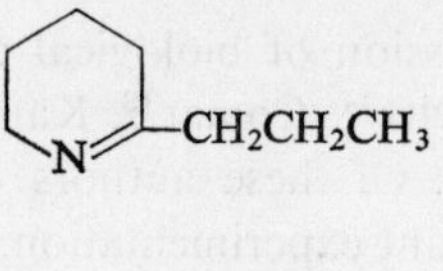

(1) 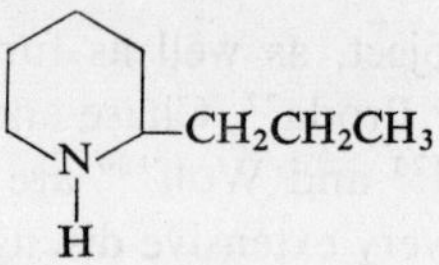   (2)

possibility of a reconversion of coniine into γ-coniceine as part of some oxidation–reduction process.

Kasprzyk and Fonberg-Broczek[99] estimated triterpenic monols and diols of *Calendula officinalis* from the seedling stage to that of seed production using thin-layer chromatography and a colorimetric test. They observed that the monols taraxasterol (3, R = H) and ψ-taraxasterol (4, R = H) were formed 3–4 days earlier than the diols arnidiol (3, R = OH) and faradiol

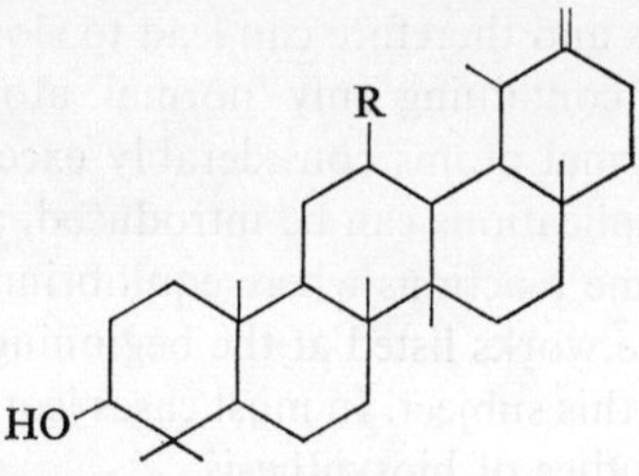

(3) 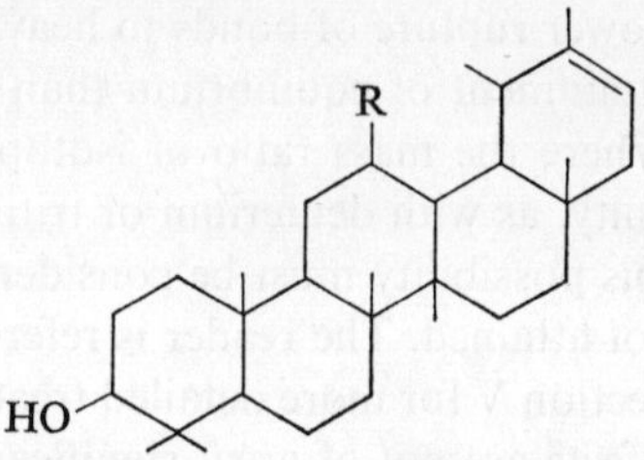   (4)

(4, R = OH), with significant synthesis of the monols being detectable in the leaves surrounding the flower bud before any trace of diol could be found. The synthesis of monols was also found to cease in the flowers before that of the diols. Both facts are, of course, indicative of diol formation through hydroxylation of the corresponding monol.

Indicative as this kind of study may be, it is not a particularly rigorous approach, since it does not usually permit distinction between sequential formation of compounds on a single pathway and independent formation on parallel pathways with a fortuitous time relationship. If the observed sequence persists through several experiments under varying conditions the latter possibility is minimized. But by and large, studies of seasonal variation serve best as forerunners of the more rigorous methods which will now be described.

V.   EXPERIMENTS WITH ISOTOPIC TRACERS

In the ensuing section a reasonable familiarity on the reader's part with the basic principles of isotopes and radioactivity is assumed. For information on this subject, as well as further discussion of biological applications, treatises by Broda[22], Chase and Rabinowitz[34], Comar[38], Kamen[98], Wang and Willis[173] and Wolf[180] are suggested. Of these authors only Comar offers any very extensive discussion of plant experimentation.

A.   Basic Assumptions of the Method

The use of isotopic tracers is, of course, based on the principle that the introduction into a molecule of atoms differing from the normal atomic weight permits the fate of such atoms to be followed, either by measurement of mass differences or, in the case of unstable nuclei, by emitted radiation. It is assumed that the isotopic atoms, having identical electronic configurations to those of normal atoms, behave chemically in the same way. This assumption must be qualified to allow for isotope effects, which result in slower rupture of bonds to heavier atoms and therefore can lead to slower attainment of equilibrium than in those containing only 'normal' atoms. Where the mass ratio of isotopic to normal atoms considerably exceeds unity, as with deuterium or tritium, complications can be introduced, and this possibility must be considered in some reactions where equilibrium is not attained. The reader is referred to the works listed at the beginning of Section V for more detailed treatment of this subject. In most cases isotope effects are not of great significance in studies of biosynthesis.

B.   Isotopes Used in Biosynthetic Investigation

It would appear that isotopes of virtually every atom known to occur in plant components have been used at some time for biosynthetic studies. Since the ready availability of commercial, reactor-produced radioisotopes, their use has been favoured over that of stable isotopes, e.g. ^{3}H over ^{2}H. A fundamental advantage of radioisotopes is the much higher sensitivity of measurement associated with their use. Balanced against this is the fact that some radioisotopes can constitute an appreciable health hazard, although this need not act as a deterrent in a suitably equipped laboratory where appropriate care is exercised. And there are instances, as is well known, notably in tracing nitrogen and oxygen, when a stable isotope must be used because no suitable radioisotope exists. For an exhaustive discussion of isotopes available for biological research, consult the monograph of Kamen[98], or of Comar[38].

1. Isotopes of carbon

By far the most commonly applied isotope in biosynthetic investigations, both in plants and other organisms, is ^{14}C. Its relative cheapness, very long half-life (5770 years), and low hazard have rendered ^{14}C a highly suitable tool for widespread use. The weakness of the β-radiation which facilitates its handling also poses problems of detection, but these have been greatly mitigated by newer instrumentation employing such sensitive techniques as gas-phase counting and scintillation spectrometry[20, 148], which permit counting efficiencies as high as 80–90%. ^{14}C has supplanted the ^{11}C used in some of the very early tracer experiments, whose applications were drastically limited by its short half-life of 20.5 minutes. The stable isotope ^{13}C has been employed, especially in double labelling (a technique to be treated in Section V.C.6) and has recently attracted attention as a result of experiments designed to locate its position in a molecule from n.m.r. signals without chemical degradation[47, 161]. This potentially facile method has been employed in certain microbiological studies of biosynthesis, but its sensitivity is low and the large isotope dilutions by preformed material almost always encountered in plant experiments would appear to restrict its applications in this field in the foreseeable future. Nevertheless, the combination of high isotopic enrichment and a 100-megacycle n.m.r. spectrometer, used in conjunction with a computer to eliminate fluctuations in the spectrum due to background noise, may well permit the use of this technique in certain cases where high incorporations can be achieved.

2. *Isotopes of hydrogen*

Second only to carbon isotopes in the present context have been those of hydrogen. Deuterium has been employed as a tracer for decades, but it has been increasingly superseded by tritium[54] which, being radioactive, has the advantage of much greater sensitivity of detection. Tritium emits only a very weak β-particle (0.02 MeV), and until the advent of scintillation spectrometry[20, 148] practical difficulties of counting tended to discourage its use. With this technique, counting efficiencies of the order of 60% are now obtainable, quite adequate in view of the much higher isotope enrichment possible relative to ^{14}C. While the low radiation energy of tritium contributes to the low hazard of its use it also renders adequate monitoring extremely difficult, and the best defence against contamination is extreme care in handling and routine decontamination. Another characteristic of tritium labelling, which can be advantageous or otherwise, is the relative ease of hydrogen exchange in organic molecules. By the use of appropriate catalytic techniques compounds can be extensively tritiated by exchange with high-specific-activity tritiated solvent such as $^{3}H_2O$, thus making feasible the labelling of some substances, e.g. polypeptides, whose preparation with ^{14}C-labelling is not practical and which are not readily tritiated in specific positions. Exchange will also occur in the absence of a catalyst in the case of acidic hydrogen, such as that in a carboxyl group. But labelling of such hydrogens for tracer purposes is not feasible because the label is again lost by exchange when the compound is exposed to a hydroxylic solvent. Compounds tritiated by catalytic exchange are normally treated with such a solvent to remove easily exchangeable tritium prior to use.

As biosynthetic pathways become better understood in plants, and attention is increasingly directed toward reaction mechanisms, the use of deuterium and tritium for specific and stereospecific labelling[40] will continue to gain prominence (cf. Section V.C.6).

3. Other isotopes

Of the remaining isotopes, ^{15}N has probably had the most widespread application, especially in biosynthetic studies involving purines and pyrimidines, amino acids, and compounds metabolically derived from these. The universal importance of phosphorus in metabolism has resulted in extensive use of the β-emitting ^{32}P. Others, such as the stable ^{18}O, and the radioactive nuclides ^{35}S and ^{36}Cl, both soft β-emitters, have had more limited and specialized use. Although these less-used isotopes will not be extensively mentioned in succeeding sections, the techniques discussed in connexion with ^{14}C and tritium are in general applicable to them as well.

C.   Some Theoretical Considerations of Tracer Methods

1. Chemical synthesis and degradation

While one cannot overemphasize the importance of adequate procedures for the labelling of compounds with isotopes, and for degradative procedures to locate the positions of labelling in a molecule, it is not our intention to discuss them in detail here. Such a discussion would encompass a significant part of the field of organic reactions, and space does not permit this. The increasing variety of commercially available labelled compounds has caused the problems of synthesis to recede in importance in recent years, and this trend will certainly continue. When a suitable labelled compound cannot be purchased, reference to its synthesis can sometimes be found in the original literature. When entering completely virgin territory the reader should consult the many treatises and monographs on synthetic organic chemistry for procedures which can be adapted to his purpose. The treatise by Murray and Williams[123], although now somewhat outdated, remains a satisfactory starting point for a literature search.

In isotopic syntheses radiochemical purity and radiochemical yield are all-important. Rigorous purification, usually involving chromatography and crystallization, must be employed to ensure purity. To maximize yield it is essential that the isotope be introduced at the latest feasible stage of the synthesis, and chemical yield up to that point is of secondary concern. Since the syntheses are almost invariably on the scale of a few millimoles at most, good yields require avoidance of such procedures as distillation which are difficult to adapt to a small scale, in favour of chromatographic separations.

Procedures for the degradation of organic compounds to locate labelled atoms are given in a monograph by Simon and Floss[152].

2. Problems of transport and permeability

If meaningful results are to be obtained, any labelled compound administered to a plant must obviously reach the synthetic sites where enzymes concerned with its metabolism may act upon it. Under natural conditions a higher plant absorbs carbon dioxide through its photosynthetic organs, and most species absorb inorganic salts in aqueous solution through the roots. Labelled carbon dioxide, water, and inorganic salts can therefore be given to a plant with great confidence that, once absorbed, they will have no difficulty in reaching any synthetic site. But as soon as organic substances are to be studied, problems immediately appear. Although many compounds, amino acids for example, pose few solubility problems, the insolubility of others, such as steroids, has proved a considerable challenge. We shall treat the practical aspects of this in Section V.D.2.

Once an organic compound has entered a plant it is faced with many potential permeability barriers. The question of compartmentation of metabolic pools in plants, which is the subject of a recent review by Oaks and Bidwell[128], is of great significance in this context. It is evident from the results of many experiments on a wide range of compounds fed to plants that these substances do reach active synthetic sites at least in part, and are metabolized there, presumably, in many cases at least, as part of endogenous pools of intermediates. But difficulties in obtaining incorporation of label from certain compounds, such as phosphate esters and mevalonate[141] point up the dangers one must guard against in this respect. For this reason negative findings from a tracer experiment must always be treated with caution, especially if the compound in question is of a type known to encounter permeability barriers in the tissue.

3. Choice of the appropriate time and system

In initiating experiments on a previously untested system the investigator should establish at the outset that it can synthesize the compound being studied. Since nicotine, for example, is synthesized in the roots of the tobacco plant[45], the futility of introducing potential precursors into cut shoots will be apparent, whereas sterile root cultures can be satisfactorily used[46]. Further examples of such localization are known. Variations in time, as well as locale in the plant, must also be taken into account. Instances are known where synthesis occurs over long periods of time or virtually throughout the life of the plant, as in the case of such fundamental metabolites as sugars, protein amino acids, and cell-wall components. But the formation of many derived (secondary) products is restricted to a short period of the plant's development[19, 99, 168]. In instances of the latter type an investigator could easily be deceived if he were to conduct his experiments during any other period. In cases of extended synthesis convenience usually dictates the choice of young plants because of greater ease in handling, and seedlings have been employed in a great many experiments for this reason.

In practice it may not be essential to carry out systematic kinetic studies before precursors are tested, and in fact this seems to have been done quite infrequently. If a trial feeding of a likely precursor gives good incorporation of isotope one concludes that extensive synthesis occurs under the prevailing conditions, although these may not be optimal. Otherwise it is well to establish beyond question that these conditions are conducive to synthesis before even tentative negative conclusions are drawn. This can be done by a comparison of incorporation from labelled carbon dioxide or glucose to the same organs at the same stage of development. If this gives negative results, feedings at other stages of growth should be tried.

4. *Comparison feedings*

Even when an apparently optimal set of conditions for feeding has been established, one must beware of uncontrolled variables such as climatic fluctuations which could affect rates of biosynthesis. When a new precursor is tested it is therefore quite essential to administer some reference compound in a parallel experiment; this can be carbon dioxide or glucose, or some compound which has been shown to be better utilized than these. Failure to do this invites potentially serious misinterpretations. As just implied, a useful initial approach in assessing suspected precursors is direct comparison in a series of parallel experiments. Success clearly depends on good guesses about which compounds to test. But in view of expanding knowledge of biosynthetic pathways the experimenter who has given reasonable thought to structural and metabolic relationships doesn't often come away entirely empty-handed, and frequently obtains good indications about the direction subsequent feeding experiments should take. Such a 'shotgun approach' is easier now that a large selection of labelled compounds (especially with ^{14}C) is available from commercial sources, many at quite moderate cost[154].

5. Criteria for assessment of precursor effectiveness

The question of what criteria to use in assessing the effectiveness of a precursor now requires examination. Those that have been used are dilution value (or occasionally its reciprocal, specific incorporation), degree of incorporation, and extent of label randomization.

Dilution value is defined as A_1/A_2, where A_1 = specific activity of precursor and A_2 = specific activity of the product (on a molar basis). To appreciate the significance of the dilution value let us consider the simplest case, a biosynthetic pathway in which precursor A is converted via several intermediates to E:

$$A \rightarrow B \rightarrow C \rightarrow D \rightarrow E$$

Let us assume that a large enough pool of E exists in the plant for isolation and analysis. The specific activity of the administered precursor, if transformed to E, will be diluted in proportion to the size of this pool of E. If there also exist pools for any of the intermediates between A and E, i.e. if the rate of their formation exceeds that of their consumption at the time of the experiment, then further dilution of specific activity will occur at each intermediate stage.

A more complex case is one in which the precursors of E have more than one metabolic fate:

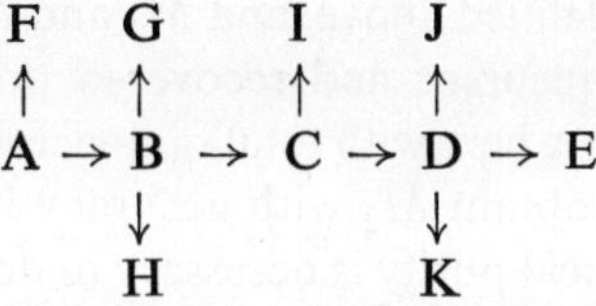

In such a case the degree of dilution at each stage in the formation of E will be greater than in the first instance, since fewer moles of each intermediate will be available to enter the next pool in the direction of E, so that the ratio, endogenous intermediate: labelled intermediate will be higher. It can be seen that the effect of both these factors will be to give a higher dilution value the further the precursor is removed sequentially from the product.

A practical advantage to the use of dilution values is that the product need not be recovered quantitatively from the plant. Since one need not be concerned with maximizing recovery, a greater range of purification techniques becomes available for use, and purification is facilitated. Only radiochemical purity is essential in applying the criterion of dilution value, if the labelled compound can be selectively assayed, although it is usually much preferable to aim at full chemical purity as well.

A disadvantage of dilution values becomes apparent when biological variability results in discrepancies of pool sizes between plants. To take an extreme case, if A is fed to w g of plant 1, containing x mmoles of E, and B is fed to w g of plant 2, containing $10x$ mmoles of E, with other factors constant, the dilution value for A will be lower than that for B, probably by a factor approaching 10, in spite of the fact that B is closer to E on the metabolic pathway. Fortunately this *is* an extreme example, and differences in comparisons of pools between plants grown under the same environmental conditions are normally much smaller, but differences of 300–400% should not be considered extraordinary. A related difficulty with dilution values arises during comparison of the efficiency with which a precursor is incorporated into different compounds in the same plant. Here, large differences in pool size can greatly limit the interpretation of dilution values. If intermediate D is converted to both E and J, which accumulate, and if synthesis has previously favoured E, administration of D will lead to a greater dilution value for E than for J, even though more J than E may have been synthesized during the actual period of the feeding.

In theory these disadvantages can be overcome by application of the second criterion, degree of incorporation, which is defined as

$$\frac{A_2 M_2}{A_1 M_1} \times 100$$

where A_1 and A_2 are as defined above, and M_1 and M_2 are the molar quantities of administered precursor and recovered product, respectively. In other words, we are dealing here with total radioactivity. This is easily calculated for $A_1 M_1$, but to obtain M_2 with accuracy implies quantitative recovery, while radiochemical purity is necessary to determine A_2. It is recognized that in practice, for complex mixtures such as those found in plant extracts, accurate values for A_2 and M_2 can be mutually exclusive; in fact, it is owing to this that techniques of isotope dilution analysis have been so extensively developed[38, 173]. If rigorous purity is achieved, per cent incorporation figures must be regarded only as minimal values as a consequence of losses in purification. If rigorous purification has not been achieved they may, like dilution values under the same circumstances, be rendered meaningless. Nevertheless, in cases where only a few, relatively clear-cut purification steps with near-complete recovery need be undertaken, and adequate purity thereby achieved, or where a good selective assay is available, per cent incorporation figures can be of great value, especially for comparison of label incorporation into related compounds, and can yield information not forthcoming from dilution data alone. We feel, however, that the inherent limitations of this approach have not always been sufficiently considered in the past.

A crucial factor that must be taken into account when one compares efficiencies of labelled compounds as precursors is the dose. During studies on the formation of quercetin in *Fagopyrum tataricum*, Watkin and Neish[174] compared variations in dilution value and in per cent incorporation when

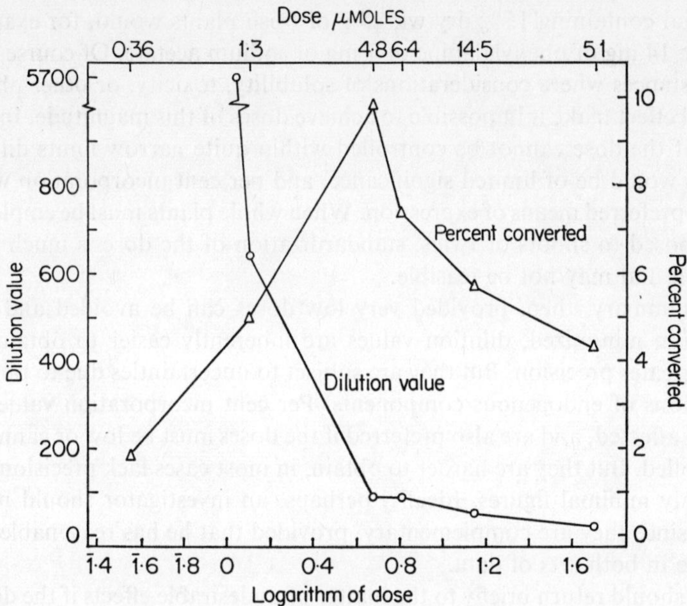

Figure 1. Effect of variations in the dose on dilution value and per cent incorporation in feedings of labelled phenylalanine to *Fagopyrum tataricum* (from Watkin and Neish[174]). Reproduced by permission of the National Research Council of Canada from the reference cited.

the doses of several known quercetin precursors were varied over as much as two orders of magnitude. Three aromatic metabolites, L-phenylalanine, cinnamic acid, and L-phenyllactic acid, and two aliphatics, glucose and acetic acid, were investigated. The results, one set of which is reproduced graphically in Fig. 1, showed enormous variation in the dilution value once the dose had fallen below a critical value, about 5 μmoles/g dry wt for phenylalanine and 20 for acetate, but a relatively constant dilution value at higher doses. In contrast, although peaking occurred in per cent incorporation as dose varied, the changes were much smaller. A good theoretical discussion of the implications of these findings, included in this paper, is recommended. Unfortunately the behaviour of other compounds over wide dose ranges is not well understood from this standpoint. But it would seem mandatory, if it is intended to compare dilution values, to ensure that doses of the order of 20 μmoles/g dry wt are used, and in addition, they should be kept as nearly constant as possible in parallel experiments whenever conditions permit. At the suggested doses, variation is unlikely to exceed normal biological variation as the result of unavoidable dose fluctuation. With plant material containing 15% dry wt, 25 g of fresh plants would, for example, receive 14 mg of phenylalanine or 7 mg of sodium acetate. Of course there are instances where considerations of solubility, toxicity, or other physiological effect make it impossible to achieve doses of this magnitude. In such cases if the dose cannot be controlled within quite narrow limits dilution values would be of limited significance, and per cent incorporation would be the preferred means of expression. When whole plants must be employed, as opposed to shoots or twigs, standardization of the dose is much more difficult and may not be feasible.

In summary, then, provided very low doses can be avoided and dose variation minimized, dilution values are inherently easier to obtain and have greater precision. But they are subject to uncertainties due to random variations of endogenous components. Per cent incorporation values are not so affected, and are also preferred if the doses must be low or cannot be controlled. But they are harder to obtain, in most cases lack precision, and are only minimal figures. Ideally, perhaps, an investigator should report both, since they are complementary, provided that he has reasonable confidence in both sets of data.

We should return briefly to the matter of undesirable effects if the dose is too high. It has long been known[38] that emitters of even moderately energetic radiation, such as ^{32}P, can inflict radiation damage on plants when administered in solution, especially in rapidly growing tissue during longer experiments[144]. Fears have been expressed that excessive doses, even if not resulting in actual damage, might produce undesirable physiological or metabolic effects such as enzyme induction. While it is difficult to find documentation of such effects in the literature they must be borne in mind as possibilities, and a wise course in the absence of specific information is to administer no larger a dose than necessary.

The third criterion of importance in the assessment of precursors is the degree of randomization or redistribution of label during conversion of precursor to product. Again ideally, if a precursor is converted to a product by a direct route, if there are no side-reactions leading to *in vivo* degradation, and if complications such as exchange in the case of labelled hydrogen do not intervene, the position of the label in the product should be entirely in accord with predictions on the basis of structural and metabolic considerations (provided, of course, that these correspond to reality!). For example, the steroid biosynthetic intermediate squalene formed after administration of [^{14}C] methyl-labelled acetate would be labelled only in alternate carbons. Such an ideal situation is rarely if ever encountered. One common reason for this is that the precursor undergoes degradative reactions, e.g. acetate in the tricarboxylic acid cycle, which lead to a randomization of label in the precursor pool. In the example just used, a small amount of ^{14}C appears in the carboxyl-carbon of acetate, and the carboxyl-derived carbons of squalene acquire some label as well[40]. Clearly the degree of randomization will tend to be a function of time, until the pool is eventually flushed free of label by the products of photosynthesis.

Another cause of redistribution of label is the existence in a biosynthetic pathway of a symmetrical intermediate. For instance, the pathway to the pyrroline ring of nicotine (5) in tobacco is believed[110] to involve putrescine (6), which is transformed with loss of the elements of ammonia to the residue 7. Feeding of [2-^{14}C]-ornithine to tobacco would yield

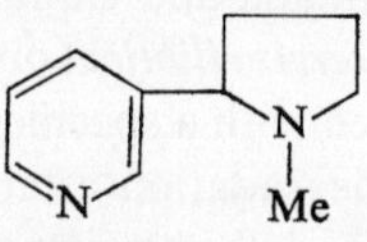

$$\begin{array}{ccc} \text{(5)} & \begin{array}{c} H_2C-CH_2 \\ | \quad\quad | \\ H_2C \quad CH_2 \\ | \quad\quad | \\ H_2N \quad NH_2 \end{array} & \text{(7)} \\ & \text{(6)} & \end{array}$$

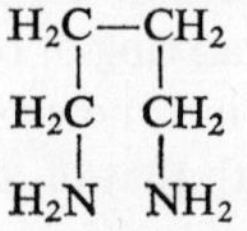

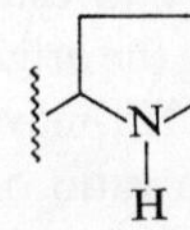

$\overset{*}{H_2}N(CH_2)_3CH_2NH_2$, but owing to the symmetrical nature of putrescine either terminal carbon has the same chance of being linked to the pyridine ring. Thus we obtain an equimolar mixture of 8 and 9, which is indistinguishable from 10 by any degradation[105].

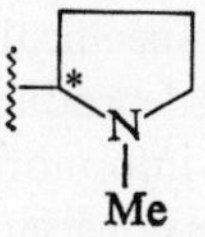

$$\text{(8)} \qquad\qquad \text{(9)} \qquad\qquad \text{(10)}$$

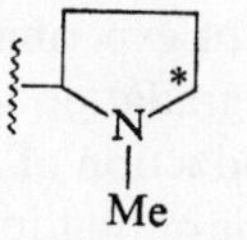

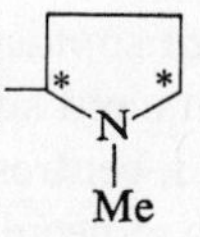

In theory no tracer experiment is complete until the product has been degraded to establish the position of the label. However, when degradation is very difficult and time-consuming, or not practical at all because of low isotope recovery, it is understandable that many workers have drawn conclusions about pathways on the assumption that randomization has not occurred. Whether this assumption is justified must be assessed from the nature of the experiment, and judgment can only be on the basis of much experience. In general it may be considered that in feeding metabolically active compounds, particularly those of simple structure and on or near the metabolic mainstream, which can have various fates, especially catabolic, one is unwise to draw conclusions without product degradation. On the other hand, once a complex carbon skeleton has been elaborated and one is administering to a plant a potential precursor, labelled with a carbon isotope, which need undergo only minor changes involving substitution or a change in oxidation level for conversion to the product under study, few would insist on degradation. There are many intermediate cases where results obtained on the basis of dilution values or per cent incorporation data alone have been accepted, but where degradative studies would have removed doubts. In deciding this question we should consider the isotope dilution which has occurred. With very low dilutions the chance that significant degradation of precursor has taken place, followed by resynthesis, is negligible, since low dilutions imply passage through few pools. But where high dilution or low incorporation is observed, localization of the label is necessary to establish the role of a precursor. Hamada and Chubachi[82], studying the origin of the furan ring of rotenone in *Derris elliptica*, observed that [2-^{14}C]-mevalonic acid lactone was incorporated with a specific incorporation ratio only one-tenth that of acetate, but degradation established that 78 % of the carbon-14 was located in carbons 7' and 8', showing a direct incorporation with intact carbon skeleton. But it should be emphasized that this type of result does not exclude the chance that only a minor pathway is being investigated. Note, too, that redistribution of label due to the participation of a symmetrical intermediate is quite compatible with low isotopic dilutions.

6. *Double and multiple labelling*

A more sophisticated type of experiment, in which specifically labelled precursors and subsequent degradation of the recovered product are both employed, centres in the introduction of isotopic label into more than one atom of a molecule. This double or multiple labelling can consist of the same isotope in different positions, or combinations of isotopes. Fortunately for the practicality of this method it is unnecessary that one given molecule bear more than one isotopic atom. Samples bearing a single label in, for example, two different positions are combined in known proportions. This mixture would be indistinguishable by degradation from a sample of equal total enrichment bearing half the number of labelled molecules each with two labelled atoms.

In general terms the advantage of the multiple labelling technique is that it permits detection of *in vivo* cleavage anywhere between the labelled atoms. Suppose we wish to study the incorporation of a substance A–B, where A and B represent two moieties, into some plant component. We could prepare two samples, one with a labelled atom in A and the other in B, feed them to different plants, and in each case recover the component in question after a suitable metabolic period, with a specific activity that invariably represents some isotopic dilution of the precursor. We might well find that there has been incorporation of label from both $\overset{*}{A}$–B and A–$\overset{*}{B}$, but since we were obliged to use two plants for this experiment, biological variability would introduce discrepancies in the per cent incorporation or dilution values of up to several hundred per cent in the two experiments. This approach thus doesn't permit us rigorously to distinguish a cleavage of A–B, with subsequent separate incorporation of the A and B moieties, from the incorporation of the A–B as an intact unit. What we need is an internal standard. If we now combine the two samples $\overset{*}{A}$–B and A–$\overset{*}{B}$ to obtain $\overset{*}{A}$–$\overset{*}{B}$, with the ratio of specific activities in the two moieties equal to x, and feed this sample to a plant, any marked deviation from x in the specific activity ratio of the corresponding atoms in the product will indicate that bond cleavage has occurred between A and B. To understand more fully the scope inherent in double and multiple labelling experiments let us examine several representative instances where the technique has been successfully used.

In a study of the biosynthesis of a thioglucoside, gluconasturtiin (12), Underhill[162] wished to examine its possible formation from γ-phenylbutyrine (2-amino-4-phenylbutyric acid, 11) in *Nasturtium officinale*:

$$\langle\bigcirc\rangle\!-\!CH_2\!-\!CH_2\!-\!CHNH_2\!-\!COOH$$

(11)

↓

$$\langle\bigcirc\rangle\!-\!\overset{3}{C}H_2\!-\!\overset{2}{C}H_2\!-\!\overset{1}{C}\!\!\underset{N-SO_4^-K^+}{\overset{S-C_6H_{11}O_5}{<}}$$

(12)

By means of separate experiments with γ-phenylbutyrine labelled with ^{14}C at C-2 or C-3, followed by degradation, he had obtained good evidence that the carbon skeleton of γ-phenylbutyrine, except the carboxyl-carbon, was incorporated intact into the aglycone of gluconasturtiin. He then prepared [2-^{14}C,^{15}N]-L-γ-phenylbutyrine and [G-^{14}C,^{15}N]-L-phenylalanine (also known to be a fairly efficient precursor) with ^{14}C/^{15}N ratios of 1.82 and 2.19, respectively, and administered them separately to *N. officinale* plants. Degradation and isotope analysis of the aglycone from the recovered gluconasturtiin gave respective ^{14}C/^{15}N ratios (for C-1) of 2.19 and 8.80. While in the case of phenylalanine this differed by a factor of 4 from that of the precursor, the two ratios were quite close in the case of phenylbutyrine. These results demonstrated that, in contrast to phenylalanine, little C— N bond cleavage had occurred during the transformation of L-γ-phenyl-butyrine to the aglycone of gluconasturtiin.

Caspi and Lewis[32], seeking evidence about the origin of cardenolides in *Digitalis lanata*, showed by means of tritiated precursors that both pregne-nolone (13) and progesterone (14) are converted to these products. Earlier animal studies had suggested the sequence:

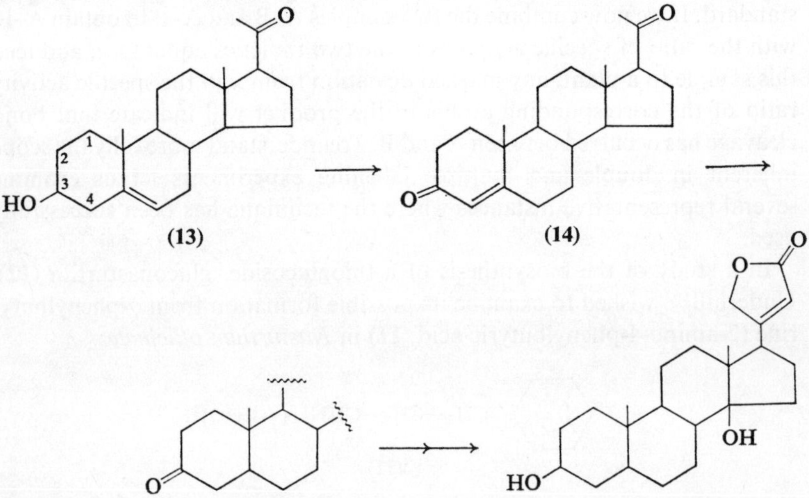

Formation of the 3-keto intermediates 14 and 15 obviously requires loss of hydrogen at the 3-position. To test this mechanism Caspi and Hornby[31] prepared [3α-^{3}H, 4-^{14}C]-pregnenolone (17) with a ^{3}H/^{14}C ratio of 13.3 and fed it to *D. lanata* leaves. The recovered digitoxigenin (16) had a ^{3}H/^{14}C ratio of 1.15, corresponding to a tritium retention of about 9 %. Since *in vitro* oxidation to digitoxigenone (the 3-ketone) proceeded without further loss of tritium, the total absence of tritium at C-3 of digitoxigenin was established, indicating cleavage of the C—H bond at position 3 and participation of a 3-ketone in the biosynthesis.

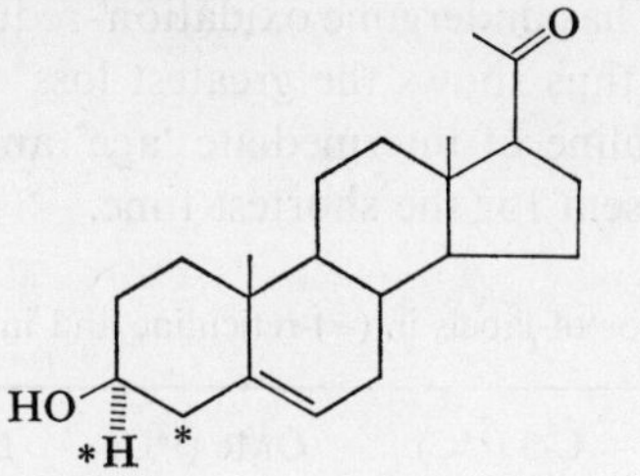

(17)

Extensive use of multiple labelling techniques has been made by Battersby and his associates in their studies on the biosynthesis of the morphine alkaloids. Battersby, Foulkes, and Binks[13] prepared four samples of reticuline, three of them labelled specifically with ^{14}C and one with tritium, and combined them to form the quadruply labelled reticuline (18) which was then resolved into its enantiomers. The (+)- and (−)-forms were administered in separate feedings to *Papaver somniferum*. The (−)-isomer (18) was well incorporated into the major opium alkaloids, and from these thebaine (19, $R^1 = R^2 = Me$) was selected for degradation. The ratios shown in

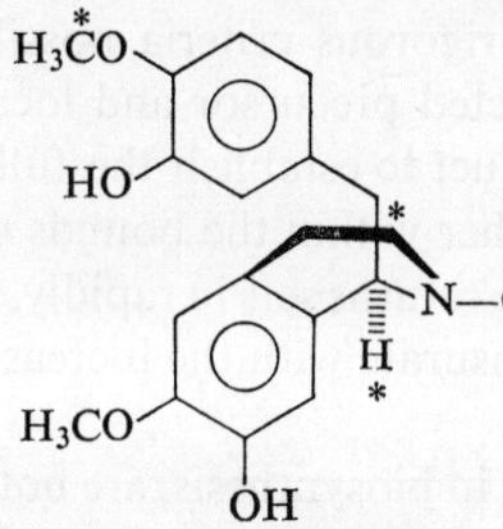

(18) 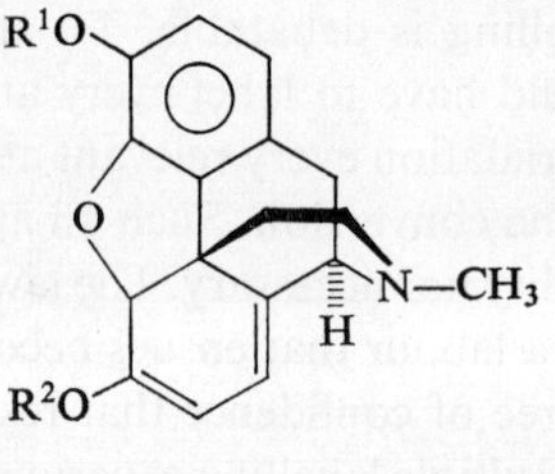   (19)

Table 1 were found. These results clearly show that no significant bond cleavage between the ^{14}C-labelled parts of the molecule had occurred, but that there was appreciable loss of tritium relative to the labelled carbon. (+)-Reticuline was utilized with comparable efficiency but there was much greater loss of tritium. The results were explained by the hypothesis that the enantiomers of reticuline were being interconverted by a redox mechanism probably involving a dehydro intermediate. Examination of the ratios in codeine and morphine, into which thebaine is progressively converted (cf.

2

Section V.C.8), showed a progressive increase in tritium retention. The authors' interpretation was that the isolated thebaine is mainly drawn from (−)-reticuline which has undergone oxidation–reduction in the plant for the longest period and thus shows the greatest loss of tritium, while codeine represents (−)-reticuline of intermediate 'age' and morphine mainly that which has been present for the shortest time.

Table 1. Ratios of labels in (−)-reticuline and in isolated thebaine[13]

	C-3 (^{14}C)	OMe (^{14}C)	NMe (^{14}C)	C-1 (^{3}H)
(−)-Reticuline	0.76	0.13	0.11	2.01; 100[a]
Thebaine	0.77	0.11	0.12	0.64; 32[a]

[a] Expressed as a percentage of the ratio ^{3}H:^{14}C (at C-3) in the reticuline fed to the plants. Reproduced by permission of the Chemical Society.

Although the few examples just given touch on only a small part of the total scope of double and multiple labelling, they should at least convey some idea of the power of this technique in the elucidation of biosynthetic mechanisms. It is especially useful where tritium is employed to study reactions involving carbon–hydrogen bond cleavage, a virtually universal biosynthetic reaction. The extent to which it is worth while to carry multiple labelling is debatable. To apply the most rigorous criteria possible one would have to label every atom of a suspected precursor and localize by degradation every relevant atom of the product to establish the full nature of the conversion. Such an approach is neither within the bounds of practicality nor necessary. The law of diminishing returns sets in rapidly, and the extra labour that ensues becomes incommensurate with the increase in the degree of confidence that results.

Multiple labelling experiments, like others in biosynthesis, are not devoid of pitfalls. A striking example is provided by the observations of Austin and Meyers[4] in the course of their research on the origin of skimmin (7-hydroxy-coumarin glucoside, **20**) in *Hydrangea macrophylla*. They fed to this plant *trans*-4-β-D-glucosyloxycinnamic acid labelled generally in the glucose moiety and in C-2 of the aglycone side-chain (**21**), and obtained the

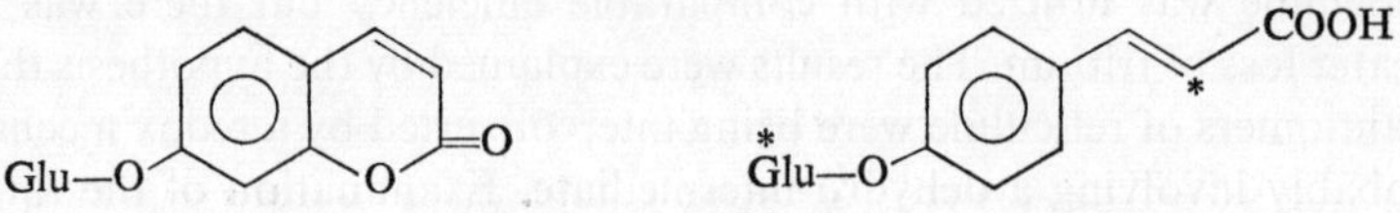

Glu—O   Glu—O

(**20**)   (**21**)

results reproduced in Table 2. After one day's metabolism the amount of glucose-^{14}C incorporated relative to aglycone-^{14}C was only about 7%, but after 5 days the ratio approached unity. The short-term results plainly indicated that a cleavage of the glycosyl linkage had occurred, and the authors rationalized the long-term data on the basis of differential uptake of the products into skimmin which, after 5 days, fortuitously yielded the gross effect of an *en bloc* utilization of the glucoside! The following comments of Austin and Meyers merit quoting verbatim:

'This interpretation of the 5-day result is directly contrary to the tacit assumption of multiple-label experimentation—that an observed retention of the initial isotopic proportion constitutes sufficient proof of the incorporation of the fed precursor *as such*. This assumption is justified only when at least one of the endogenous metabolic pools of the precursor breakdown

Table 2. Percentage incorporations into skimmin[a][4]

	1 day	3 days	5 days
2-[^{14}C] inc.	3.05	4.12	4.28
U-[^{14}C] glucose inc.	0.20	2.51	3.94
Relative inc. glu/chain	6.7%	61.0%	92.2%

[a] Dose = 12 μmoles to each plant, 2-[^{14}C]-activity = 1.07×10^6 c.p.m./mmole, U-[^{14}C]glucose activity = 1.04×10^6 c.p.m./mmole.

Reproduced by permission of the Chemical Society.

products is large enough to cause a significant dilution of activity in the event of recombination . . . If this is not the case, or if the rates of uptake of the breakdown products are sufficiently different during the metabolic period, then the observed overall dilutions of the products may be similar. Recombination at some stage will therefore give rise to a misleading result.'

The inference from these findings is that in cases where retention of the isotopic ratio of an administered precursor is observed in the product it would be prudent, where possible, to repeat the experiment on a different time scale and to conduct *in vitro* enzyme studies, to establish the existence of the phenomenon on the most rigorous basis.

Another caveat in double labelling, and for that matter single labelling studies, concerns isotope fractionation during separation procedures, which can amount to 0.5% or more in some chromatographic separations, and has been noted among such varied compounds as steroids[103], amino acids[77] and sugars[116]. To minimize the effect single fractions should be avoided and as much of the peak as possible collected for analysis.

7. *Isotope trapping and isotope competition*

The methods outlined thus far in Section V are suitable for the identification of precursors of a plant component under study, but they do not establish that a precursor, however efficiently utilized without isotope randomization, is an intermediate on the pathway in question—that is, according to our terminology, a compound both formed and converted under identical conditions. Only the second criterion, conversion to the product, has been shown. In many instances the first criterion is readily satisfied by actual isolation from the plant of a compound chemically identical to the precursor. In fact, such natural occurrence is frequently the first fact to suggest testing a compound as a precursor, but in other cases the discovery of a precursor–product relationship stimulates a successful search for a previously undiscovered plant component, which is then recognized as an intermediate.

However, an intermediate in a metabolic pathway accumulates only if it is formed during any given period faster than it can be further metabolized. If it is consumed as fast as it can be formed, means other than direct isolation must be found to demonstrate its role. For this purpose the technique of *in vivo* isotope trapping has been used. Let us assume we wish to determine whether compound Y is an intermediate on the biosynthetic pathway to Z. We select a compound X, which we have already established as a precursor of Z, presumably via Y, label it with a convenient isotope, and feed it to an appropriate plant that is forming Z. At the same time we also feed to the plant a quantity of non-isotopic Y. With the advent of the added Y the pool of Y in the plant, which may have approached zero magnitude under natural conditions, is vastly augmented. Two processes now operate in an idealized case if Y is indeed an intermediate between X and Z. Molecules of labelled Y formed from the administered X enter the pool, while molecules of Y leave the pool during further metabolism to Z and possibly to other products. Eventually the pool of Y will again approach zero magnitude as the steady state is once more established. But in the interval an isotopically labelled pool of Y will exist temporarily in the plant and will be isolable. Some trial and error is needed to select suitable doses, and it may be necessary to experiment with several metabolic periods. In some cases where enough endogenous Y accumulates, *in vitro* trapping is feasible by addition of Y during working-up.

A positive result in such a case, especially if supported by data from degradations, demonstrates that Y is formed from X, but a negative result does not eliminate Y as an intermediate in the biosynthesis of Z; it could still be on another pathway to Z not involving X. If this is so we could still expect labelled Y to act as a precursor of Z in a direct feeding experiment. But even if Y is on the X → Z pathway the result could still be negative if the augmented pool of Y brings about feedback inhibition of an earlier reaction in the pathway. In such a case, if $X \to Y \to Z$ is the sole biosynthetic route, Z would also have very low or zero specific activity. If there is another pathway not involving Y, the specific activity of Z formed from labelled X could exceed that of Y. In trapping experiments it is, therefore, important to determine the specific activities of the end-products as well as of the trapped intermediate and, if possible, the recovered precursor.

The role of marmesin (**23**) in the biosynthesis of furanocoumarins (**24**) was investigated by Brown, El-Dakhakhny and Steck[28] in a typical application of isotope trapping. Earlier studies had shown that umbelliferone (**22**, $R = H$) is an intermediate in the elaboration of furanocoumarins, and that tritiated marmesin is a precursor, but although marmesin as a glycoside is a component of *Ammi majus* it had not been detected in

(**22**)

(**23**) (**24**)

Ruta graveolens. [2-^{14}C]-Umbelliferone (163 μCi/mmole) was therefore administered to this latter species along with inactive marmesin and allowed to undergo metabolism for 2 days. Extraction of the plants yielded radioactive marmesin (12 and 37 μCi/mmole). The direct feeding and trapping experiments together had thus established both the formation and further conversion of marmesin by *Ruta*, and demonstrated its status as an intermediate in spite of its non-accumulation in the plant.

For studies in intact plants or organs, isotope trapping is of little use when substances with relatively central metabolic roles are under consideration. For example, to trap label in an amino acid while the plant is synthesizing protein leaves us no further ahead in an investigation of alkaloid formation. But as we move further along the pathway from this amino acid to the alkaloid, the potential for the use of trapping increases. This is especially true if we can use cell-free systems of more restricted synthetic capability than an intact plant.

Closely related to isotope trapping is the technique of isotope competition, and the two can be combined in the same experiment. In the latter approach it is argued that the addition of the pool-diluting non-radioactive intermediate Y together with the labelled precursor X will result in a product Z of lower specific activity by comparison to a parallel experiment in which only the labelled precursor X is fed.

Wong[181] has resorted to isotope competition methods in an attempt to decide whether the chalcone isoliquiritigenin (**25**) or the flavanone liquiritigenin (**26**) is the more immediate precursor of flavonoids in *Trifolium subterraneum*. As the two compounds are biochemically interconvertible,

           (25)   (26)

direct feeding experiments of one or the other, labelled, would be difficult to interpret. Instead, parallel experiments were carried out with seedlings of *Trifolium* in which each compound, labelled with ^{14}C, was fed together with a non-radioactive sample of the other. In all cases the feeding with labelled **25** and unlabelled **26** yielded flavonoid and isoflavonoid products with higher specific activities, by factors of about two, than did labelled **26** and unlabelled **25**. From this it was deduced that the chalcone **25**, not having to be converted to the products in question via the augmented pool of the flavanone **26**, was the more direct precursor.

A limitation of this technique is the fact that the differences in observed specific activities are usually within normal limits of biological variation, so that its rigorous application requires extensive replication and statistical analysis, which are frequently impractical. Cases are also known[97, 160] where isotope incorporation has been increased in the presence of a competitor. Swain[160] thinks it probable that in these experiments the inactive compound stimulated the production of enzymes which catalysed the individual steps in the biosynthesis and so increased the formation of the product, a phenomenon which clearly could complicate interpretation. Isotope trapping experiments are also difficult to interpret in some instances. Readers conversant with metabolic control mechanisms will be aware of possible complications of both techniques in addition to those already discussed. Problems of interpretation may also be posed by differences in translocation rates between the administered compounds. Both these methods call for a good deal of caution when conclusions are being drawn from their results alone.

8. Kinetic experiments

Because of the truism that the formation of the first compounds in a biosynthetic pathway must chronologically precede that of compounds at the end, studies of the time sequence by which a series of metabolites is formed have yielded valuable data about biosynthesis. The outstanding example of this method has been the pioneering research of Calvin and his group[9] on the metabolic path of carbon fixed as carbon dioxide by photosynthesizing green algae. In view of the important findings which have attended its application to higher plants it is rather surprising that its use has been relatively limited in this area.

The principle of the technique is to administer a labelled precursor to a series of plants or plant organs, harvest them after increasing metabolic periods, and follow the incorporation of label into the compound of interest and into related substances which could act as intermediates in its formation. The usual procedure has been to administer one 'pulse' of the precursor followed by a metabolic period in a normal atmosphere, so that the system is thereafter flushed by unlabelled metabolites formed photosynthetically from $^{12}CO_2$. Sequential studies can be undertaken after administration of any labelled compound, but investigators in practice have favoured carbon dioxide which, in addition to being the universal precursor, is also the only one whose absorption is essentially instantaneous, thus permitting sequential studies as short as a few minutes. (With algae, Calvin's group was able to reduce the period to seconds.)

Possibly the best-known application to higher plants has been the work of Rapoport and his associates on opium and tobacco alkaloids. Rapoport, Stermitz and Baker[133] studied the interrelationship among the alkaloids morphine (**19**, $R^1 = R^2 = H$), codeine ($R^1 = Me$, $R^2 = H$) and thebaine ($R^1 = R^2 = Me$) of the opium poppy *Papaver somniferum*. After administration of $^{14}CO_2$ for 1.8 hours followed by 0.3 hours in normal air the ratio of morphine:codeine:thebaine specific activities (M:C:T) was $5:77:>300$ after prior removal of the methyl groups. After an 8-day experiment during which the plants were kept in a chamber and allowed to reabsorb their expired $^{14}CO_2$ the M:C:T was 463:346:200. The interpretation of these results was that thebaine is successively transformed by O-demethylation to codeine and morphine. The authors argued that the specific activities of the three alkaloids would at first rise with increasing formation from $^{14}CO_2$-derived intermediates, and then fall as the flushing with $^{12}CO_2$ proceeded. If thebaine were the earliest of these alkaloids to be formed its specific activity would peak first and then diminish first, with a concomitant increase in the specific activity of morphine, which appears to be a metabolically in-active storage product. The observed data are in accord with this inter-pretation. Additional extensive studies with similar techniques on the alka-loids of *P. somniferum* and *Nicotiana glutinosa* have since been reported (for references see Martin, Warren and Rapoport[117]).

Other investigations employing similar principles have been carried out. Dietrich and Martin[48] have extended the *P. somniferum* studies to obtain information on the pathways of alkaloid biosynthesis in *Conium maculatum*. Battaile and Loomis[11] and later Hefendehl, Underhill and von Rudloff[86] have used $^{14}CO_2$ feedings for periods as short as 5 minutes to acquire evi-dence for the biosynthetic sequence piperitone $\rightarrow$ *l*-menthone $\rightarrow$ *l*-menthol in *Mentha piperita*. Brown[24] has used the data from $^{14}CO_2$ feeding to *Hierochloe odorata* to show that the glucoside **27** is formed earlier than

$$\text{COOH}$$

$$\text{O—Glu}$$

(27)

coumarin, providing evidence in support of an *ortho*-hydroxylation mecha-nism of coumarin formation.

A more elaborate extension of kinetic principles concerns their applica-tion to the measurement of turnover rates, which have an obviously great relevance to biosynthesis. It is beyond the scope of the present chapter to deal more than superficially with the considerable body of theory in this area, which has been intensively treated by Reiner[134, 135]. The experimental application of this theory has been quite infrequent, and even a cursory reading of Reiner's discussion makes the reason plain, as the complexities inherent in the living experimental systems pose formidable if not frightening difficulties.

In his discussion Reiner[134] concludes that to calculate the turnover of a substance it is necessary in tracer experiments to know not only the immediate precursor and its specific activity, but also the concentration of the substance itself and its specific activity. Since concentration measure-ments require an analytical determination or a quantitative isolation, the problems to be faced in dealing with complex mixtures are all too evident. In general, too, calculation of the turnover rate of a compound involved in a metabolic network demands that the concentration and activity of each compound in that network be ascertained, while additional complications are introduced for any reactions with higher than first-order kinetics. Reiner further concludes that there is no simple criterion for ascertaining that one compound is the *immediate* precursor of a second except in the special case when only one precursor exists.

While it is hard to see in the light of the above how studies of metabolic turnover can ever be definitive in the elucidation of biosynthetic pathways, this aspect of kinetic studies can certainly be a useful adjunct to other methods already discussed, and merits greater attention than it has received to date. As examples of recent investigations using this approach we shall close this section with a brief reference to papers by Fritig and his associates describing work with tobacco cell cultures (see also Section VI).

In the first paper[17] the interrelationships among 4,4-dimethylsterols, 4-methylsterols, and sterols were studied, with particular attention to the previously proposed pathway:

cycloartenol → methylene-24 → cycloeucalenol → obtusifoliol → methylene-24
               cycloartenol   lophenol
   (I)   (II)   (III)   (IV)   (V)

Experimentally, generally ^{14}C-labelled acetate was fed to the cultures for periods of 5, 30 and 150 minutes. Analyses revealed that IV became labelled less rapidly than I but more rapidly than II or III. Furthermore, although the specific activity ratios A_I/A_{II} and A_{IV}/A_V decreased with time, as expected from theory if the pathway were as above, A_{II}/A_{III} and A_{III}/A_{IV} increased with time. It was thus impossible to account for the results by this simple linear scheme, and the authors proposed instead a more complex, branched pathway from I in which the more or less parallel routes converge at IV, and which agrees with the kinetic results. From this approach a working hypothesis has been devised for subsequent testing.

The second paper[64] deals with the formation of the coumarin scopoletin (**22**, R = OMe). Labelled phenylalanine was fed as a precursor to the cultures for periods up to 6 hours, and isotopic analyses during this period showed the turnover rate of free scopoletin to be much higher than that of its glucoside, scopolin, in the same culture. The deduction was made that the main pathway leading to scopoletin does not involve scopolin as an intermediate. Turnover rates of the free forms of cinnamic and ferulic acids were also demonstrably lower than those of their bound forms, implying that the free forms of the cinnamic acids are involved as intermediates between phenylalanine and the coumarin. As stated above, this type of observation cannot conclusively establish a metabolic pathway, but when supported by other observations, as in this case, its contributory value can be quite considerable.

9. *Expression of results of radioisotope analysis*

Before we leave the subject of isotopic studies it is worth emphasizing that all results of isotope analysis should be expressed in absolute units. This is no problem with stable isotopes, as the use of atom per cent excess is universally accepted. But too many papers in the literature contain results of radioisotope analyses expressed only as counts per minute (c.p.m.). If the counting efficiency is not given (and it frequently is not) there is no way of comparing directly figures obtained on different counters. The counting efficiency should always be determined and the results expressed as disintegrations per minute (d.p.m.). Even better, and in accord with a growing trend, is to divide d.p.m. by 2.22×10^6 to obtain the results in microcuries (μCi), a convenient unit for most phytochemical research. In many experiments when specific activities are required it is also convenient to express results in μCi/mmole.

D.   Administration of Compounds

While this discussion is included for convenience in Section V on isotopic tracers, the reader will appreciate that it is equally applicable to the administration of non-labelled compounds.

1. *Carbon dioxide*

We have referred earlier in this review to problems encountered in ensuring that an administered compound has access to the biosynthetic site(s) of the product under study. With carbon dioxide the problem is minimal, and in administering label in this form it is necessary only to enclose the plant or plant organ in some sort of chamber where it can metabolize the CO_2 fed, under controlled conditions. Chambers of glass[121, 175], rigid plastic[85, 133], and polyethylene film[145] have been described, and the choice is largely one of convenience. Sunlight, artificial light, or both in combination can be used. The chamber can be designed so as to permit generation of the CO_2 within it or in a small connected chamber by acidification of barium carbonate or a solution of soluble carbonate. In the case of external generation some means is provided for sweeping the gas into the main chamber in a current of air, as by maintaining a small negative pressure inside the chamber and drawing outside air through the reaction mixture. A non-volatile acid, usually perchloric or sulphuric, is employed in the generator, and a fan can be used to assure rapid circulation of the CO_2. Although absorption of the carbon dioxide can be followed by an infrared analyser[133], a simpler but effective method, when ^{14}C is used, involves insertion of the probe of a laboratory monitor not pointed at any nearby part of the plant, and following uptake of the CO_2 by the decrease in the count rate[24].

2. Compounds in solution

For the feeding of compounds in solution to plants, uptake through the ends of stems or petioles cut under water[113] is one of the most convenient methods if usable[21, 102, 130, 177]. In a few cases it is ruled out by the nature of the plant, as in species which, on cutting, exude a latex or other substance that blocks the transportation system to the entry of water (e.g. *Typha latifolia*[23]). As mentioned above it is also inapplicable in cases where the organs cut off do not elaborate the product. There may also be a limitation imposed by the length of time a cutting will survive and metabolize normally, or at least without evident deterioration. Temporary wilting appears to be of little importance but longer-term effects may be observed as well. Some species can be kept for weeks without noticeable ill effect while inserted into sand or expanded mica kept moist, as gardeners have found while propagating plants by the rooting of cuttings. The use of Vermiculite moistened with Hoagland's or other suitable nutrient[91] has proved satisfactory for metabolic periods of the order of a week after feeding[27], and this time could undoubtedly be extended. Wilting can be controlled to a degree by enclosure of the shoot in a transparent or translucent chamber of high humidity to reduce transpiration[27], but this must be balanced against possible slowing of the rate of uptake if the shoot is enclosed during feeding. A difficulty frequently encountered is excessively slow uptake even in the open air, and there seems to be no way of predicting when this will occur or of avoiding it entirely. In general, bright light and, if necessary, a current of air[113] will usually ensure a reasonable rate of uptake.

Insolubility in water of many organic compounds has been overcome in various ways. Acids and bases can be administered as their salts. In theory some organic acids might be expected to be liberated from their salts at the acid pH of plant saps, but in practice the salts are usually well translocated and appear to reach the synthetic sites, at least in part, without undue difficulty. In view of the almost universal distribution of β-glucosidases in plants insoluble phenols can be administered as their glycosides[28], which are then cleaved after absorption and translocation. Coumarins have been fed to plants after opening of the lactone ring with alkali, as salts of the corresponding acid[28]. Barz and Grisebach[8], wishing to administer the phenol 4-methoxy-2′,4′-dihydroxychalcone to *Cicer arietinum*, studied the degree of uptake of this compound from several solutions. Although feeding in the presence of a 20% excess of NaOH led to uptake of 82% of 1 mg from 10 ml of solution within 31 hours, even better uptake (96%) was obtained from a solution made up of 9.5 ml of $Na_4P_2O_7$/HCl, 0.1 M, pH 8.5, and 0.5 ml of glycol monomethyl ether. A very good incorporation (5.3%) of ^{14}C from this compound into the isoflavone formononetin was observed with the latter method.

For compounds such as hydrocarbons and steroids which defy solubilization by the aforementioned techniques, different approaches have been successfully used. Non-ionic surface-active agents have been used to disperse insoluble compounds in a form suitable for absorption by plants[42, 132, 149, 150]. Sonication used in conjunction with the surface-active agent can be effective in dispersion[96].

Methods useful for feeding of compounds to cut shoots are also applicable, of course, to certain other kinds of feeding. These include feeding to sterile cell cultures (Section VI), absorption through roots[58, 61, 112, 130], administration of solutions to leaf discs or sections[3, 106] or root or tuber sections[41, 83, 139], uptake of solutions by seeds and seed pods[5, 12, 79, 132, 169], infiltration into fruits[62] and administration to isolated organelles such as chloroplasts[42, 147]. The technique of vacuum infiltration merits special mention in the case of leaf discs, when the amount of plant material is relatively small[3, 43]. The plant material in contact with the solution to be administered is subjected to moderate vacuum for a few minutes to remove interstitial gases, which are replaced by the solution as atmospheric pressure is restored. This technique also has application to larger organs[62].

The use of non-aqueous solvents for feeding water-insoluble compounds to plants does not appear to have been very widely explored. Diglyme (bis-(2-methoxyethyl) ether)[33], dimethylsulphoxide (DMSO)[28] and sunflower seed oil[75], have been used to dissolve certain compounds for feeding, not always with unqualified success. Further experimentation along these lines may yield technical advances of considerable importance, but caution is indicated in view of the demonstration that DMSO, at least, can alter plant metabolism, as evidenced by quantitative changes in alkaloid formation in several species of *Datura*[151]. A variant of this approach that has been of particular value is the application of non-aqueous solutions to leaf surfaces followed by evaporation of the solvent. Bennett and Heftmann[16] have found that in the administration of squalene to *Pharbitis nil* seedlings by this method absorption was accelerated by subsequent application of a thin film of silicone oil by spraying as a solution in light petroleum. The silicone, carrying most of the squalene with it, was absorbed within a few hours. The technique has been extended to the administration of steroids to other species[32, 71]. Simply spraying an aqueous solution of a water-soluble compound on the leaf has also proved an effective means of administration[111].

A good many additional techniques have been reported, some of them primarily of value in special circumstances, for getting solutions into plants. If the plant has a hollow stem or leaf, considerable volumes of solutions can easily be injected with a syringe into the interior, where they are readily absorbed; this technique has been used for onion[75], wheat[29] and parsnip[26] plants. To a more limited degree direct injection into 'solid' stems[59] or even leaves[137, 155] can be used to administer solutions. Wick-feeding is an approach which has found favour and has been used successfully with a variety of plants[38, 146, 177]. Several pieces of non-mercerized cotton thread are passed through the stem of the plant and so arranged that both ends of the thread emerge from the stem close together. The ends are then allowed to dip into a small beaker, conical centrifuge tube, or other convenient container holding the solution to be administered, which passes into the stem by capillary action. Some workers have employed a similar principle by inserting fine glass capillary tubes into plants with the other end in the solution[60]. Feeding to trees poses unique problems which have been met with some ingenuity[137]. Solutions have been administered to spruce trees through the end of a twig from which the needles have been cut, but absorption is slow[63]. Mentzer[119] and his associates[57] administered solutions through twigs from which the ends had been cut. They observed a suction which exhibits a seasonal variation but which can, under favourable circumstances, draw several hundred millilitres of solution into the plant. Sectioning under water is again recommended, and the cut end can be joined to a reservoir by rubber tubing. For smaller quantities another satisfactory procedure has been to remove small pieces of bark to expose the cambium, and then implant the precursor, or inject a solution of it into the cavity from a syringe[63, 84]. The solution is absorbed within a few hours. All these methods possess the common advantage that they are applicable to plants rooted in soil or other support without further disturbing the plant, and are thus of special value for long-term feeding experiments.

Before leaving the subject of tree feeding we should mention the possibilities presented by the Japanese bonsai (dwarfed) plants. Hillis and Ishikura[95], studying biosynthesis of polyphenolic compounds in eucalypts, found the standard technique of floating leaf discs on solutions of radioactive compounds inapplicable, since oils and waxes in the leaves rendered the periphery of the disc impermeable to the aqueous solution. Bonsai plants, while they resemble normal plants in polyphenol composition, have leaves of only one-tenth normal size. Stems carrying several of these leaves were placed in the solution and entire leaves removed at intervals for isotopic analysis.

A related subject is the use of plants of normally small size. Among the Lemnaceae (duckweeds) are found the smallest flowering plants, and species of this family grown in sterile culture have been used for biosynthetic studies on flavonoid biosynthesis[15, 138, 172]. The plantlets of *Lemna minor* float on the surface of the culture medium, and since they reproduce asexually in a few days increase in weight is relatively rapid. Suspected flavonoid precursors can be introduced into the medium by filtration sterilization.

Mention of sterile cultures raises the subject of microbial contamination, with which we shall close this section. Early workers in the field of plant biosynthesis were quite concerned with the possibility of microbial degradation of compounds in solution before their absorption by the plant, especially when absorption required long periods of time. But with most of the techniques described here, when absorption is complete within a few hours, it poses no problem in practice provided that normal precautions are taken. In some 20 years' experience in this field we have never encountered complications of this nature, although in some special cases the possibility undoubtedly can exist. Administered solutions should be kept free of suspended material, and if no turbidity or mycelium is seen the possibility of significant microbial contamination can be neglected.

VI. EXPERIMENTS WITH CELL AND ORGAN CULTURES

Even though many of the associated techniques are identical with those used in other types of plant biosynthetic research, there are enough basic differences to warrant brief discussion of this field in a separate section. A vast literature has grown up around sterile cultures of plant cells and organs, but it is beyond the scope of this chapter to consider the specialized techniques developed to initiate and maintain them. The reader may consult works of Carew and Staba[30], Hildebrandt[94], White and Grove[178] and Willmer[179] for detailed treatment of these questions.

From the standpoint of biosynthetic investigation the use of cell cultures, in particular, offers a number of potential advantages over that of organized plants or differentiated plant organs. Cell suspensions offer the ease of handling long associated with bacterial and algal cultures by microbial biochemists. More rapid growth is sometimes observed in these than with cells of an organized plant, with the potential of more rapid metabolite formation. Problems of permeability may remain, but as difficulties of translocation are avoided, greater ease of getting a potential precursor to a synthetic site should be attainable. In many cases growing an organized plant for biosynthetic study poses formidable if not insurmountable practical problems, as in the case of a large tropical tree to be cultivated in the temperate zone. In such cases cell cultures initiated from germinated seeds may be obtainable and may elaborate the desired product. Because suspension cultures are grown already enclosed in flasks, collection of evolved carbon dioxide, if desired, is facilitated[131], as is investigation under anaerobic conditions[64]. Finally, the axenic nature of these cultures eliminates any complications that might result from microbial contamination*.

Balanced against all this are several drawbacks. A cell or organ culture is not a plant in the normal sense, and its metabolism may exhibit important quantitative and even qualitative differences, so that only experiment can enable a decision to be made whether a culture will serve a given purpose. In some instances, as in the case of some alkaloids and volatile oils, a culture has failed to elaborate a desired product or has formed it only in reduced amounts[30]. With certain species, notably among monocotyledons, it may prove very difficult even to initiate a culture. And growing a given weight of plant material in sterile culture is more prodigal of time and space than is the case for an organized plant, with consequently greater expense.

In comparison with the great volume of work published in the field as a whole few investigations have yet taken place on the detailed pathways of biosynthesis in sterile plant cell and organ cultures. But most of the principles and techniques discussed in this chapter with reference to organized plants are equally applicable to cultures, *mutatis mutandis*. For this reason we intend to discuss in this section only the methods of administering compounds to cultures, and by the use of relevant references to indicate the more important approaches that have been used in biosynthetic studies with this kind of plant material.

The principal additional factor to be considered in the administration of substrates to these cultures is the desirability of their being supplied in sterile solution, although in experiments of short duration this is not strictly necessary[92]. Sterile plant material is originally obtained by surface sterilization of seeds or stem sections. Hypochlorite, bromine or peroxide solutions are most commonly employed but antibiotics have been used[159]. Heat-stable additives can be autoclaved with the culture medium or sterilized by heat in a separate solution and added aseptically to a sterile medium. This procedure normally ensures sterility but in some special cases antibiotics have been added after the heat treatment[159]. If the compound to be administered is labile to heat one can use the alternative technique of sterilizing a solution by passage through ultra-fine glass frits or cellulose ester membranes such as Millipore† filters. Administration of the substrate is often made coincident with transfer of the culture to fresh medium.

---

* It should be noted that organized plants of considerable size can also be grown in sterile culture on a defined medium after seed sterilization or from organized suspension cultures[156].

† Millipore Filter Corp., Bedford, Mass., U.S.A.

Solubility problems appear less acute with cell cultures than in many studies with organized plants. Precursors are usually added to a volume of medium in the range 25–100 ml, an order of magnitude greater than commonly employed in feeding organized plants, but in addition the need for dissolving relatively large weights of labelled substrate has often been avoided by the use of high specific activities. As explained in Section V.C.5, the resulting low dose levels in these cases would indicate the use of per cent incorporation, as opposed to dilution value, as a criterion of precursor effectiveness in comparative studies with isotopes. In general even the feeding of highly hydrophobic substrates seems feasible, as witness the successful administration of cholesterol, in 0.2 ml of 70% ethanol, to *Dioscorea* cultures suspended in 100 ml of aqueous medium[158].

Compounds whose pathway of formation has been studied in cell or organ culture have included proteins[51, 52, 107, 131], alkaloids[74, 122, 142, 153, 159], phenols and derived products[35, 50, 64, 67, 92, 176] and steroid precursors[17, 158]. While cell culture techniques at best will remain supplementary to those involving organized plants, additions to this list can be anticipated as the potential of cell culture methods in relation to biosynthesis is more extensively explored.

VII.   EXPERIMENTS WITH ENZYMES

As described in earlier sections of this chapter many of the postulated biosynthetic sequences proposed for plant products are obtained by the administration of radioactive or inactive compounds to higher plants. The detection and isolation of enzymes catalysing individual steps of a proposed pathway are often required as proof for that pathway. In other words the existence of the enzymes gives added credibility to the proposed pathway although it does not rule out the possibility that alternate routes may exist in the plant. Richards and Hendrickson[136] express it in this way, 'It must, of course, always be recognized that the biogenetic inferences derived by structural examination can never be more than hypothetical in any case and must ultimately be supported by direct experiment, using such tools as radioactive tracers or direct enzyme isolation.' Conn and Butler[39] in a review in which they discuss the biosynthesis of cyanogenic glucosides make a similar statement. However Swain[160] cautions that it should not be assumed that the isolation of a single enzyme is necessarily indicative of the normal biosynthetic pathway. Therefore the detection and subsequent isolation of an enzyme may not be absolute proof for the existence of a sequence. But it is nevertheless convincing evidence favouring the proposed route.

We have made no effort to deal here with all the excellent work that has been done in this field. For the reader that would like to acquaint himself more thoroughly with enzymes and how to use them there are many sources of information; perhaps the most detailed is the series edited by Colowick and Kaplan[37]. The reader requiring general knowledge related to enzymes should consult the works of Dixon and Webb[49], Gutfreund[78] and Mahler and Cordes[114]. For more detailed information regarding enzyme assays Bergmeyer[18] is recommended. Various authors discuss a variety of problems related to enzyme methodology in *Modern Methods of Plant Analysis*[129]. The following examples selected from many in the literature will serve to demonstrate the value of enzyme investigations in verifying postulated pathways.

The biosynthetic pathway from phosphoenol pyruvate and D-erythrose-4-phosphate to the aromatic amino acids *via* shikimic acid as elucidated by Davis and coworkers[44] in microorganisms and by Neish[127] in plants is an excellent example of the value of enzyme investigations. In microorganisms much of the sequence prior to shikimic acid has been verified by the isolation of enzymes. Similar enzymes have also been isolated from various higher plant materials, thus substantiating the pathway in higher plants. Balinsky and Davies[7] using cauliflower buds have purified the enzyme, dehydroquinase, which interconverts 5-dehydroquinic acid and 5-dehydroshikimic acid. The same investigators[6] have isolated dehydroshikimic acid reductase, the next enzyme in the sequence, from etiolated pea epicotyls. Others have demonstrated that a similar enzyme is present in *Phaseolus aureus* seedlings[125] and in immature bamboo[93]. Quinic acid, which is common in higher plants, is not on the direct pathway from 3-deoxy-D-arabinoheptulosonic acid-7-phosphate to shikimic acid, but is derived from 5-dehydroquinic acid which is on the pathway. Gamborg[66] demonstrated that an enzyme, quinate dehydrogenase, which interconverts quinic acid and 5-dehydroquinic acid, is present in the cell suspension cultures of *Ph. aureus*. The same author[68] later reported that appropriate enzymes found in other plant cell suspension cultures interconvert shikimic acid and quinic acid.

The conversion of shikimic acid to the aromatic amino acids, phenolic acids and lignin has been reviewed by Neish[127] and by Brown[25]. The enzymes responsible for the conversion of shikimic acid to 5-phosphoshikimic acid, 3-enolpyruvylshikimic acid-5-phosphate, chorismic acid and then to prephenic acid have not as yet been demonstrated in higher plants. Prephenate dehydrogenase, which converts prephenic acid to 4-hydroxyphenylpyruvic acid, has been isolated from *Ph. vulgaris* cotyledons[69]. Whether the same enzyme produces phenylpyruvate from prephenic acid is not known at the present time, but it has been demonstrated that both phenylalanine and tyrosine are synthesized from prephenic acid by an enzyme prepared from 3-week-old *Ph. aureus* plants[70]. An aminotransferase which may be involved in the production of aromatic amino acids has been isolated from *Ph. aureus* shoots[65]. Two enzymes which convert tyrosine and phenylalanine to *p*-coumaric acid (4-hydroxycinnamic acid) and cinnamic acid respectively have been isolated and investigated in some depth. Tyrosine ammonia-lyase was detected by Neish[126] in various species belonging to the Gramineae and purified from barley stems. Koukol and Conn[104] isolated phenylalanine ammonia-lyase from barley stems. The conversion of *trans*-cinnamic acid to *p*-coumaric acid by the enzyme, cinnamic acid hydroxylase, was first reported by Nair and Vining[124] who isolated it from spinach leaves. Later Russell and Conn[143] reported the isolation of a similar enzyme from pea seedlings. The hydroxylation of *p*-coumaric acid to caffeic acid has been accomplished by employing an enzyme isolated from *Beta vulgaris* leaves[170]. Hess[87] reported on the methylation of caffeic acid to ferulic acid by an enzyme obtained both from wheat seedlings and the young leaves of *Petunia hybrida*. He later showed that an enzyme preparation obtained from various plant materials could further methylate 5-hydroxyferulic acid to sinapic acid[88].

Radioactive tracers have played a very important role in the elucidation of the biosynthetic pathway of glucosinolates in higher plants[53, 166]. These investigations demonstrate that glucosinolates are derived from amino acids, for example allylglucosinolate (sinigrin) from homomethionine[36] and 2-phenylethylglucosinolate (gluconasturtiin) from γ-phenylbutyrine[162]. The two amino acids are derived from methionine and phenylalanine by a chain-lengthening mechanism which involves condensation with acetate[166]. An enzyme complex, 2-keto-4-phenylbutyrate synthase, prepared from *Nasturtium officinale* leaves by Kindl and Wetter[101] carries out the condensation of acetate with 2-keto-3-phenylpropionic acid yielding 2-keto-4-phenylbutyric acid.

Biosynthetic studies with double-labelled compounds led Underhill and Chisholm[165] and Underhill[162, 163] to postulate that intermediates between the amino acid (**28**) and the corresponding glucosinolate (**33**) must be compounds in which the carbon to nitrogen bond remains intact (see Section V.C.6). The compounds from **28** to **32** have been administered to plants and shown to be incorporated into glucosinolates. *N*-Hydroxyphenylalanine (**29**, $R = C_6H_5CH_2$) is incorporated into benzylglucosinolate[100], as is phenylacetaldehyde oxime (**30**, $R = C_6H_5CH_2$)[164], phenylacetothiohydroximate (**31**, $R = C_6H_5CH_2$), and desulphobenzylglucosinolate (**32**, $R = C_6H_5CH_2$)[167]. Kindl and Underhill[100] have partially purified an enzyme complex which they obtained from the leaves of *Tropaeolum majus*, *Sinapis alba* and *Nasturtium officinale* which catalyses the transformation of N-hydroxyphenylalanine (**29**) to phenylacetaldehyde oxime (**30**). The enzyme which catalyses the formation of desulphobenzylglucosinolate (**32**) by glucosyl transfer from UDP glucose to phenylacetothiohydroximate (**31**) has been isolated from the leaves of *Tropaeolum majus* and

purified 20-fold by Matsuo and Underhill[118]. They also found the enzyme in cell-free extracts of *Sinapis alba, Nasturtium officinale* and *Armoracia lapathifolia*.

Conn and his group in their studies of the cyanogenic glucosides have also indicated that conclusive evidence for a biosynthetic pathway is often best provided by enzymic studies[39]. The purification and properties of a UDP glucose: ketone cyanohydrin β-glucosyltransferase isolated from *Linum usitatissimum* confirmed a previous postulate that the substrate, acetone cyanohydrin, is glucosylated by UDP glucose to form the cyanogenic glycoside, linamarin (**34**)[80]. In a later report[81] the same workers suggest

that a single glucosyltransferase appears to be responsible for the glucosylation of acetone cyanohydrin and butanone cyanohydrin respectively, a fact that can only be demonstrated by the isolation of the appropriate enzyme.

A final example of the importance of enzymes in establishing biosynthetic pathways in plants is found in the field of terpenoid biosynthesis. Experiments performed on intact plants showed that the incorporation of 2-^{14}C-mevalonate is very low, and in some cases it appeared that the carbon labelling was introduced *via* [14]CO_2. These results suggest that the major problem is one of translocation of the precursor to the site of monoterpene biosynthesis[76] and point up the need for investigating biosynthetic pathways in a cell-free system which can eliminate problems of translocation and compartmentation[10, 128] (see Section V.C.2).

VIII. ACKNOWLEDGEMENTS

The authors wish to thank their colleagues for their helpful discussions and advice during the preparation of this manuscript.

Acknowledgement must be made to the Chemical Society and the National Research Council of Canada for permission to reproduce material that has already appeared in print.

IX. REFERENCES

1. E. A. Adelberg, *Bacteriol. Rev.*, **17**, 253 (1953).
2. R. E. Alston, in *Biochemistry of Phenolic Compounds* (Ed. J. B. Harborne), Academic Press, London, 1964, pp. 171–204.
3. K. Asada, K. Saito, S. Kitoh and Z. Kasai, *Plant Cell Physiol. (Tokyo)*, **6**, 47 (1965).
4. D. J. Austin and M. B. Meyers, *Chem. Commun.*, 125 (1966).
5. D. Baisted, E. Capstack, Jr. and W. R. Nes, *Biochemistry*, **1**, 537 (1962).
6. D. Balinsky and D. D. Davies, *Biochem. J.*, **80**, 292 (1961).
7. D. Balinsky and D. D. Davies, *Biochem. J.*, **80**, 300 (1961).
8. W. Barz and H. Grisebach, *Z. Naturforsch.*, **22b**, 627 (1967).
9. J. A. Bassham and M. Calvin, *The Path of Carbon in Photosynthesis*, Prentice-Hall, Englewood Cliffs, N.J., 1957.
10. J. Battaile, A. J. Burbott and W. D. Loomis, *Phytochemistry*, **7**, 1159 (1968).
11. J. Battaile and W. D. Loomis, *Biochim. Biophys. Acta*, **51**, 545 (1961).
12. A. R. Battersby, R. Binks, J. J. Reynolds and D. A. Yeowell, *J. Chem. Soc.*, 4257 (1964).
13. A. R. Battersby, D. M. Foulkes and R. Binks, *J. Chem. Soc.*, 3323 (1965).
14. G. W. Beadle and E. L. Tatum, *Proc. U.S. Nat. Acad. Sci.*, **27**, 499 (1941).
15. E. Beck and O. Kandler, *Z. Pflanzenphysiol.*, **55**, 71 (1966).
16. R. D. Bennett and E. Heftmann, *Phytochemistry*, **4**, 475 (1965).
17. P. Benveniste, M. J. E. Hewlins and B. Fritig, *Eur. J. Biochem.*, **9**, 526 (1969).
18. H.-U. Bergmeyer, *Methods of Enzymatic Analysis*, Academic Press Inc., New York, 1963.
19. T. Beyrich, *Pharmazie*, **21**, 365 (1966).
20. J. B. Birks, *The Theory and Practice of Scintillation Counting*, Pergamon, Oxford, 1964.
21. D. E. Bland and A. F. Logan, *Phytochemistry*, **6**, 1075 (1967).
22. E. Broda, *Radioactive Isotopes in Biochemistry*, Elsevier, Amsterdam, 1960.

23. S. A. Brown, *Can. J. Botany*, **39**, 253 (1961).
24. S. A. Brown, *Can. J. Biochem. Physiol.*, **40**, 607 (1962).
25. S. A. Brown, *BioScience*, **19**, 115 (1969).
26. S. A. Brown, *Phytochemistry*, **9**, 2471 (1970).
27. S. A. Brown, unpublished results.
28. S. A. Brown, M. El-Dakhakhny and W. Steck, *Can. J. Biochem.*, **48**, 863 (1970).
29. S. A. Brown and A. C. Neish, *Can. J. Biochem. Physiol.*, **32**, 170 (1954).
30. D. P. Carew and E. J. Staba, *Lloydia*, **28**, 1 (1965).
31. E. Caspi and G. M. Hornby, *Phytochemistry*, **7**, 423 (1968).
32. E. Caspi and D. O. Lewis, *Science*, **156**, 519 (1967).
33. E. Caspi, D. O. Lewis, D. M. Piatak, K. V. Thimann and A. Winter, *Experientia*, **22**, 506 (1966).
34. G. D. Chase and J. L. Rabinowitz, *Principles of Radioisotope Methodology*, Burgess, Minneapolis, 1967.
35. M. Chen, S. J. Stohs and E. J. Staba, *Lloydia*, **32**, 339 (1969).
36. M. D. Chisholm and L. R. Wetter, *Can. J. Biochem.*, **44**, 1625 (1966).
37. S. P. Colowick and N. O. Kaplan, *Methods in Enzymology*, Vols. 1–7, Academic Press Inc., New York, 1955–1964.
38. C. L. Comar, *Radioisotopes in Biology and Agriculture*, McGraw-Hill, New York, 1955.
39. E. E. Conn and G. W. Butler, in *Perspectives in Phytochemistry* (Eds. J. B. Harborne and T. Swain), Academic Press, London, 1969, p. 47.
40. J. W. Cornforth, *Quart. Rev. Chem. Soc.*, 125 (1969).
41. C. Costes, *Ann. Physiol. Vegetale*, **7**, 25 (1965).
42. C. Costes, *Ann. Physiol. Vegetale*, **7**, 105 (1965).
43. C. Costes, *Phytochemistry*, **5**, 311 (1966).
44. B. D. Davis, *Advan. Enzymol.*, **16**, 247 (1955).
45. R. F. Dawson, *Am. J. Botany*, **29**, 66 (1942).
46. R. F. Dawson, D. R. Christman, A. D'Adamo, M. L. Solt and A. P. Wolf, *J. Am. Chem. Soc.*, **82**, 2628 (1960).
47. D. Desaty, A. G. McInnes, D. G. Smith and L. C. Vining, *Can. J. Biochem.*, **46**, 1293 (1968).
48. S. M. C. Dietrich and R. O. Martin, *Biochemistry*, **8**, 4163 (1969).
49. M. Dixon and E. C. Webb, *Enzymes*, 2nd ed., Longmans, Green and Co., London, 1964.
50. D. K. Dougall, *Australian J. Biol. Sci.*, **15**, 619 (1962).
51. D. K. Dougall, *Plant Physiol.*, **40**, 891 (1965).
52. D. K. Dougall, *Plant Physiol.*, **41**, 1411 (1966).
53. M. G. Ettlinger and A. Kjaer, in *Recent Advances in Phytochemistry*, Vol. 1 (Eds. T. J. Mabry, R. E. Alston and V. C. Runeckles), Appleton-Century-Crofts, New York, 1968, p. 59.
54. E. A. Evans, *Tritium and its Compounds*, Van Nostrand, London, 1966.
55. J. W. Fairbairn and S. B. Challen, *Biochem. J.*, **72**, 556 (1959).
56. J. W. Fairbairn and P. N. Suwal, *Phytochemistry*, **1**, 38 (1961).
57. J. Favre-Bonvin, M. Massias, C. Mentzer and J. Massicot, *Phytochemistry*, **7**, 1555 (1968).
58. F. Ferron, J. Van Assche and C. Costes, *Ann. Physiol. Vegetale*, **9**, 245 (1967).
59. H. G. Floss, H. Guenther and L. A. Hadwiger, *Phytochemistry*, **8**, 585 (1969).
60. H. G. Floss and U. Mothes, *Phytochemistry*, **5**, 161 (1966).

61. H. G. Floss and H. Paikert, *Phytochemistry*, **8**, 589 (1969).
62. C. Frenkel, I. Klein and D. R. Dilley, *Phytochemistry*, **8**, 945 (1969).
63. K. Freudenberg, H. Reznik, W. Fuchs and M. Reichert, *Naturwissenschaften*, **42**, 29 (1955).
64. B. Fritig, L. Hirth and G. Ourisson, *Phytochemistry*, **9**, 1963 (1970).
65. O. L. Gamborg, *Can. J. Biochem.*, **43**, 723 (1965).
66. O. L. Gamborg, *Biochim. Biophys. Acta*, **128**, 483 (1966).
67. O. L. Gamborg, *Can. J. Biochem.*, **45**, 1451 (1967).
68. O. L. Gamborg, *Phytochemistry*, **6**, 1067 (1967).
69. O. L. Gamborg and F. W. Keeley, *Biochim. Biophys. Acta.*, **115**, 65 (1966).
70. O. L. Gamborg and F. J. Simpson, *Can. J. Biochem.*, **42**, 583 (1964).
71. A. M. Gawienowski and C. C. Gibbs, *Phytochemistry*, **8**, 2317 (1969).
72. T. A. Geissman and E. Hinreiner, *Botan. Rev.*, **18**, 77 (1952).
73. T. A. Geissman and E. Hinreiner, *Botan. Rev.*, **18**, 165 (1952).
74. M. R. Gibson and G. A. Danquist, *J. Pharm. Sci.*, **54**, 1526 (1965).
75. T. N. Godnev and R. M. Rotfarb, *Dokl. Akad. Nauk SSSR*, **147**, 962 (1962); Consultants Bureau translation, p. 1215.
76. T. W. Goodwin, in *Biosynthetic Pathways in Higher Plants* (Eds. J. B. Pridham and T. Swain), Academic Press, London, 1965, p. 57.
77. J. P. Greenstein and M. Winitz, *Chemistry of the Amino Acids*, Vol. 2, Wiley, New York, 1961, p. 1479.
78. H. Gutfreund, *An Introduction to the Study of Enzymes*, Blackwell Scientific Publications, Oxford, 1965.
79. L. A. Hadwiger, H. G. Floss, J. R. Stoker and E. E. Conn, *Phytochemistry*, **4**, 825 (1965).
80. K. Hahlbrock and E. E. Conn, *J. Biol. Chem.*, **245**, 917 (1970).
81. K. Hahlbrock and E. E. Conn, *Phytochemistry*, **10**, 1019 (1971).
82. M. Hamada and M. Chubachi, *Agr. Biol. Chem. (Tokyo)*, **33**, 793 (1969).
83. K. R. Hanson, *Phytochemistry*, **5**, 491 (1966).
84. M. Hasegawa and M. Shiroya, *Botan. Mag. (Tokyo)*, **79**, 595 (1966).
85. M. D. Hatch and C. R. Slack, *Biochem. J.*, **101**, 103 (1966).
86. F. W. Hefendehl, E. W. Underhill and E. von Rudloff, *Phytochemistry*, **6**, 823 (1967).
87. D. Hess, *Z. Naturforsch.*, **19b**, 447 (1964).
88. D. Hess, *Z. Pflanzenphysiol.*, **53**, 460 (1965).
89. D. Hess, *Z. Pflanzenphysiol.*, **60**, 348 (1969).
90. D. Hess, *Z. Pflanzenphysiol.*, **61**, 286 (1969).
91. E. J. Hewitt, *Sand and Water Culture Methods Used in the Study of Plant Nutrition*, Commonwealth Agricultural Bureaux, Farnham Royal, Bucks, 1966, 187 ff.
92. T. Higuchi, *Can. J. Biochem. Physiol.*, **40**, 31 (1962).
93. T. Higuchi and M. Shimada, *Plant Cell Physiol. (Tokyo)*, **8**, 61 (1967).
94. A. C. Hildebrandt, in *Modern Methods of Plant Analysis*, Vol. 5 (Eds. H. F. Linskens and M. V. Tracey), Springer-Verlag, Berlin, 1962, p. 383.
95. W. E. Hillis and N. Ishikura, *Phytochemistry*, **9**, 1517 (1970).
96. C. Hitchcock, L. J. Morris and A. T. James, *Eur. J. Biochem.*, **3**, 419 (1968).
97. A. Hutchinson, C. D. Taper and G. H. N. Towers, *Can. J. Biochem. Physiol.*, **37**, 901 (1959).
98. M. D. Kamen, *Isotopic Tracers in Biology*, Academic Press, New York, 1957.
99. Z. Kasprzyk and M. Fonberg-Broczek, *Physiol. Plantarum*, **20**, 321 (1967).
100. H. Kindl and E. W. Underhill, *Phytochemistry*, **7**, 745 (1968).

101. H. Kindl and L. R. Wetter, presented at a meeting of Pacific Slope Biochemical Conference, Davis, California, June 15–17, 1967, p. 115.
102. M. Kito, H. Kokura, J. Izaki and K. Sasaoka, *Phytochemistry*, **7**, 599 (1968).
103. P. D. Klein, in *Advances in Tracer Methodology*, Vol. 2 (Ed. S. Rothchild), Plenum Press, New York, 1965, p. 145.
104. J. Koukol and E. E. Conn, *J. Biol. Chem.*, **236**, 2692 (1961).
105. B. L. Lamberts, L. J. Dewey and R. U. Byerrum, *Biochim. Biophys. Acta*, **33**, 22 (1959).
106. G. Lamoureux, R. H. Shimabukuro, H. R. Swanson and D. S. Frear, *J. Agr. Food Chem.*, **18**, 81 (1970).
107. D. T. A. Lamport, *Nature*, **202**, 293 (1964).
108. J. Langridge, *Australian J. Biol. Sci.*, **11**, 58 (1958).
109. J. Langridge and R. D. Brock, *Australian J. Biol. Sci.*, **14**, 66 (1961).
110. E. Leete, *J. Am. Chem. Soc.*, **80**, 2162 (1958).
111. E. Leete, H. Gregory and E. G. Gros, *J. Am. Chem. Soc.*, **87**, 3475 (1965).
112. E. Leete, L. Marion and I. D. Spenser, *Can. J. Chem.*, **32**, 1116 (1954).
113. F. Loewus and M. M. Baig, *Methods Enzymol.*, **18**, 22 (1970).
114. H. R. Mahler and E. H. Cordes, *Biological Chemistry*, 2nd ed., Harper and Row, New York, 1971.
115. J. Marmur, *J. Mol. Biol.*, **3**, 208 (1961).
116. L. M. Marshall and R. E. Cook, *J. Am. Chem. Soc.*, **84**, 2647 (1962).
117. R. O. Martin, M. E. Warren and H. Rapoport, *Biochemistry*, **6**, 2355 (1967).
118. M. Matsuo and E. W. Underhill, *Phytochemistry*, **10**, 2279 (1971).
119. C. Mentzer, *C. R. Acad. Sci.*, **245**, 2354 (1957).
120. M. W. Miller, E. D. Garber and P. D. Voth, *Nature*, **195**, 1220 (1962).
121. T. Minamikawa, S. Yoshida and M. Hasegawa, *Plant Cell Physiol. (Tokyo)*, **10**, 283 (1969).
122. S. Mizusaki, T. Kisaki and E. Tamaki, *Agr. Biol. Chem. (Tokyo)*, **29**, 714 (1965).
123. A. Murray III and D. L. Williams, *Organic Syntheses with Isotopes*, Interscience, New York, 1958.
124. P. M. Nair and L. C. Vining, *Phytochemistry*, **4**, 161 (1965).
125. M. Nandy and N. C. Ganguli, *Arch. Biochem. Biophys.*, **92**, 399 (1961).
126. A. C. Neish, *Phytochemistry*, **1**, 1 (1961).
127. A. C. Neish, in *Biochemistry of Phenolic Compounds* (Ed. J. B. Harborne), Academic Press, London and New York, 1964, p. 295.
128. A. Oaks and R. G. S. Bidwell, *Ann. Rev. Plant Physiol.*, **21**, 43 (1970).
129. E. Paech and M. V. Tracey, in *Modern Methods of Plant Analysis*, Vol. 6 (Eds. H. F. Linskens and M. V. Tracey), Springer-Verlag, Berlin, 1963, pp. 295–488.
130. H. J. Perkins and D. W. A. Roberts, *Biochim. Biophys. Acta*, **58**, 486 (1962).
131. J. K. Pollard and F. C. Steward, *J. Exp. Botany*, **10**, 17 (1959).
132. K. H. Raab, N. J. de Souza and W. R. Nes, *Biochim. Biophys. Acta*, **152**, 742 (1968).
133. H. Rapoport, F. R. Stermitz and D. R. Baker, *J. Am. Chem. Soc.*, **82**, 2765 (1960).
134. J. M. Reiner, *Arch. Biochem. Biophys.*, **46**, 53 (1953).
135. J. M. Reiner, *Arch. Biochem. Biophys.*, **46**, 89 (1953).
136. J. H. Richards and J. B. Hendrickson, *The Biosynthesis of Steroids, Terpenes, and Acetogenins*, W. A. Benjamin, Inc., New York, 1964, p. 15.
137. W. A. Roach, 'Plant injection for diagnostic and curative purposes', *Imperial Bureau of Horticulture and Plantation Crops Technical Communication No. 10*, East Malling, Kent, England, 1938.

138. R. M. Roberts, R. H. Shah and F. Loewus, *Plant Physiol.*, **42**, 659 (1967).

139. R. M. Roberts, R. Shah and F. Loewus, *Arch. Biochem. Biophys.*, **119**, 590 (1967).

140. R. Robinson, *Structural Relations of Natural Products*, Oxford University Press, Oxford, 1955.

141. L. J. Rogers, S. P. J. Shah and T. W. Goodwin, in *Biochemistry of Chloroplasts*, Vol. 2 (Ed. T. W. Goodwin), Academic Press, London, 1967, pp. 283–292.

142. A. Romeike and O. Aurich, *Phytochemistry*, **7**, 1547 (1968).

143. D. W. Russell and E. E. Conn, *Arch. Biochem. Biophys.*, **122**, 256 (1967).

144. R. S. Russell and R. P. Martin, *Nature*, **163**, 71 (1949).

145. K. Saito and Z. Kasai, *Plant Cell Physiol.* (*Tokyo*), **9**, 529 (1968).

146. K. Saito and Z. Kasai, *Phytochemistry*, **8**, 2177 (1969).

147. M. Sato, *Phytochemistry*, **5**, 385 (1966).

148. E. Schram, *Organic Scintillation Detectors*, Elsevier, Amsterdam, 1963.

149. K. E. Schulte and S. Foerster, *Tetrahedron Letters*, 773 (1966).

150. K. E. Schulte, G. Rücker, W. Meinders and W. Herrmann, *Phytochemistry*, **5**, 949 (1966).

151. L. A. Sciuchetti, *Ann. N.Y. Acad. Sci.*, **141**, 139 (1967).

152. H. Simon and H. G. Floss, *Bestimmung der Isotopenverteilung in markierten Verbindungen*, Springer, Berlin/München, 1967.

153. M. L. Solt, R. F. Dawson and D. R. Christman, *Plant Physiol.*, **35**, 887 (1960).

154. J. L. Sommerville, *The Isotope Index*, Vol. 8, Scientific Equipment Co., Indianapolis, 1967.

155. W. E. Splittstoesser, *Plant Cell Physiol.* (*Tokyo*), **10**, 87 (1969).

156. E. J. Staba, in *Recent Advances in Phytochemistry*, Vol. 2 (Eds. M. K. Seikel and V. C. Runeckles), Appleton-Century-Crofts, New York, 1969, p. 75.

157. W. Steck and B. K. Bailey, *Can. J. Chem.*, **47**, 2425 (1969).

158. S. J. Stohs, B. Kaul and E. J. Staba, *Phytochemistry*, **8**, 1679 (1969).

159. R. J. Suhadolnik, A. G. Fischer and J. Zulalian, *Biochem. Biophys. Res. Commun.*, **11**, 208 (1963).

160. T. Swain, in *Biosynthetic Pathways in Higher Plants* (Eds. J. B. Pridham and T. Swain), Academic Press, London, 1965, p. 9.

161. M. Tanabe and G. Detre, *J. Am. Chem. Soc.*, **99**, 4515 (1966).

162. E. W. Underhill, *Can. J. Biochem.*, **43**, 179 (1965).

163. E. W. Underhill, *Can. J. Biochem.*, **43**, 189 (1965).

164. E. W. Underhill, *Eur. J. Biochem.*, **2**, 61 (1967).

165. E. W. Underhill and M. D. Chisholm, *Biochem. Biophys. Res. Commun.*, **14**, 425 (1964).

166. E. W. Underhill and L. R. Wetter, in *Biosynthesis of Aromatic Compounds. Proceedings of the 2nd Meeting of the Federation of European Biochemical Societies* (Ed. G. Billek), Pergamon Press, Oxford, 1966, p. 129.

167. E. W. Underhill and L. R. Wetter, *Plant Physiol.*, **44**, 584 (1969).

168. E. W. Underhill and H. W. Youngken, Jr., *J. Pharm. Sci.*, **51**, 121 (1962).

169. R. T. van Aller, H. Chikamatsu, N. J. de Souza, J. P. John and W. R. Nes, *J. Biol. Chem.*, **244**, 6645 (1969).

170. P. F. T. Vaughan and V. S. Butt, *Biochem. J.*, **113**, 109 (1969).

171. R. P. Wagner and H. K. Mitchell, *Genetics and Metabolism*, Wiley, New York, 1964, pp. 288–320.

172. J. W. Wallace, T. J. Mabry and R. E. Alston, *Phytochemistry*, **8**, 93 (1969).

173. C. H. Wang and D. L. Willis, *Radiotracer Methodology in Biological Science*, Prentice-Hall, Englewood Cliffs, N.J., 1965.
174. J. E. Watkin and A. C. Neish, *Can. J. Biochem. Physiol.*, **38**, 559 (1960).
175. L. H. Weinstein, C. A. Porter and H. J. Laurencot, Jr., *Contrib. Boyce Thompson Inst.*, **20**, 121 (1959).
176. F. Weygand and H. Wendt, *Z. Naturforsch.*, **14b**, 421 (1959).
177. T. A. Wheaton and I. Stewart, *Phytochemistry*, **8**, 85 (1969).
178. P. R. White and A. R. Grove, *Proceedings of an International Conference on Plant Tissue Culture*, McCutchan, Berkeley, 1965.
179. E. N. Willmer, *Cells and Tissues in Culture*, Vol. 3, Academic Press, London, 1966. Chaps. 8–10.
180. G. Wolf, *Isotopes in Biology*, Academic Press, New York, 1964.
181. E. Wong, *Phytochemistry*, **7**, 1751 (1968).

The Biosynthesis of Ergot Alkaloids

R. THOMAS DEPARTMENT OF CHEMISTRY,
UNIVERSITY OF SURREY, GUILDFORD, SURREY

and

R. A. BASSETT DEPARTMENT OF BIOCHEMISTRY,
IMPERIAL COLLEGE OF SCIENCE AND
TECHNOLOGY, LONDON, S.W.7

I. INTRODUCTION

The ergot alkaloids constitute the largest known group of nitrogenous fungal metabolites[1] and to date approximately fifty alkaloids have been isolated and chemically characterized. The incentive for this considerable micro-biological and chemical undertaking, stemmed primarily from a recognition of the important pharmacological properties of these natural substances, some of which were utilized as long ago as the sixteenth century while other novel properties have only very recently been discovered.

There are few drugs with as intriguing a history as ergot[2], which is the name originally given to the sclerotium which develops on cereal crops and grasses infected by the parasitic fungus *Claviceps purpurea*, a member of the Hypocreaceae family. From the early middle ages, consumption of ergot-ized rye has been held responsible for epidemics of gangrene and convul-sions resulting in widespread loss of life. The disease was variously known as 'St. Anthony's fire', 'holy fire' or simply 'ergotism', and serious outbreaks have been reported as recently as 1927 in Russia and 1950 in France[3]. The ergot of wild grasses still remains a menace to grazing livestock.

The medicinal value of ergot preparations was first recorded in a German herbal published in 1582, prior to the recognition of ergot as the cause of ergotism, which followed a century later. The crude drug was formally introduced into medicine in the United States early in the nineteenth century and was subsequently prescribed in many other countries.

Today the pure alkaloids and their semi-synthetic derivatives find a variety of medical applications. These include ergometrine which, because of its oxytoxic property, is of value in obstetrics, and ergotamine which relieves migraine[3, 4]. The semi-synthetic product lysergic acid diethyl-amide, better known as LSD, is currently not only used as a psychothera-peutic agent[5] but is also abused as a potent hallucinogenic drug. Agro-clavine, an alkaloid of the clavine group, has very recently been shown to possess novel pharmacological properties, some of which indicate that it may be a potential post-coital contraceptive drug[6].

Until 1960 the only known source of the ergot alkaloids was a few species of the genus *Claviceps*[4] growing either parasitically on grasses or cereals, or saprophytically on a laboratory culture medium. In recent years, however, ergot alkaloids have been found in other fungi including *Penicillium* species[7, 8], *Aspergillus* species[4, 9] and in *Rhizopus arrhizus*[9]. Abe and coworkers[10, 11] in an extensive survey of fungi, based on their ability to pro-duce Van Urk positive materials in submerged and static culture, have ob-tained chromatographic evidence for the production of ergot alkaloids, mainly elymoclavine, agroclavine, festuclavine and the fumigaclavines by

all the major fungal groups. The distribution of ergot alkaloids is not confined to the fungi, however, as they are also found in higher plants belonging
to the genera *Ipomea*[12, 13], *Rivea*[12] and *Argyreia*, and the alkaloids produced
by these plants are in general clavines and simple lysergic acid derivatives;
these findings were summarized in the review by Der Marderosian[14].

There are numerous excellent reviews[3, 4] covering the mycology, pharma-
cology and chemistry of the ergot alkaloids. The present chapter is
concerned primarily with the biosynthesis of these alkaloids and their
associated interconversions, and with relevant considerations of alkaloid
production in submerged cultures; recent reviews on these biosynthetic
aspects include those of Agurell[15], Ramstad[16], Kelleher[17] and Voigt[18].

II. CHEMISTRY

Nearly 100 years have passed since the isolation and crystallization of the
first ergot alkaloid from crude ergot, and in the interim period about fifty
additional alkaloids have been characterized in the course of numerous
chemical studies, which included a total synthesis of ergotamine[19]. On
structural grounds, the ergot alkaloids can be grouped into distinct classes,
namely the classical lysergic acid peptide alkaloids, the clavine alkaloids,
both of which contain the tetracyclic ergoline skeleton (**1**, R = H), the tricyclic chanoclavines such as chanoclavine-I (**2**), and a new group with at
present a single representative, clavicipitic acid (**3**).

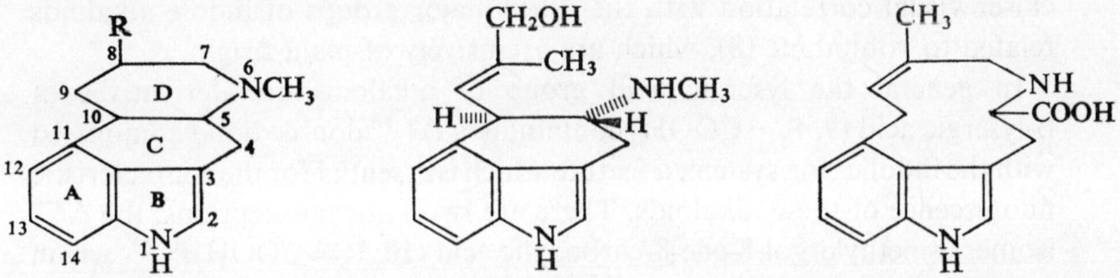

Figure 1.

In view of the long and diversified background of chemical studies of the
ergot alkaloids, it is not surprising that differing structural conventions have
evolved. Thus in addition to the formulation represented in **1**, two-dimen-
sional shapes such as **4**, **5** and **6** are currently in use by other authors.

These alternative conventions are equally acceptable, but in this article
structures based on **1** will be used throughout, as it is felt that this affords a
more direct pictorial association with the prevalent representation of the

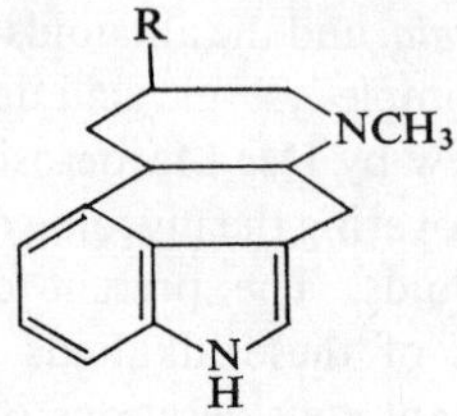

Figure 2.

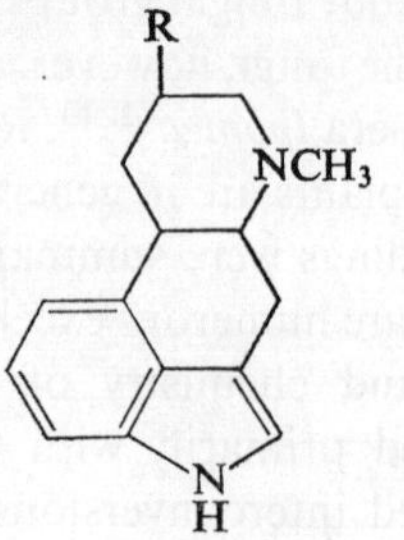

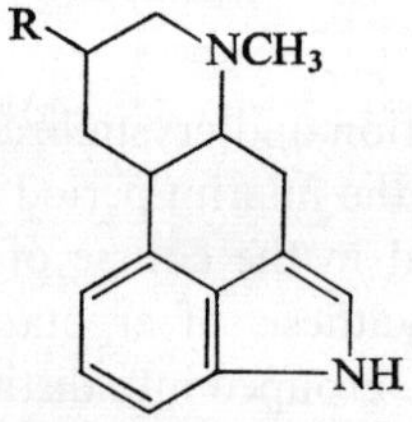

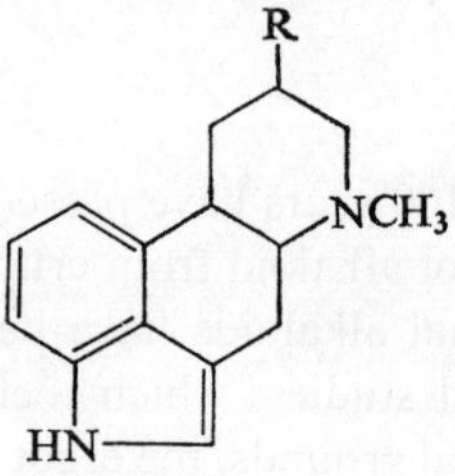

indolic ring of the established precursor tryptophan (7). This also permits a closer visual correlation with the other major groups of indole alkaloids related to yohimbine (8), which are exclusively of plant origin.

In general the lysergic acid group of alkaloids are derivatives of D-lysergic acid (9, R = COOH) containing a $\Delta^{9,10}$ double bond conjugated with the indolic ring system, a feature which is essential for the characteristic fluorescence of these alkaloids. There are two known exceptions, the $\Delta^{8,9}$ isomer 6-methylergol-8-ene-8-carboxylic acid (10, R = COOH)[20,21], which

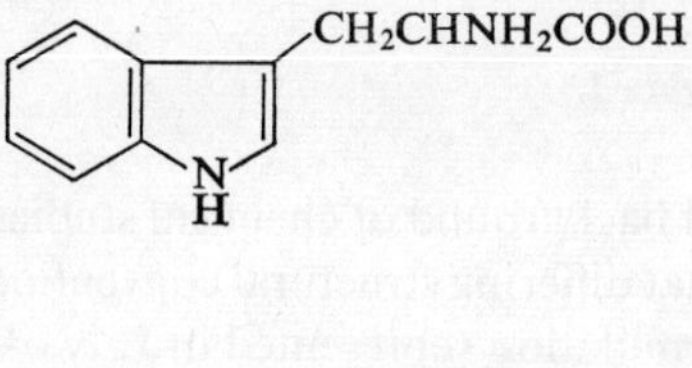

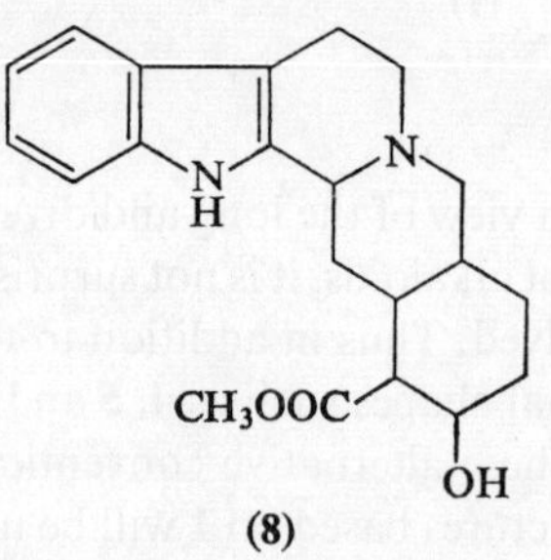

Figure 3.

(9) **(10)**

Figure 4.

coexists with D-lysergic acid in *C. purpurea*, and the saturated ergoline derivative dihydroergosine[22] containing the ergoline skeleton (1).

The peptide alkaloids can be further divided into two subgroups: (i) the simple amides such as lysergic acid amide (9, R = CONH$_2$) and the methylcarbinolamide (9, R = CONHCHOHCH$_3$), (ii) the more complex so-called peptide alkaloids of which the archetype is ergotamine (11) in which the lysergic acid moiety is linked to a cyclic tripeptide via an amide bond.

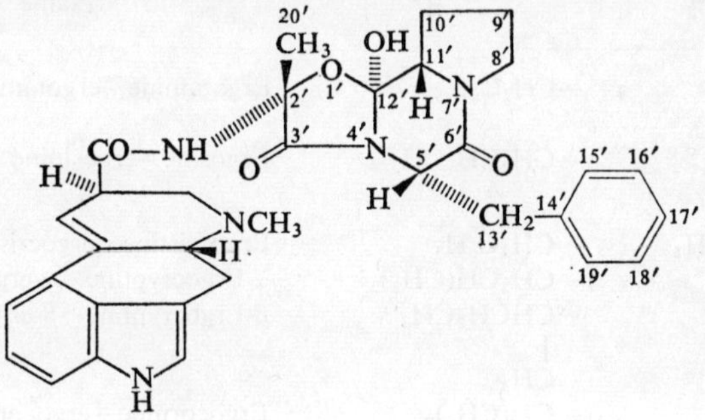

(11)

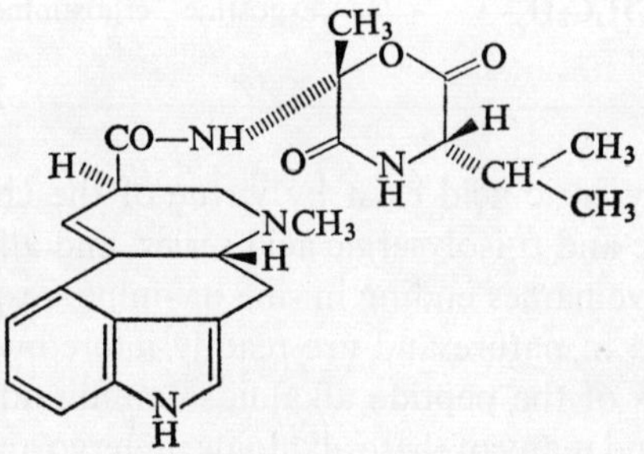

(12)

Figure 5.

Although the majority of the peptide alkaloids are of the complex tripeptide type, a dipeptide alkaloid ergosecaline (**12**) is also known. Examples of the tripeptide alkaloids are classified in Table 1.

Table 1. Classification of the tripeptide alkaloids

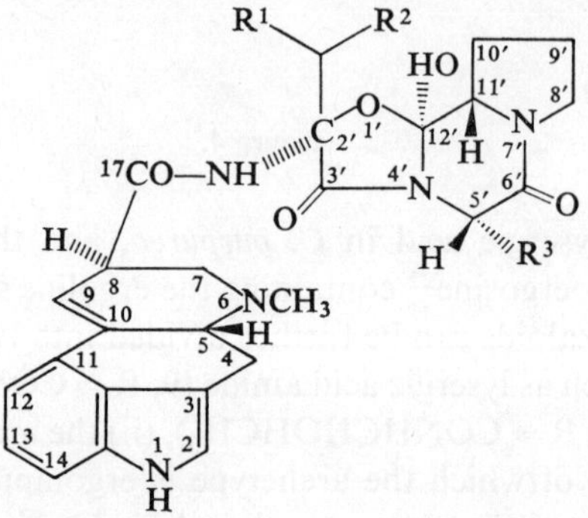

Family	R^3	Name
Ergotamine $R^1 = R^2 = H$	$—CH_2C_6H_5$	Ergotamine—ergotaminine
	$—CH_2CH(CH_3)_2$	Ergosine—ergosinine
Ergotoxine $R^1 = R^2 = CH_3$	$—CH_2C_6H_5$	Ergocristine—ergocristinine
	$—CH_2CH(CH_3)_2$	α-Ergocryptine—α-ergocryptinine
	$—CHCH_2CH_3$ $\quad\vert$ $\quad CH_3$	β-Ergocryptine—β-ergocryptinine
	$—CH(CH_3)_2$	Ergocornine—ergocorninine
Ergostine $R^1 = H$ $R^2 = CH_3$	$—CH_2C_6H_5$	Ergostine—ergostinine

Two isomers of D-lysergic acid exist by virtue of the chirality at C-8; these are the D-lysergic and D-isolysergic acid series, and alkaloids derived from these isomers have names ending in -ine or -inine, respectively. Both types have been found in nature and are readily interconverted in dilute alkali. The side-chains of the peptide alkaloids contain additional asymmetric carbon atoms and many of these alkaloids undergo an acid-catalysed reversible isomerization to the so-called aci-series, for example aci-ergotamine. These differ from the parent in the configuration at C-2′ of the

Table 2. Examples of the clavine alkaloids[a]

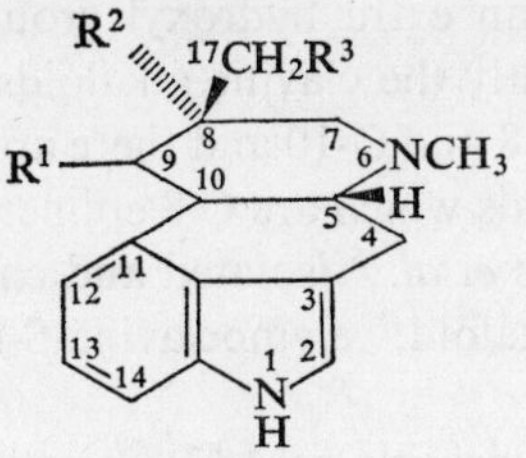

Double bond	Name	R¹	R²	R³	C-10 chirality
$\Delta^{8,9}$-Ergolene alkaloids	Agroclavine	H	None	H	10 R
	Elymoclavine	H	None	OH	10 R
	Elymoclavine-β-D-fructoside	H	None	—O—fructoside	10 R
	Molliclavine	OH	None	OH	Unknown
$\Delta^{9,10}$-Ergolene alkaloids	Penniclavine–isopenniclavine	H	OH	OH	
	Setoclavine–isosetoclavine	H	OH	H	
	Lysergol–isolysergol	H	H	OH	
	Lysergine–isolysergine	H	H	H	
$\Delta^{9,10}$-, $\Delta^{8,17}$-	Lysergene	H	None	None	
Ergoline alkaloids (saturated ring D)	Festuclavine	H	H	H	10 R
	Pyroclavine (C-8 epimer of festuclavine)	H	H	H	10 R
	Costaclavine	H	H	H	10 S
	α-Dihydrolysergol	H	H	OH	10 R
	Fumigaclavine A	CH₃COO	H	H	Unknown
	Fumigaclavine B	OH	H	H	Unknown

[a] The isoclavine series (including pyroclavine) are C-8 epimers of the parent clavines.

side-chain[23]. Ultraviolet irradiation of aqueous acid solutions of lysergic acid derivatives results in the addition of water across the $\Delta^{9,10}$ double bond, giving rise to the non-fluorescent lumialkaloid series[24].

In the tetracyclic clavine alkaloids, examples of which are listed in Table 2, C-17 is present as a methyl or hydroxymethyl group and all of the clavines of this group possess the same configuration at C-5 as do the lysergic acid alkaloids. The double bond may be $\Delta^{8,9}$ or $\Delta^{9,10}$ or may be reduced. Some clavine alkaloids carry an extra hydroxyl group in ring D, for example molliclavine. Consequently the clavine alkaloids may contain one or two asymmetric centres at C-8 and C-10 and there are several pairs of naturally occurring clavine alkaloids which are C-8 epimers, for example setoclavine and isosetoclavine. Floss *et al.*[25] isolated and characterized a new type of substituted clavine alkaloid, elymoclavine-β-D-fructoside from ergot cultures.

Chanoclavine and clavicipitic acid (3) are atypical in that they do not possess the tetracyclic ergoline skeleton. Of the chanoclavines, four isomers are known to occur naturally[26], and of the clavicipitic acid type only one member has been reported[27]. Recently, three novel clavines have been described, cycloclavine (**29**) and rugulovasins A and B, which are stereo-isomers of structure **30** (see addendum).

III.   CLAVICEPS FERMENTATION

A.   Cultivation of Ergot

Until the 1960's the ergot alkaloids were obtained commercially from rye ergot, by artificial or spontaneous infection of the rye plant generally referred to as parasitic culture. No real progress was made in the isolation of a viable alkaloid-producing fungus which could be cultured on a synthetic medium, until Abe[28,29] reported the production of clavine alkaloids by this so-called saprophytic culture of ergot. In the following years, a number of other clavine alkaloids were similarly isolated by Abe and coworkers[30-32] and Stoll, Hofmann *et al.*[33,34] Little additional work has been done on the clavine alkaloids produced in submerged culture since the field was comprehensively reviewed in 1964 by Abe[30], but improved strains of *C. fusiformis* have recently been developed by Mantle[35] producing about 6000 μg/ml of agroclavine. In view of the newly discovered pharmacological properties of agroclavine[6], and the fact that they provide suitable model systems for biosynthetic studies, one may anticipate a renewed interest in submerged culture fermentations with clavine-producing strains.

In contrast to the clavine alkaloids, the well-known pharmacological properties of the classical lysergic acid alkaloids have provided more impetus for their study. Attempts to produce these alkaloids in submerged culture were not successful[36] until Arcamone *et al.*[37,38] reported a method for the large scale fermentation of the simple alkaloid lysergic acid methylcarbinolamide by *C. paspali*. Subsequently other groups reported successful alkaloid production in submerged culture. *C. paspali* fermentations were described in some detail by Pacifici *et al.*[39] and Gröger and Tyler[40]. Kobel *et al.*[20] and later Castagnoli and Mantle[21], succeeded in isolating $\Delta^{8,9}$ D-lysergic acid, the latter investigators from *C. purpurea*. Various aspects of these fermentations were recently reviewed by Kelleher[17].

The production of the more complex peptide alkaloids in low yield by a surface culture of *C. purpurea* was achieved in the late 1950's by several groups[41–44] and the progress of these and similar fermentation studies was reviewed by Abe[30]. In 1966 the production of peptide alkaloids was achieved in yields in excess of 1000 μg/ml using submerged culture fermentations[45, 46]. These strains of *C. purpurea* which yielded primarily ergotamine, produced no conidia and alkaloid accumulation was shown to be favoured by high sugar concentrations (e.g. 30% sucrose). Amici *et al.*[47] observed that 90% of the total alkaloid was present in the mycelium, whereas with the *C. purpurea* strain of Tonolo[45], the alkaloids were chiefly present in the culture filtrate. Recently, Amici *et al.*[48] have reported the production, by strains of *C. purpurea*, of high yields of members of the ergotoxine family, in submerged culture.

B. Time-course of Peptide Alkaloid Production

Essentially, this discussion will be restricted to the more recently isolated high yielding alkaloid strains of *Claviceps*. Amici *et al.*[46, 47] in a physiological study of ergotamine production, described in detail the course of fermentation, the results of which compared favourably with those subsequently reported for high yielding ergotoxine[48] strains of *C. purpurea*. The fermentations were continued for 14–16 days and could be divided into several phases. The first phase is characterized by rapid utilization of certain constituents, namely phosphate, nitrogen (ammonia), citric acid and sucrose. During this phase there is rapid growth and low production of alkaloids. The second phase which begins between 4 and 6 days, is characterized by slower growth and by a marked tendency to accumulate lipids, sterols and to produce alkaloids in the mycelium. The course of an ergotamine-producing fermentation[57] is shown in Figures 6(a) and (b).

The early work of Taber[49–51] and others[52, 53] has shown that in alkaloid-producing fermentations, there is a phase of 'true growth' generally referred to as the trophophase[54, 55], which is normally terminated on exhaustion of a limiting nutrient from the medium. Alkaloid biosynthesis occurs subsequently in the so-called idiophase[54–56]. In recent investigations with relatively high yielding strains of *C. purpurea* producing ergotamine[57] and the ergotoxine group of alkaloids[48], it appears that although a two-phase fermentation pattern can be distinguished, the trophophase and idiophase become less distinct. In strains of *C. fusiformis*[35], producing agroclavine in very high yields (e.g. 6000 μg/ml), the phase of true growth and alkaloid production are nearly synchronous and therefore such means of describing the course of these high yielding fermentations appear to be of limited operational value.

However, Amici *et al.*[48] considered that the passage from the phase of rapid growth and phosphate depletion, to the alkaloid-producing phase, could be initiated by the exhaustion of phosphate or magnesium, both of which can cause the accumulation of lipids[58]. The importance of the initial inorganic phosphate levels for alkaloid production has been demonstrated by Taber and Vining[59] with low yielding strains of *C. purpurea*, and Arcamone *et al.*[60] who used high yielding strains of *C. purpurea* producing ergotamine. The latter authors reported that alkaloid production is favoured by growth-limiting phosphate concentrations, and concluded that during the alkaloid-producing phase the reduction of protein synthesis may make available the simple nitrogenous precursors, namely amino acids, required for the synthesis of peptide alkaloids.

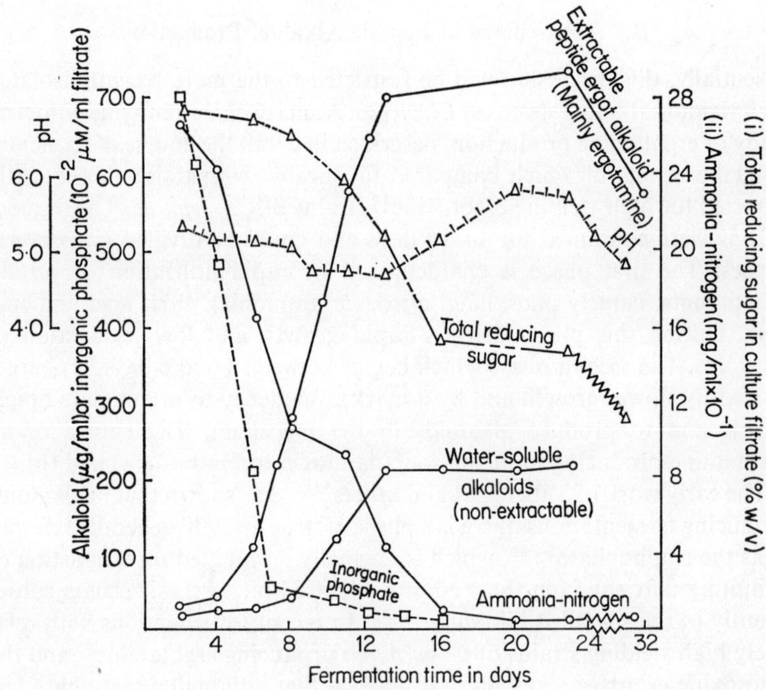

Figure 6(a)

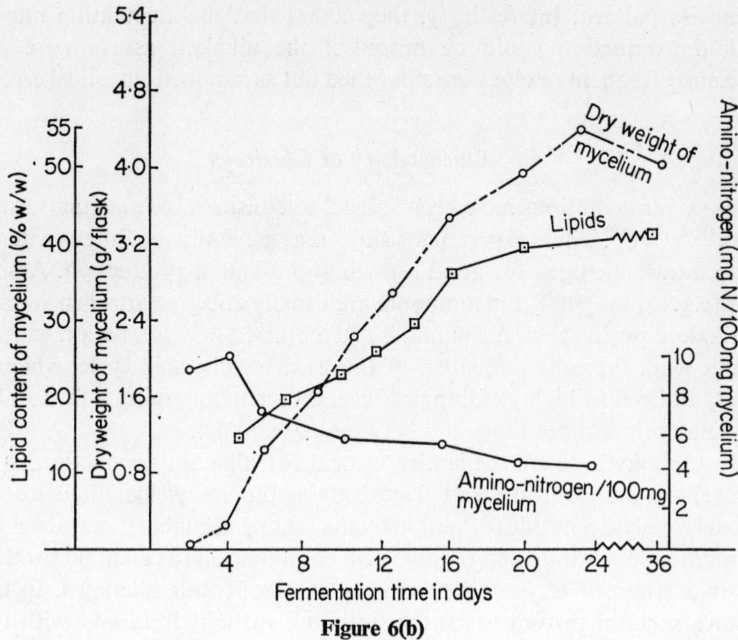

Figure 6(b)

Figures 6(a) and (b). Alkaloid production and metabolism of *C. purpurea*. These figures show the course of a typical fermentation of a relatively high ergotamine-producing strain of *C. purpurea* (Fr) Tul (IC/114/1) grown in submerged culture, in a 30% (w/v) sucrose-containing production stage medium. The initial uptake of inorganic phosphate and ammonia-nitrogen can be readily seen (but this phenomenon is *not* restricted to alkaloid producers), followed at 4–8 days by the initiation of rapid alkaloid production, which decreases at 14–16 days. Unlike the Amici strain of *C. purpurea*, this produces alkaloid mainly in the filtrate. At fermentation times of more than three weeks the alkaloid titre falls. Mycelial protein is produced most rapidly during the early period of the fermentation, corresponding in this strain to the period of low alkaloid synthesis. In agreement with Amici and coworkers, high lipid accumulation was observed in the mycelium during the latter phase of the fermentation in high alkaloid-yielding *Claviceps*.

Similarly, Rosazza *et al.*[61] carried out a detailed study of the inorganic requirements for both growth and alkaloid production. Their results indicated that the requirement for peak growth was essentially the same as that for maximal alkaloid production with respect to potassium and magnesium levels, and in general their findings for phosphate-dependence confirmed those of Amici *et al.*[48] and Arcamone *et al.*[60] In relation to the kinetic course of the fermentation, it is interesting to note that several workers, notably, Bu'Lock and Barr[53] reported that the course of alkaloid production in the high yielding clavine strains examined, is not linear in time but rather follows a stepwise pattern. Interestingly, they stated that this fluctuating rate of alkaloid production could be missed if the alkaloid assays were not sufficiently frequent or else were smoothed out as assumed statistical errors.

C.   Biochemistry of *Claviceps*

Claviceps fermentations have been subject to considerable medium formulation[38, 47, 48, 62]. These experiments show that generally the fungus will not utilize nitrate-nitrogen for good growth and alkaloid production. Amino acids (e.g. asparagine), and ammonia are usually effective nitrogen sources for alkaloid production. Ammonia is often utilized in combination with an organic acid, typically a member of the tricarboxylic acid cycle, which is rapidly utilized in high yielding producers, thus maintaining a favourable pH. The optimum pH range lies between 5.0 and 5.4.

The carbohydrate source is also critical for alkaloid production, with sucrose, mannitol, sorbitol and glucose giving the best yields, often accompanied by polysaccharide formation by the fungus. Kelleher[17] examined the optimum carbohydrate and organic acid combination for alkaloid production in a strain of *C. paspali* producing simple peptide alkaloids. In this instance sorbitol proved to be the optimum carbohydrate and with this sugar, the yields of alkaloid decreased with change of organic acid substrate in the order fumarate, succinate, malate, tartrate and citrate. With mannitol and maltose, the order was similar with respect to the acids, but the overall yield of alkaloid was considerably lower. Sucrose, glucose and lactose gave little or no alkaloid. Peak production was observed to be delayed in media containing sorbitol or fumarate, the overall fermentation time requiring 30 days. As pointed out by Kelleher[17] this could be reduced by modification of the seed stage medium to a 20-day fermentation.

In an analogous examination[47, 63] of the relationship of alkaloid production and the ability to utilize sucrose and citric acid, marked strain variations were observed. Three strains were isolated from sectors of a giant colony of *C. purpurea* 275F1 and were named V (low producer), C and W (non-producers). In terms of sucrose and citric acid utilization, strain 275F1 consumed large amounts of sucrose and citric acid, strain V utilized large amounts of sucrose, but small amounts of citric acid, strain C utilized large amounts of citric acid and small amounts of sucrose, while strain W utilized only limited quantities of both. Furthermore, alkaloid synthesis was shown to be associated with the accumulation of large quantities of lipids and sterols in the mycelium.

Although no recent investigations have been reported concerning the pathways of carbohydrate metabolism in high yielding strains of *Claviceps*, studies of low yielding strains of *C. purpurea* have been reported by McDonald *et al.*[64, 65] It was shown that glucose metabolism operated via the normal pentose and glycolytic Krebs pathways. Enzymatic and chromatographic evidence for the complete pentose pathway was presented leading through pentose and sedoheptulose phosphates to hexose monophosphate. Furthermore, pentose and sedoheptulose phosphates were reformed from fructose-6-phosphate if NADPH was provided. Phosphorylated intermediates such as hexose phosphates, 6-phosphogluconate and ribose-5-phosphate were not able to permeate the cell wall of whole cells, but it was suggested by McDonald[64] that these intermediates could be synthesized intracellularly by the action of hexokinase which was shown to be present. Cell-free extracts readily oxidized glucose-6-phosphate, fructose-6-phosphate, glucose-1-phosphate, 6-phosphogluconate and fructose-1,6-diphosphate when triphenyltetrazolium chloride was used as an electron acceptor. Calculations based on various specifically ^{14}C-labelled forms of glucose, indicated that 90% of the glucose substrate was oxidized via the glycolysis–tricarboxylic acid route and only 10% via the pentose pathway.

It was also found[65] that the succinic dehydrogenase of *C. purpurea*, (E.C. 1.3.99.1), which in yeast and animals is particle bound[66], was more readily made soluble, and that under the conditions of assay phenazine methosulphate alone functioned as an electron acceptor, in contrast to the other succinic dehydrogenase acceptors. Fumarate, pyrophosphate and malonate were competitive inhibitors of the dehydrogenation, with dissociation constants (K_i) for their respective enzyme–inhibitor complexes of 0.9, 0.42 and 0.03 mM respectively.

The relation between high alkaloid production and rapid phosphate uptake by *Claviceps* led Kim *et al.*[67] to undertake a series of inhibition experiments with arsenate, which can interfere with various aspects of phosphate metabolism. It appeared from these studies that the molar ratio of phosphate and arsenate was more important than the absolute concentration of arsenate, in controlling the levels of alkaloids biosynthesized. At certain ratios a 100% increase in alkaloid was obtained without causing a significant reduction in mycelial dry weight. It was also noted that arsenate decreased the rate of phosphate uptake during the course of the fermentation period. In the course of an investigation of the role of arsenate, several respiratory inhibitors were examined. In addition one uncoupler of oxidative phosphorylation was tested, namely 2,4-dinitrophenol which was found to give a significant increase (about 50%) in alkaloid production. Clearly further studies concerning carbohydrate metabolism could yield valuable results both in increasing alkaloid yield and in enhancing the understanding of the underlying control processes relating alkaloid biosynthesis and primary metabolism.

Zahid and Baxter[68] have reported that phenobarbitone stimulates an increase in the incorporation of ^{14}C-tryptophan into total alkaloid, and in the light of the established effect of phenobarbitone on microsomal enzyme induction[69], Ambike, Baxter and Zahid[70] have subsequently demonstrated the presence of cytochrome P-450 in the microsomal fraction of *C. purpurea* and also observed a stimulation of the level of this cytochrome by phenobarbitone. The polycyclic hydrocarbon 3-methylcholanthrene exhibited a similar stimulatory effect, which in both instances was paralleled by an increase in alkaloid synthesis. Cytochrome P-450 is a widely distributed component of many mixed-function oxidases, responsible for the aerobic hydroxylation and resulting detoxification of drugs and foreign substrates[71] in animal tissues and some bacteria. Its apparent association with alkaloid synthesis in *C. purpurea* may be significant in relation to the recent report of Hsu and Anderson[72] that cell-free preparations of this fungus contain an NADPH-dependent hydroxylase which effects oxidative conversion of agroclavine (**10**, $R = CH_3$) to elymoclavine (**10**, $R = CH_2OH$). This hydroxyl group had been previously shown not to originate from water, although its direct derivation from molecular oxygen was not demonstrated. This would be consistent with a pathway based on a coupled redox system requiring the NADPH reduction of cytochrome P-450, which is the oxygen-activating component of the microsomal mixed-function oxidase, although additional intermediate electron-transporting coenzymes may be involved.

D. Amino Acid Metabolism

The characterization of the precursors of the ergoline ring, and the side-chain of the peptide alkaloids, has prompted several studies of precursor feeding in an effort to increase alkaloid yields[38, 53, 73, 74]. Arcamone *et al.*[38] tested several amino acids and found that DL-tryptophan stimulated a substantial increase in the production of lysergylmethylcarbinolamide when added at the start of the fermentation. Pacifici *et al.*[73] also observed increased production of alkaloids on addition of tryptophan. Perhaps one of the most interesting observations was that reported by Bu'Lock and Barr[53] who carried out a detailed study of tryptophan uptake and alkaloid synthesis. With their strain of *Claviceps*, it was demonstrated that in order to ensure substantial alkaloid production, exogenous tryptophan was required at the very beginning of the fermentation during the early phase of growth, when no alkaloid was being synthesized. In addition, it was established that in this strain, the requirement for exogenous tryptophan was lost in the final phase

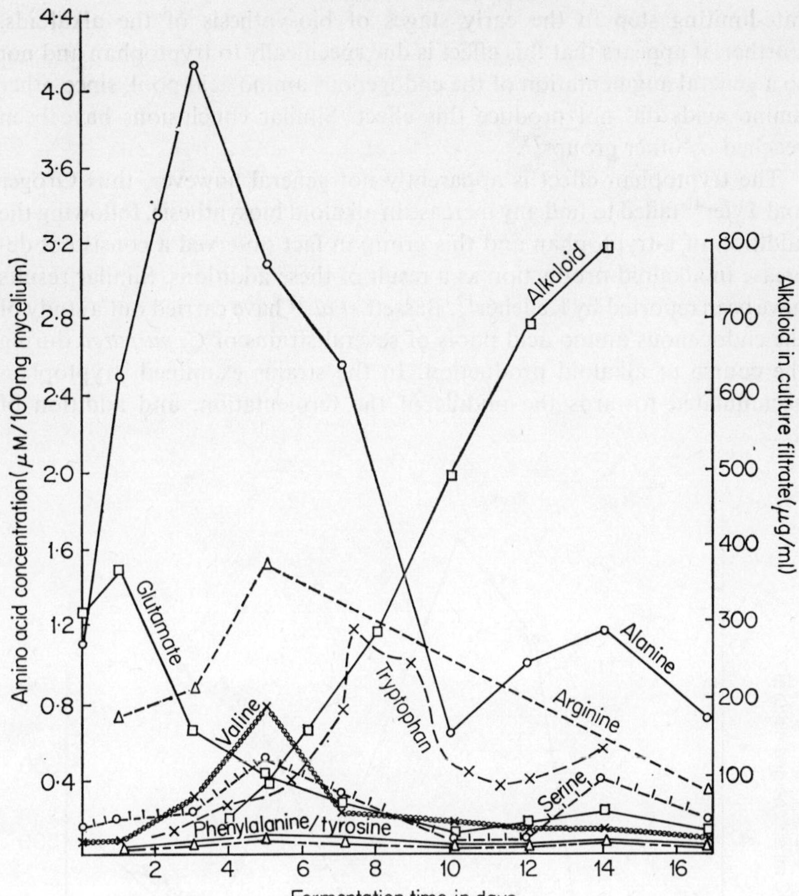

Figure 7. Extractable amino acids of *C. purpurea* (IC/114/1) mycelium. This figure shows the levels of extractable amino acids recovered from the mycelium of *C. purpurea* producing ergotamine as the principal peptide alkaloid component, when grown in submerged culture. For the sake of clarity proline, leucine, isoleucine and glycine are not shown, but their levels in the mycelium were comparable with those of phenylalanine and tyrosine. In this strain and the one discussed later (Fig. 8), the high level of extractable alanine and tryptophan is seen to fall at the onset of ergotamine biosynthesis. Alanine does stimulate alkaloid production if added at this time in the fermentation. It would appear from these and similar results that the fluctuation in the mycelial amino acid pools follows a two-phase pattern as discussed in the previous section.

of alkaloid production, when for the first time a high level of endogenous tryptophan was observed. From these and other results it was concluded that tryptophan induced an enzyme 'synthetase', which was involved in a rate-limiting step in the early stages of biosynthesis of the alkaloids. Further, it appears that this effect is due specifically to tryptophan and not to a general augmentation of the endogenous amino acid pool, since other amino acids did not produce this effect. Similar conclusions have been reached by other groups[75].

The tryptophan effect is apparently not general however, thus Gröger and Tyler[40] failed to find any increase in alkaloid biosynthesis, following the addition of L-tryptophan and this group in fact observed a consistent decrease in alkaloid production as a result of these additions. Similar results have been reported by Kelleher[17]. Bassett *et al.*[57] have carried out a study of the endogenous amino acid pools of several strains of *C. purpurea* during the course of alkaloid production. In the strains examined, tryptophan accumulated towards the middle of the fermentation, and addition of

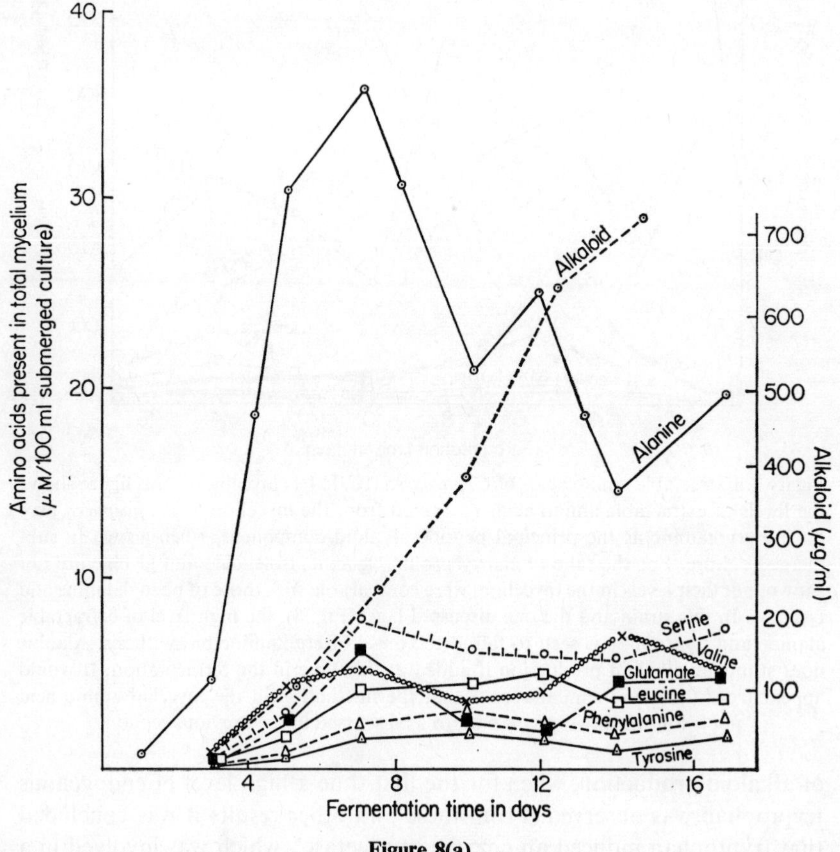

Figure 8(a)

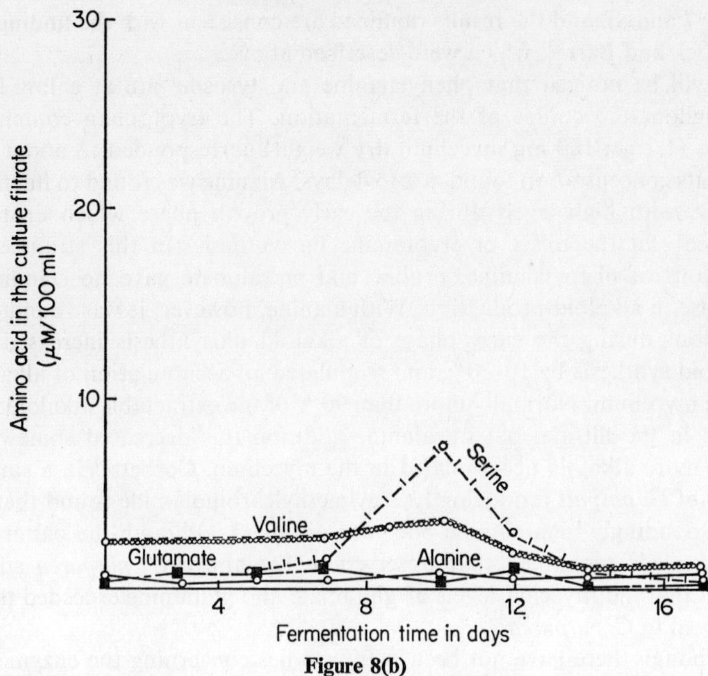

Figures 8(a) and (b). Amino acid levels in *C. purpurea* (IC/123). The amino acid pool sizes are shown during the course of a typical alkaloid fermentation on a chemically defined medium. Mycelial alanine concentration showed the widest fluctuation, the level dropping at the onset of ergotamine biosynthesis. The amino acid content of the culture filtrate after 17 days was very low, indicating little autolysis.

tryptophan to the cultures at both the early and late stages of fermentation produced a small increase (10–20%) in alkaloid production in all of these strains. Mantle[76] has observed that with a dihydroergosine-producing strain of *Claviceps*, alkaloid yield actually dropped on the addition of tryptophan.

With *Claviceps* producing clavines or simple lysergic acid alkaloids, Gröger and Tyler[40], Bu'Lock[53] and Pacifici[73] found that addition of other amino acids effected either no increase or only a small increase of the order of 10% in alkaloid production.

There have been no reports in the literature concerning the general pattern of amino acid metabolism during the course of a fermentation by *Claviceps*. In the case of relatively high yielding ergotamine strains of *C. purpurea*, Bassett *et al.*[57] examined both mycelial and culture filtrate amino acids periodically during the course of a 20-day submerged culture fermentation

(Figs. 7 and 8), and the results obtained are consistent with the findings of Bu'Lock and Barr[53], which were described above.

It will be noticed that phenylalanine and tyrosine are at a low level throughout the course of the fermentation. The tryptophan content at 7 days (1.2 μM/100 mg mycelium dry weight) corresponded to about five times its concentration found at 2 to 4 days. Alanine was found to fluctuate widely, with high levels during the early growth phase which dropped markedly at the onset of ergotamine biosynthesis. In this strain early additions of phenylalanine, proline and mevalonate gave no significant increase in alkaloid production. With alanine, however, it was found that additions during the early phase of alkaloid biosynthesis increased net alkaloid synthesis by 10–20% and stimulated an accumulation of alkaloid in the mycelium. Normally more than 90% of the extractable alkaloid was found in the filtrate, but on alanine addition this decreased somewhat, while more alkaloid accumulated in the mycelium. Corbett[77] in a similar study of *C. paspali* producing lysergylmethylcarbinolamide found that no correspondingly high alanine pool was obtained, although the pattern of amino acids resembled that observed in the above *C. purpurea* study, except that the mycelial levels of glutamate and glutamine exceeded those observed in *C. purpurea*.

Although there have not been many studies concerning the enzymes in *Claviceps* species responsible for the biosynthesis of the aromatic amino acids, there are clear indications that tryptophan production follows the normal pathway via chorismate and is subject to the usual feed-back controls[78, 79]. Possibly these controls may be less rigid, since tryptophan analogues can stimulate alkaloid production[75]. The steps from chorismic acid to prephenic acid, catalysed by chorismic mutase, have been examined in *Claviceps* SD58 and Pb156 in some detail by Sprossler and Lingens[80, 81], especially, in relation to the activating and inhibiting effects of tryptophan, substituted tryptophans and a range of other amino acids. At certain concentrations, L-tryptophan, L-tryptophan methyl ester hydrochloride, a dimethylfluorotryptophan and N_α-methyl-DL-tryptophan all produced similar activation effects on chorismate mutase, whereas 2-methyl-, 4-methyl-, and 6-methyl-DL-tryptophan had a lower activating effect. D-Tryptophan and some compounds with large substituents generally showed only negligible activating effects, although D-tryptophan was more effective at high concentrations. Fifteen other amino acids were tested and of these, both histidine and cysteine were shown to be strong activators of this enzyme. The chorismate mutase was found to be inhibited by L-phenylalanine and tyrosine.

Tryptophan degradation appears to follow a variety of patterns including some which are peculiar to *Claviceps*[82]. In this context Teuscher has argued[83] that there is a positive link between the rate of uptake and metabolism of L-tryptophan, with the high alkaloid yielding capacity of a range of *C. purpurea* strains.

E. Differentiation and Alkaloid Production

In a series of mycological and biochemical studies, Tonolo *et al.*[84-86] have established a relationship between the morphology of the constitutive hyphae, both in surface and submerged culture (and in the sclerotia produced on host plants), and alkaloid production by certain strains of *C. purpurea*. Mantle and Tonolo[84] showed that a non-plectenchymatic growth form (sphacelial form) produces no alkaloid, whereas a plectenchymatic or sclerotial form of the fungus was associated with alkaloid production. Furthermore, it appeared that in submerged cultures of *C. purpurea*, the sphacelial form correlated with the early growth phase, whereas in the later stages of the fermentation when the growth rate is reduced the mycelium is predominantly sclerotial. If such a change in morphology is generally related to alkaloid production in other *Claviceps* species, then the factors causing this change would be of considerable importance. To date there do not appear to have been any direct studies along these lines. However, there do appear to be at least two metabolic activities which are associated with this morphological transition, namely, lipid metabolism and polysaccharide biosynthesis.

F. Lipid Metabolism

The glyceride oil of naturally occurring *C. purpurea* sclerotia (ergot oil) has long been known to contain 25–30% of ricinoleic acid (D-12-hydroxy-*cis*-9-octadecenoic)[87, 88]. Morris and Hall[89] demonstrated that these triglycerides have unusual structures, in that the hydroxyl groups of the ricinoleic acid chains are *not* free but instead are esterified with a variety of long-chain fatty acids to form estolides. The ergot oil thus contains not only normal triglycerides, but also tetra-, penta- and hexaacid-triglycerides. The accumulation of linoleic or oleic acid when ricinoleic acid was found to be absent from the mycelial oil of non-plectenchymatic cultures of *C. purpurea*, suggested that the oleic or linoleic acids may be intermediates in the biosynthesis of ricinoleic acid. The role of linoleic acid as an intermediate was confirmed experimentally in immature parasitic sclerotia by Morris *et al.*[90] Mantle *et al.*[91] further showed that when cultures of *C. purpurea* produced a sclerotial growth form in culture, the principal component of the triglyceride oil was ricinoleic acid, whereas sphacelial growth forms (non-plectenchymatic) not only contained less total triglyceride oil, but then the ricinoleic acid component was absent and was replaced by linoleic and oleic acid. Amici *et al.*[47, 48] have commented on the importance of lipid accumulation, especially in relation to peptide alkaloid production by *C. purpurea.*

Thus the control mechanisms which determine the change from sphacelial to sclerotial growth forms in submerged culture, the latter representing the phase of alkaloid synthesis, also appear to initiate the biosynthesis of a unique estolide triglyceride, based on ricinoleic acid. The ricinoleic acid content could be used to indicate the extent of plectenchymatic (sclerotial growth) tissues of *C. purpurea* in laboratory cultures and consequently of alkaloid production in these strains.

G.   Polysaccharide Biosynthesis

The extra-cellular metabolism of carbohydrates has been studied in *Claviceps* species producing high yields of clavine and peptide alkaloids. The temporary accumulation of polysaccharides by *Claviceps* was noted by Perlin and Taber[92], who characterized the product. Similarly, Buck *et al.*[93a] investigated extra-cellular glucan produced by *C. fusiformis* when grown in submerged culture (Fig. 9). The polysaccharide was shown to be a branched glucan consisting of β-1,3-D-glucopyranosyl units linked to form the main chain, with single β-1,6-D-glucopyranosyl units substituted at intervals along the glucan chain. Branched glucans of similar structure were also detected in natural sclerotia of *C. fusiformis* and in the cell walls of mycelia in submerged culture.

From the growth curve it will be noticed that in these strains of *C. fusiformis*, the extra-cellular polysaccharide level remains high even in the later phase of the fermentation. Recently, Dickerson *et al.*[93b] reported a novel strain of *C. fusiformis* which, while producing glucan early in the fermentation as a product of *sphacelial* fungal tissue, was subsequently able to autolyse the glucan to glucose. The autolytic activity was attributed to a constitutive β-1,3-glucanase and β-glucosidase which could be detected as soon as the fungal hyphae differentiated from the sphacelial to the sclerotial growth form, indicating that perhaps the glucan was a synthetic product of the sphacelial tissue. Glucanase production (Fig. 10) followed a sigmoid pattern, reaching a high level within 12 days and the liberated glucose contributed to apparent renewed growth towards the end of the fermentation. Interestingly, this activity maintained a minimum viscosity during the later stages of the fermentation which permitted adequate aeration, thereby facilitating ergot alkaloid production by the fungus.

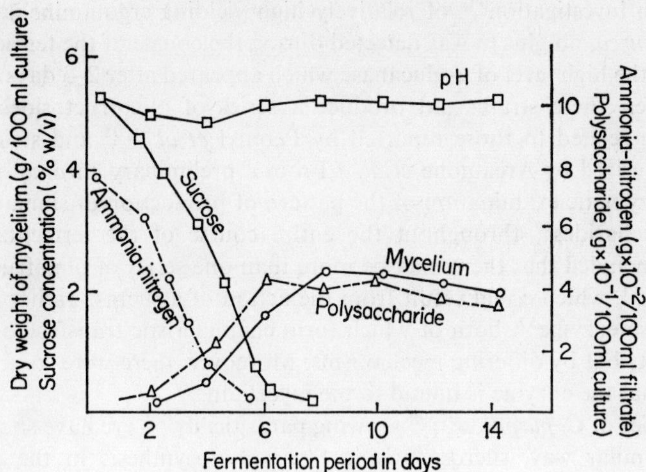

Figure 9. Course of *C. fusiformis* fermentation. *C. fusiformis* grown in submerged culture on a defined sucrose-containing medium, showing an accumulation of polysaccharide (glucan).

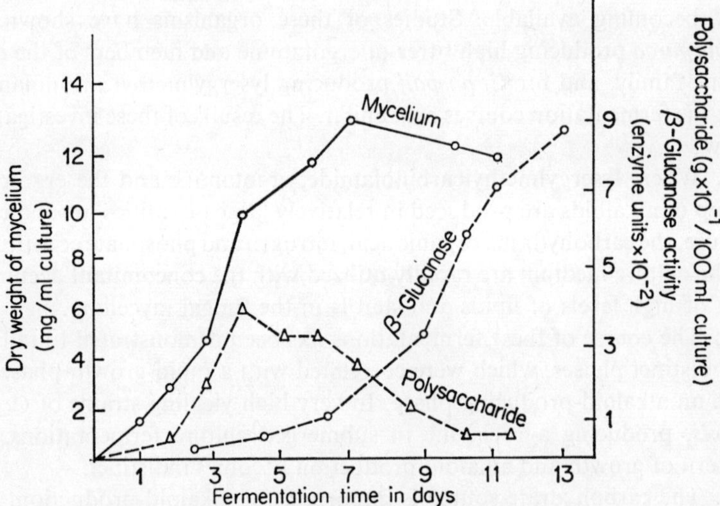

Figure 10. Course of fermentation of a β-glucanase-producing strain of *C. fusiformis*. The *C. fusiformis* strain when grown in submerged culture on a sucrose medium, initially produces a viscous β-glucan and subsequently a β-glucanase, which decreases the viscosity of the culture medium in the later stages of the fermentation.

In an investigation[93b] of relatively high yielding ergotamine strains of *C. purpurea*, no glucan was detected during the course of the fermentation due to the high level of β-glucanase which appeared after 2–3 days growth. However, these strains did produce a series of oligofructosides, which may be related to those reported by Perenyi *et al.*[94, 95] and structurally characterized by Arcamone *et al.*[96] From a preliminary chromatographic and enzymatic examination of the pattern of production of similar filtrate polysaccharides[97] throughout the entire course of the fermentation, it was concluded that there may be more than one series of oligofructosides produced, which could result from the action of a sucrase rather than an invertase enzyme[57], both of which form characteristic transfructosylation products but by differing mechanisms. Moreover, there were some indications that the enzyme is bound to the mycelium.

Studies of *C. purpurea*[97, 98] growing parasitically on rye have shown that in a similar way, sucrose arising from photosynthesis in the plant is metabolized to a series of oligofructosides by the sphacelial tissues of the developing parasitic fungus, and these are chromatographically similar to those produced in submerged culture.

In summarizing this section, it is apparent that due to the ingenious efforts of many microbiologists, strains of *Claviceps* producing the full range of the peptide ergot alkaloids, in submerged culture, on a synthetic medium are now becoming available. Studies of these organisms have shown, for *C. purpurea* producing high titres of ergotamine and members of the ergotoxine family, and for *C. paspali* producing lysergylmethylcarbinolamide, that the fermentation courses are similar. The results of these investigations show:

1. Where lysergylmethylcarbinolamide, ergotamine and the ergotoxine group of alkaloids are produced in relatively large quantities in submerged culture, the carbohydrate, organic acid, nitrogen and phosphate constituents of the culture medium are rapidly utilized with the concomitant accumulation of high levels of lipids and sterols in the fungal mycelium.

2. The course of these fermentations has been demonstrated to fall into two distinct phases, which were correlated with a rapid-growth phase and with an alkaloid-producing phase. In very high yielding strains of *C. fusiformis*, producing agroclavine in submerged culture fermentations, this pattern of growth and alkaloid production becomes indistinct.

3. The carbohydrate source is important for alkaloid production, with sucrose, mannitol, sorbitol and glucose yielding the best alkaloid titre, often when used in combination with a readily utilized Krebs-cycle acid. Moreover polysaccharides are frequently produced by these fungi, when grown in submerged culture, and so-called parasitic culture.

4. Amino acids and ammonia typically provide efficient nitrogen sources for good growth and alkaloid production.

5. Studies of the utilization and metabolism of amino acids have given a wide range of results. Tryptophan additions can produce a diversity of effects, from a substantial stimulation of alkaloid accumulation, to a marked decrease in alkaloid production, depending upon the strain of *Claviceps* used. Other amino acids have been found to produce a less marked effect.

6. In *C. purpurea*, a correlation between mycelial morphology and alkaloid yield has been demonstrated. Furthermore, lipid and polysaccharide biosynthesis have been shown to be coincidental with these morphological changes. As pointed out, such biochemical and morphological changes may provide valuable markers for those engaged in strain selection aimed at enhanced alkaloid production.

7. Although studies of intermediate metabolism of *Claviceps* are still rudimentary, initial results indicate that such lines of investigation will probably be of value, both in increasing alkaloid titre and in gaining an understanding of the control processes relating primary and secondary metabolism in this fungus.

IV. PATHWAYS OF ERGOT ALKALOID BIOSYNTHESIS

A. Primary Precursors of the Ergoline Moiety

A wide range of modern chemical, microbiological and biochemical techniques have been utilized in an attempt to elucidate the mode of biosynthesis of these compounds. As recently pointed out by Kelleher[17], these studies have already led to more than 100 publications on the subject. The effort, indeed, has proved to be most fruitful although many details remain to be clarified. On the basis of structural analysis, Robinson[99] proposed that tryptophan or a biogenetically related species was involved in the biosynthesis of the ergoline ring system. Early experiments designed to test this hypothesis, however, led to some confusion, probably because strains of *Claviceps* producing satisfactory amounts of alkaloids in submerged culture were not, at that time, available. Thus Suhadolnik *et al.*[100] concluded from their experiments that tryptophan did not appear to function as a precursor of ergometrine. They used sclerotia growing parasitically on rye plants, and then injected the precursors into the internodal region of the plant, prior to extracting the alkaloids from the maturing sclerotia. Their results illustrate some of the difficulties involved in studies using parasitic culture conditions (Table 3), although Mothes *et al.*[101], also using parasitic cultures, succeeded in obtaining significant incorporation of ^{14}C-tryptophan into ergot alkaloids.

Many experiments using saprophytic culture conditions subsequently confirmed the role of tryptophan as a primary precursor of both the clavine and lysergic acid alkaloids.

The very wide range of incorporation efficiencies reported by various workers in the field is particularly noteworthy. Understandably, comparatively poor incorporation of added tryptophan has been reported in strains producing small amounts of ergot alkaloids[102, 103, 104], but Gröger *et al.*[103] demonstrated that increased tryptophan incorporation could take place in the presence of added pyridoxal phosphate. However, even in relatively high yielding strains some observers have obtained only poor incorporation of tryptophan[105, 106], whereas others[107–109] have reported incorporation values approaching nearly 40% of the added radioactive tryptophan.

Table 3. The results of early precursor feeding experiments, using parasitic cultures of *Claviceps*[100]

Precursor	Incorporation into ergometrine, expressed as a percentage of precursor
L-^{14}C-Tryptophan	0.012
L-^{3}H-Tryptophan	0.004
^{3}H-Anthranilic acid	0.006
2-^{14}C-Acetate	0.005
1-^{14}C-Phenylalanine	0.003

Arcamone *et al.*[74] in feeding experiments with *C. paspali* which produced high yields of lysergylmethylcarbinolamide in submerged culture, reported that 59.1% of the added DL-(3-^{14}C)-tryptophan was incorporated into the alkaloids.

Various explanations have been advanced to account for this wide range of incorporation efficiencies. Gröger *et al.*[105, 110] considered that the explanation could lie partly in the dissimilar abilities of the various fungal strains to degrade tryptophan to anthranilic acid and dihydroxybenzoic acid. Moreover Teuscher[83] has argued for a positive link between rapid tryptophan metabolism and high alkaloid-producing capacity in a range of *C. purpurea* strains. It is clear that a range of experimental conditions have been employed in these studies, such as the time of addition of tryptophan, the incubation period before harvesting labelled alkaloids and the differing strains of *Claviceps* used; any one of these parameters could account at least in part for this variation. In reports of unexpectedly low tryptophan incorporation efficiencies, no indication was given as to whether or not the fungus was able to metabolize the preformed ergot alkaloids, and the existence of a significant turnover rate would profoundly affect the observed incorporation efficiency.

A considerable literature has now accumulated concerning attempts to define more precisely the exact status of the tryptophan-utilizing pathway. In particular the work of Floss *et al.*[111] may be noted, in which they demonstrated the incorporation into elymoclavine of all the carbon, hydrogen and nitrogen atoms of the alanine side-chain of L-tryptophan, with the exception of the carboxyl carbon atom, as previously shown by Gröger *et al.*[112] The relative incorporation efficiencies of doubly labelled $^{15}N/^{14}C$- and $^{3}H/^{14}C$-tryptophan into elymoclavine were determined. Whereas the β-carbon and β-hydrogen atoms were equally efficiently incorporated, the α-hydrogen atom from DL-tryptophan was incorporated with only half the efficiency of the β-carbon. Furthermore, the α-hydrogen atom from D-tryptophan showed negligible incorporation into elymoclavine relative to the β-carbon atom. Approximately half the α-nitrogen from DL-tryptophan and almost all the α-nitrogen from D-tryptophan was lost in the course of biosynthesis. The incorporation of a ^{14}C-label from D-tryptophan, but not the α-^{3}H- or α-^{15}N-labelled amino acid, indicated a conversion of D-tryptophan to L-tryptophan, probably via indolyl-3-pyruvic acid. Plieninger and coworkers[109] demonstrated that indolyl-3-pyruvic acid was in fact a good precursor of the ergot alkaloids. One interesting consequence of the retention of the α-hydrogen atom from L-tryptophan, taken in conjunction with the absolute configuration of the C-5 bridgehead carbon in the ergoline nucleus, is that decarboxylation must be accompanied by an inversion of configuration, during the cleavage of the L-tryptophan carboxyl bond with the formation of the C-5 to C-10 bond in chanoclavine-I (**2**).

Feeding experiments with known precursors of tryptophan and various derivatives have been carried out. These have produced an understanding of some of the earliest stages in the pathway leading to the ergoline nucleus, as well as serving to substantiate the conclusions derived from the tryptophan feeding data. Although known tryptophan precursors were readily incorporated such as anthranilic acid, indole and indolyl-3-pyruvic acid[109], modifications of tryptophan have resulted in severe restriction of its precursor function. Floss *et al.*.[114, 115] did not obtain any incorporation of 4-hydroxytryptophan, nor did Agurell[15] with 4-hydroxytryptamine, or Baxter *et al.*[104] with 5-hydroxytryptophan. Plieninger *et al.*[109] in a related series of precursor feeding experiments, prepared 4-, 5- and 6-deuterated tryptophans and demonstrated that whereas 5- and 6-deuteriotryptophans were incorporated with retention of the isotopic label on using 4-deuteriotryptophan the deuterium was not incorporated. These results appear to rule out activation of the C-4 position of tryptophan by hydroxylation and confirmed the expected result that the C-4 hydrogen is eliminated, however a C-4,5 epoxide intermediate has not been excluded. Similarly, negative incorporation results were obtained with tryptamine[116, 117], N-methyltryptamine[117, 118] and N_α-methyltryptophan[118]. These observations were consistent with the view that tryptophan is neither decarboxylated nor methylated before condensation. Cavender and Anderson[119] have obtained from *C. purpurea* PRL-1980 a partially purified cell-free supernatant preparation, capable of synthesizing agroclavine, elymoclavine, chanoclavine-I and chanoclavine-II from the primary precursor tryptophan. Clearly, such a system has the potential to yield valuable results concerning these early stages.

The N-methyl group of the ergoline system appears to be derived via the usual methionine transmethylation pathway[116, 120], but since N-methyltryptophan and N-methyltryptamine were not incorporated, the stage at which methylation occurs in the biosynthetic sequence remains to be determined.

Although Robinson correctly predicted that tryptophan was involved in the biosynthesis of the ergot alkaloids, his speculation that the remaining five carbon atoms were derived from succinate and formate was not supported by subsequent labelling studies. The true biosynthetic origin of this five-carbon unit was demonstrated simultaneously in 1960 by three groups[121–123]. Gröger[124] initially supplemented saprophytic cultures with cold mevalonic acid, but found no increased production of clavine alkaloids. However, using isotopically labelled precursors, it was readily established that the isoprenoid intermediates mevalonic acid[121, 123, 125], dimethylallyl pyrophosphate and isopentenyl pyrophosphate[109, 119, 126] were incorporated into both the clavine and lysergic acid families of ergot alkaloids. Generally, the incorporation efficiencies of labelled[108, 125, 127, 128] DL-mevalonic acid (that is 3RS-mevalonate) have been lower than those reported for tryptophan, with values in the range 0.3–2% being typical for relatively good alkaloid-producing strains. Table 4 shows incorporation efficiencies obtained for mevalonate and acetate, when fed to clavine-producing strains. Not surprisingly, poor incorporation values have also been reported (such as 0.02%) in a strain of *C. purpurea* which produced a low yield of the peptide alkaloid ergosine[128]. This is probably a consequence of the diversity of the demands on the metabolic pools of mevalonate, especially in the ergot fungi where high levels of sterols accumulate in the sclerotial fungal mycelium, during the active period of alkaloid biosynthesis.

Degradation studies by the three initial groups of investigators, confirmed the specific isoprenoid origin of the non-tryptamine moiety. Birch *et al.*[123, 127] concluded from feeding (3RS)-2-^{14}C-mevalonic acid lactone, that 86% of the activity of elymoclavine was located at the C-17 hydroxymethyl substituent and 8.6% at C-7 of the ergoline nucleus. This result has been confirmed by other degradation studies which always showed some randomization between the C-17 and C-7 positions, of the order of 9:1. Baxter *et al.*[128] found that 90% of the activity of 2-^{14}C-mevalonate-derived ergosine was located in the lysergic acid moiety, and Castagnoli *et al.*[108] reported a similar distribution of activity in lysergylmethylcarbinolamide obtained from the same precursor. Floss *et al.*[129] showed that only the 3R-enantiomer of mevalonate is utilized and this has been confirmed by Seiler *et al.*[130]

Table 4. Precursors of the non-tryptamine moiety[123]

Precursor	Product	Incorporation efficiency (percentage of total precursor)
DL-(2-^{14}C)-Mevalonic acid	Agroclavine	1.0
1-^{14}C-Sodium acetate	Agroclavine	0.43
2-^{14}C-Sodium acetate	Agroclavine	0.83
DL-(2-^{14}C)-Mevalonic acid	Elymoclavine	1.4
1-^{14}C-Sodium acetate	Elymoclavine	1.0
2-^{14}C-Sodium acetate	Elymoclavine	2.1

In general the incorporation of labelled mevalonate and acetate into the clavine alkaloids, has been considerably lower than that of tryptophan; representative acetate and mevalonate incorporation values are shown.

Double-labelling experiments[121, 131] using mevalonic acid with ^{14}C and ^{3}H at selected positions, have also clarified the details of incorporation of this precursor. This appears to follow the same initial steps as in terpene biosynthesis, involving conversion to isopentenyl pyrophosphate and dimethylallyl pyrophosphate[109, 119, 126], following loss of the carboxyl group as demonstrated by Baxter *et al.*[132] The relative incorporations of mixtures of (3RS)-2-^{14}C-mevalonate and mevalonate with tritium substituted at carbon atoms 2, 4 and 5, have been examined in detail. While no relative loss of C-2 protons was observed during the conversion of 2-^{14}C,[3]H-mevalonate[15] to 17-^{14}C,[3]H-festuclavine and pyroclavine, in related experiments with 4-ditritiomevalonate and the 5-ditritio analogue[15, 131], both mixtures showed a 50% loss of tritium. The results are predictable from structural considerations and more recently the stereospecificity of the proton loss at C-4 and C-5 has been determined in a series of elegant studies.

Thus, Seiler *et al.*[130] fed a mixture of (3RS)-2-^{14}C-mevalonate and enzymatically synthesized (3RS,5R)-5-[3]H_1-mevalonate to *Pennisetum typhoideum* and observed a considerable relative loss of tritium, whereas the mixture with (3R,5S)-5-[3]H_1-mevalonate retained all of the tritium, thus establishing that the C-10 proton of the $\Delta^{8,9}$-ergolene alkaloids is specifically derived from the 5-pro-S-hydrogen of mevalonate. Floss *et al.*[133] have independently reached the same conclusion.

Table 5. Incorporation of stereospecifically labelled mevalonate
and chanoclavine-I[129,134]

Precursor	Product	Result of feeding experiment
(3R,4R)-2-^{14}C-4-^{3}H-Mevalonic acid lactone	Elymoclavine	70% [3]H retained
(3R,4S)-2-^{14}C-4-^{3}H-Mevalonic acid lactone	Elymoclavine	0.1% [3]H retained
(3R,4R)-2-^{14}C-4-^{3}H-Mevalonic acid lactone	Chanoclavine-I	100% [3]H retained
(3R,4S)-2-^{14}C-4-^{3}H-Mevalonic acid lactone	Chanoclavine-I	1.2% [3]H retained
(3R,4R)-2-^{14}C-4-^{3}H-Mevalonic acid lactone	*N*-(α-Hydroxyethyl)-lysergamide	70% [3]H retained
7-^{14}C-9-^{3}H-Chanoclavine-I	Elymoclavine	69% [3]H retained

The table shows:

1. the stereospecific elimination of the 4S-hydrogen of (3R,4S)-mevalonic acid in forming chanoclavine-I

2. the partial retention (70%) of the 4R-hydrogen from 9-^{3}H-chanoclavine-I in the biosynthetic steps leading to ring D formation.

The C-9 proton of elymoclavine had been earlier shown by Floss *et al.*[129, 134] in an analogous double-labelling study, to arise from the 4-pro-R-hydrogen of mevalonate to the extent of a 70% retention of the original precursor label (see Table 5). This established that the isomerization of the isopentenyl to the dimethylallyl pyrophosphate follows the same steric course as was observed in the extensive investigations of the biosynthesis of cholesterol intermediates by Popjak and Cornforth[143], namely that the C-2 of mevalonate becomes the methyl of dimethylallyl pyrophosphate which is *cis* to the olefinic proton. This necessitates an isomerization of the double bond during the subsequent steps forming chanoclavine-I, as will be discussed in the next section.

B. Advanced Intermediates

Our understanding of the nature of the intermediates involved in the biosynthetic sequence leading to the formation of the ergoline ring system, has advanced considerably in recent years. Plieninger[135], Agurell[136] and Baxter[131] suggested that isopentenylation of the indole moiety at the 4-position, may be an early stage in ergoline biosynthesis and this hypothesis has now been substantiated. The postulated intermediate L-4-dimethylallyltryptophan (**13**) was first synthesized in Plieninger's laboratory[137].

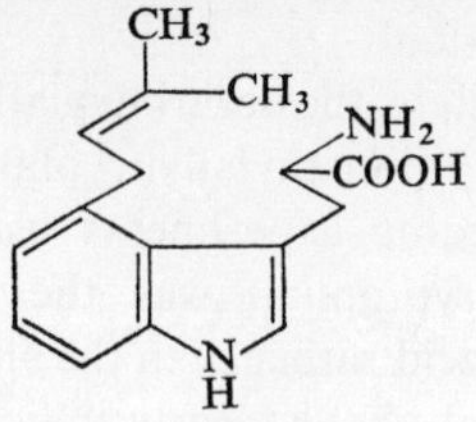

Figure 11.

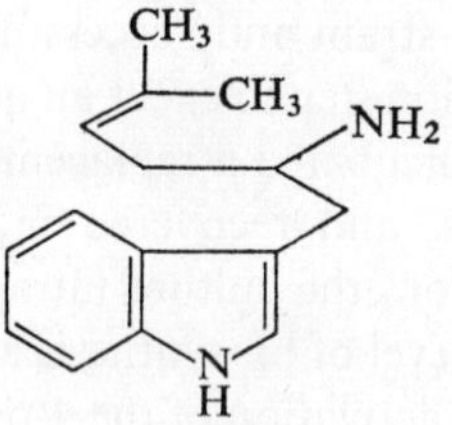

(13)

Figure 12.

(14)

An alternative hypothetical intermediate was the tryptamine side-chain-alkylated product 1-(β-indolyl)-2-amino-5-methyl-hex-4-ene (**14**) which was suggested and first synthesized by Weygand *et al.*[138, 139]

Early experiments with labelled 4-dimethylallyltryptophan[135] and 1-(β-indolyl)-2-amino-5-methyl-hex-4-ene[138], both gave rise to labelled elymoclavine. Plieninger[140] in a later series of experiments compared the incorporation efficiencies of the two intermediates, one ^{14}C-labelled and the other ^{3}H-labelled, when fed simultaneously. In all of these experiments, 4-dimethylallyltryptophan proved to be the more efficient precursor. Incorporation values obtained with 4-dimethylallyltryptophan were quite high, with 5–14% reported by Plieninger *et al.*[126, 135] and 15–20% by Floss and Robbers[141], and by Agurell[106]. Clearly, 4-dimethylallyltryptophan rivals tryptophan as an efficient precursor, and Agurell[106] in competitive feeding experiments using equimolar amounts of DL-^{14}C-4-dimethylallyl-tryptophan and DL-^{3}H-tryptophan, concluded that the former was actually 5 to 10 times more efficient than tryptophan as a precursor in the strain of *C. paspali* which he examined. Although both postulated intermediates labelled elymoclavine, Plieninger *et al.*[126] showed that doubly labelled 4-dimethylallyltryptophan was incorporated intact into agro-clavine, whereas studies by Weygand *et al.*[139] indicated that radioactive 1-(β-indolyl)-2-amino-5-methyl-hex-4-ene was actually degraded prior to incorporation into the alkaloid.

The next logical intermediate, the decarboxylation product of 4-dimethyl-allyltryptophan, namely 4-dimethylallyltryptamine was suggested by Plieninger *et al.*[126] In feeding experiments using both potential inter-mediates, 4-dimethylallyltryptophan was the more efficient precursor, suggesting that the amino acid rather than the amine was the natural inter-mediate. The evidence cited above clearly indicated that 4-dimethylallyl-tryptophan functions as an early intermediate in the biosynthesis of the ergoline ring system, and this metabolite was subsequently isolated from *Claviceps* cultures simultaneously by Agurell *et al.*[142] and Robbers *et al.*[141] Agurell used a *Pennisetum* strain and successfully utilized ethionine as an inhibitor of the N-methylation step to effect an accumulation of 4-dimethyl-allyltryptophan. Robbers incubated a replacement culture in an anaerobic environment for 48 hours, and recovered most of the accumulated 4-dimethylallyltryptophan from the culture filtrate.

Still outstanding at this level of biosynthesis are two main problems. The first concerns the nature of activation of the 4-position of the indole moiety of tryptophan prior to isopentenylation, while the other concerns the nature of ring C closure. The latter problem is of particular interest in that, as pointed out earlier (p. 71), it involves a loss of the carboxyl group of tryptophan with an inversion of configuration at the α-carbon atom during its conversion to C-5 of the ergoline nucleus. In relation to the mechanism of these early stages it may be noted that several workers have recently developed cell-free systems capable of carrying out a number of steps in the biosynthetic sequence.

Thus Cavender and Anderson[119] have obtained from *C. purpurea* PRL-1980, a partially purified cell-free supernatant preparation capable of synthesizing agroclavine, elymoclavine and chanoclavines-I and -II from the primary precursors tryptophan, isopentenyl pyrophosphate and methio-nine. Clearly such a system has the potential to yield valuable results con-cerning stages in the biosynthetic sequence, and already Hsu and Anderson[72] have been able to demonstrate the presence of an aerobic NADPH-dependent agroclavine hydroxylase.

With the discovery of chanoclavine, a tricyclic clavine with an unclosed D-ring[144], it was logical to consider that this served as a late intermediate in the biosynthesis of the tetracyclic ergoline ring system. Early experiments using labelled chanoclavine resulted however in some confusion. Agurell[15, 145] and Baxter *et al.*[131] reported negative or inconclusive results, while Mothes[146] observed good incorporation of chanoclavine into the clavine alkaloids. The problem was not resolved until it was realized that 4

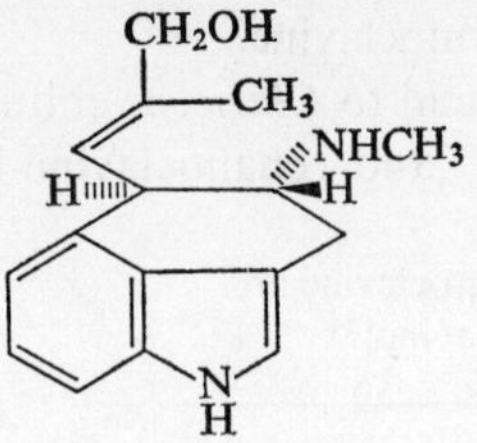

(2) 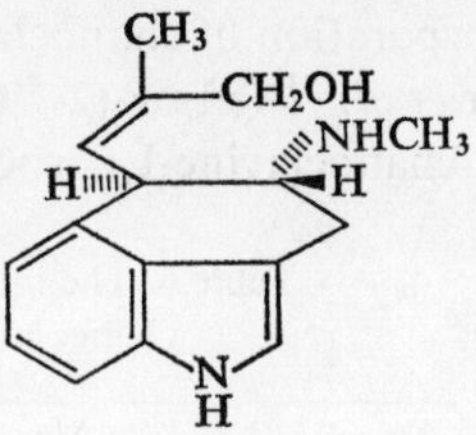   (15)

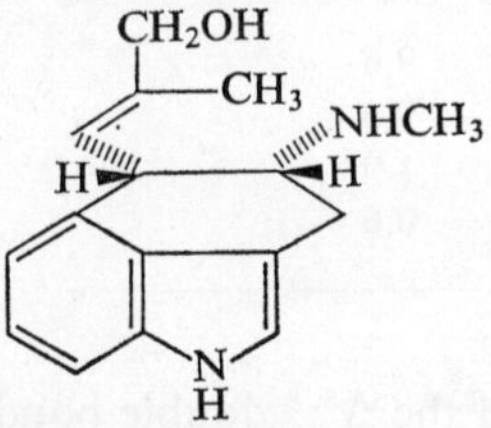

(16) 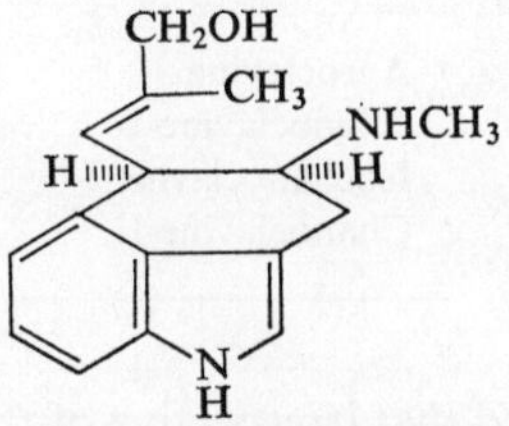   (17)

Figure 13.

isomeric chanoclavines existed in nature[26], namely chanoclavine-I **(2)**, its geometric isomer isochanoclavine-I **(15)** and (+)-chanoclavine-II **(16)**, the C-10 epimer of chanoclavine-I, together with its enantiomer (−)-chanoclavine-II **(17)**, which was isolated as the racemate (Fig. 13); the stereochemistry of structures **2** and **15** was confirmed by Acklin *et al.*[147]

From a knowledge of the lysergic acid labelling pattern obtained from DL-(2-14C)-mevalonic acid and the known stereochemistry of the chanoclavines, isochanoclavine-I would be favoured as the logical intermediate allowing the most facile closure[148] of the D-ring. However, specific labelling experiments have demonstrated that chanoclavine-I alone functions as an obligatory intermediate. Floss *et al.*[129] compared the incorporation efficiencies of the chanoclavines and agroclavine as intermediates leading to the formation of elymoclavine. The results are shown in Table 6. Arigoni[149] has reported that both desoxychanoclavine-I and nor-desoxychanoclavine-I are ineffective precursors of the ergoline nucleus. Consequently hydroxylation may occur prior to ring C formation involving an oxidative cyclization (for details see Figure 18). The oxygen atom of the C-17-hydroxymethyl group of chanoclavine-I was shown not to originate from water and was thus assumed to arise from molecular oxygen[150].

Generally, incorporation levels obtained with chanoclavine-I feeding experiments have been high, with values of up to 40%[148, 151, 152]. Ogunlana et al.[153], using cell-free systems produced from *Claviceps* cultures, obtained 20% incorporation of chanoclavine-I into elymoclavine.

Gröger et al.[151] fed DL-(2-^{14}C)-mevalonic acid to *Claviceps* cultures and isolated chanoclavine-I, isochanoclavine-I and chanoclavine-II and

Table 6. The incorporation of agroclavine and
chanoclavine into elymoclavine[129]

Intermediate	Incorporation of label into elymoclavine (%)
Agroclavine	9.6
Chanoclavine-I	9.0
Isochanoclavine-I	1.9
Chanoclavine-II	0.6

observed that irrespective of the geometry of the $\Delta^{8, 9}$ double bond, about 90% of the radioactivity of the alkaloid was located in the *C*-methyl group and 7% in the hydroxymethyl group. Floss et al.[154] then examined the incorporation of ^{14}CH$_3$-labelled chanoclavine-I into elymoclavine, in which it was demonstrated by degradation that the label appeared exclusively in the C-17-hydroxymethyl group of elymoclavine (see Fig. 14).

Fehr and coworkers[148] in an analogous series of experiments prepared (^{14}CH$_2$OH)-chanoclavine-I by reductive cleavage of 17-^{14}C-elymoclavine, prior to incorporation into 6-methylergol-8-ene-8-carboxylic acid with 2% efficiency. Degradation of the labelled acid established that 96.4% of the activity was located at C-7 and about 6.4% in the C-17-carboxyl group. (^{14}CH$_2$OH)-Isochanoclavine-I was not significantly incorporated into elymoclavine; these results are shown in Fig. 15.

Their results showed that during the normal biosynthetic conversion of chanoclavine-I to elymoclavine, a *cis–trans* isomerization of the $\Delta^{8, 9}$ double bond takes place. This is in addition to the isomerization apparently

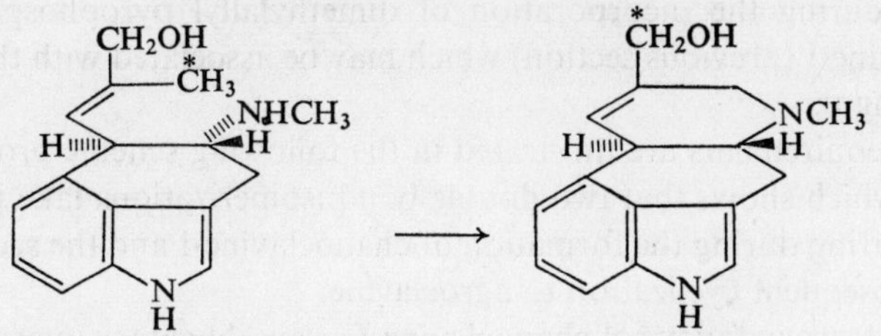

$^{14}CH_3$-Chanoclavine-I   →   Elymoclavine

Figure 14. Incorporation of chanoclavine-I into elymoclavine[154].

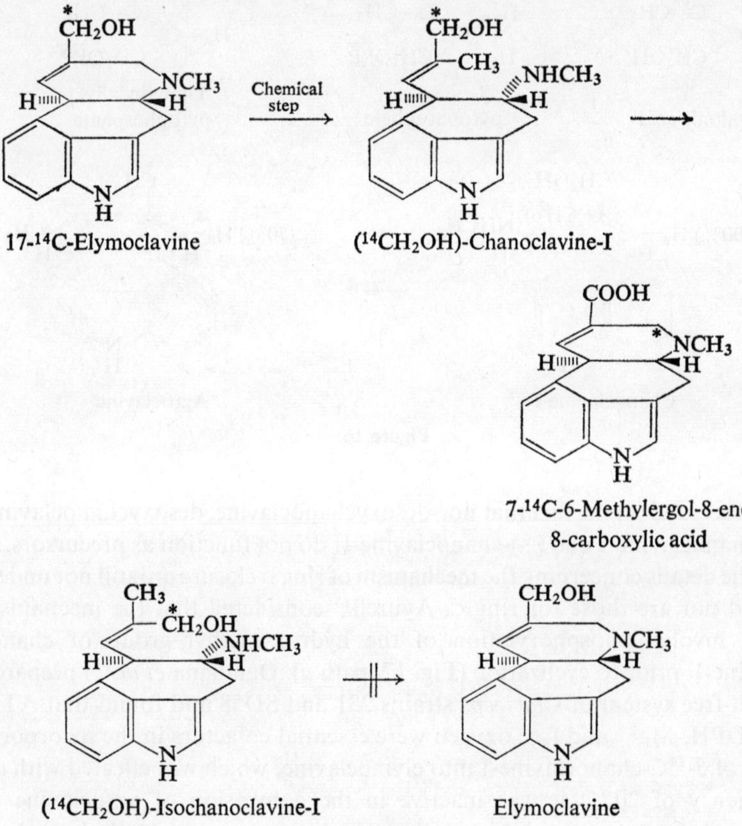

17-^{14}C-Elymoclavine   —Chemical step→   ($^{14}CH_2OH$)-Chanoclavine-I   →

7-^{14}C-6-Methylergol-8-ene-
8-carboxylic acid

($^{14}CH_2OH$)-Isochanoclavine-I   ⇏   Elymoclavine

Figure 15.

involved during the incorporation of dimethylallyl pyrophosphate into chanoclavine-I (previous section) which may be associated with the formation of ring C.

These requirements are illustrated in the following scheme proposed by Floss[155] which shows that two double bond isomerizations take place, the first appearing during the formation of chanoclavine-I and the second during its subsequent cyclization to agroclavine.

The evidence in favour of chanoclavine-I as an obligatory intermediate is based upon (a) high incorporation efficiencies which argue against degradation prior to incorporation, (b) the specific labelling pattern of the ergolene

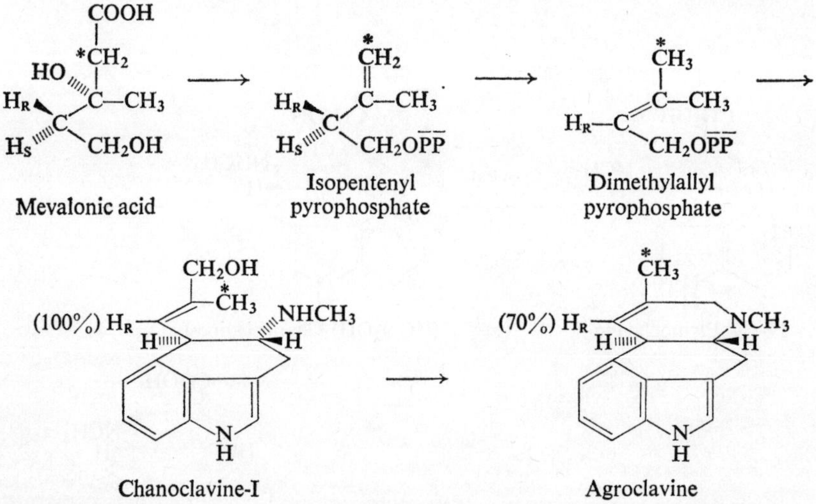

Mevalonic acid   Isopentenyl pyrophosphate   Dimethylallyl pyrophosphate

(100%) Chanoclavine-I   (70%) Agroclavine

Figure 16.

product, and (c) the fact that nor-desoxychanoclavine, desoxychanoclavine, isochanoclavine-I and ($\pm$)-chanoclavine-II do not function as precursors.

The details concerning the mechanism of ring D closure are still not understood nor are those for ring C. Agurell[15] considered that the mechanism may involve phosphorylation of the hydroxymethyl group of chanoclavine-I, prior to cyclization (Fig. 17, path a). Ogunlana et al.[153] prepared a cell-free system of *Claviceps* strains 231 and SD58 and found that ATP, NADPH, Mg^{2+} and free oxygen were essential cofactors in the incorporation of 5-^{14}C-chanoclavine-I into elymoclavine, which was effected with an efficiency of 20%; it was inactive in the conversion of agroclavine to elymoclavine. On the basis of these findings it was argued that phosphorylation of chanoclavine-I is followed by 8,17-epoxidation, providing a possible mechanism for ring D closure. However, Floss *et al.*[157] reported that one of the methylene-hydrogen atoms of chanoclavine-I was lost in closure of ring D, and extended these findings[158], again using cell-free preparations of *Claviceps* species. This latter group synthesized 17-^{3}H-chanoclavine-I-aldehyde, and found that 40% was incorporated into elymoclavine compared to an incorporation efficiency of about 10% for labelled chanoclavine-I. It was further shown by degradation of the synthesized elymoclavine, that the tritium label was as expected exclusively located on carbon-17. Alternative mechanisms for the cyclization of chanoclavine, involving addition of an enzyme[159] (Fig. 17, path *b*), or phosphate substitution[160] (Fig. 18, path *c*) would also accommodate the required inversion step, but would not readily account for the observed 70% retention of tritium at C-9 derived from 9-^{3}H-chanoclavine-I.

A further possible mechanism[161] involving initial cyclization to a pyrollidine derivative followed by ring enlargement via an aziridinium intermediate could also account for the labelling data, especially if the final step is spontaneous with a non-stereospecific elimination of a C-9 proton, the observed 70% retention representing a deviation from the predicted 50% due to a primary isotope effect (see Fig. 18, path *d*). This figure also outlines two schemes for the possible origin of ring C involving stereospecific oxidative cyclization of dimethylallyltryptophan to chanoclavine-I. It is not known whether methylation takes place before or after ring C is formed.

C. Interrelationships of the Ergot Alkaloids

Until the 1950's the lysergic acid alkaloids were the only types known to be present in ergot, but following the discovery by Abe of agroclavine the first of the clavine alkaloids, many other clavines have been subsequently isolated (Table 2) by Abe, Hofmann and Stoll and their colleagues[3, 4, 29, 30, 33, 34].

The methods used to study the interrelationships of the clavine and lysergic acid alkaloids vary, but in general they have followed the techniques first applied by Agurell[15, 162]. The potential precursor alkaloid, for example agroclavine, was labelled biosynthetically using DL-(2-^{14}C)-mevalonic acid and isolated prior to reincubating with another *Claviceps*. After a suitable period, the distribution of label among the newly isolated alkaloids of the second fungus was then determined. Unfortunately, in these studies degradations have not been carried out to ascertain the exact labelling pattern, with a single exception[163], but generally it has been considered that the high conversions recorded leave little doubt of a direct precursor function.

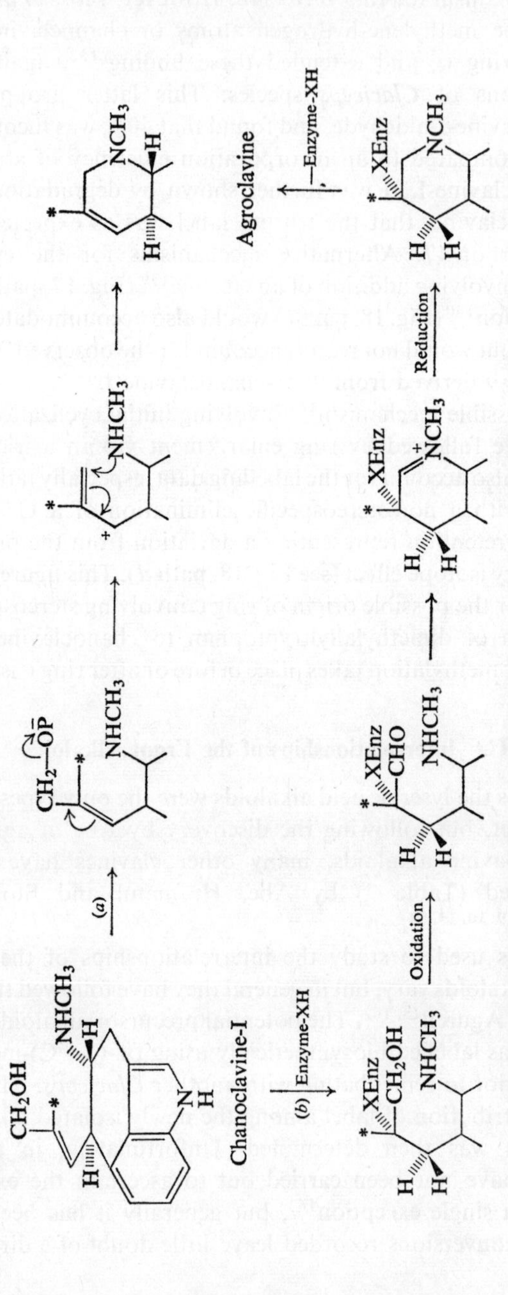

Figure 17. Hypothetical mechanisms (a)[15] and (b)[159] for the cyclization of chanoclavine-I to agroclavine.

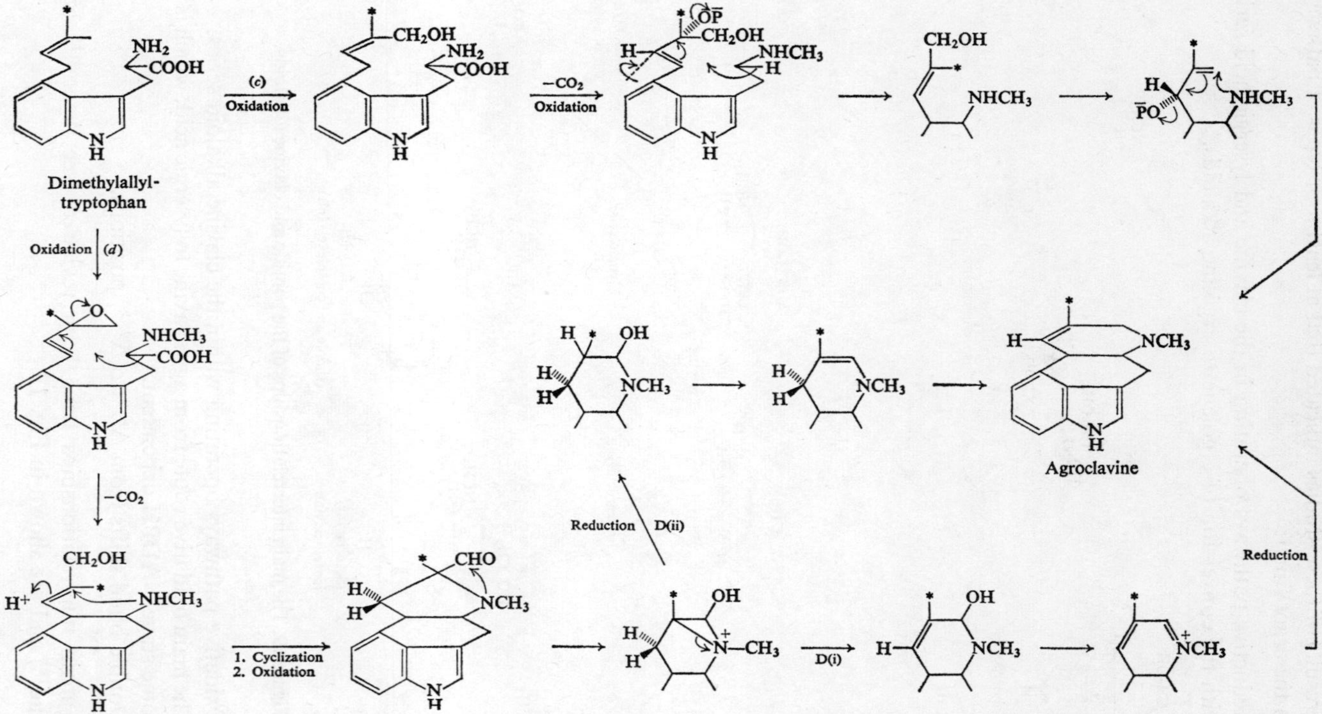

Figure 18. Hypothetical mechanisms (c)[160] and (d)[161] for the formation of rings c and d of agroclavine. (A direct ring d cyclization to elymoclavine has been postulated[153].)

The results of these studies indicated that at least four types of pathways exist in the ergot fungi:

1. Reductive pathways operating in the clavine and lysergic acid series with the conversion of ergolene to ergoline alkaloids.

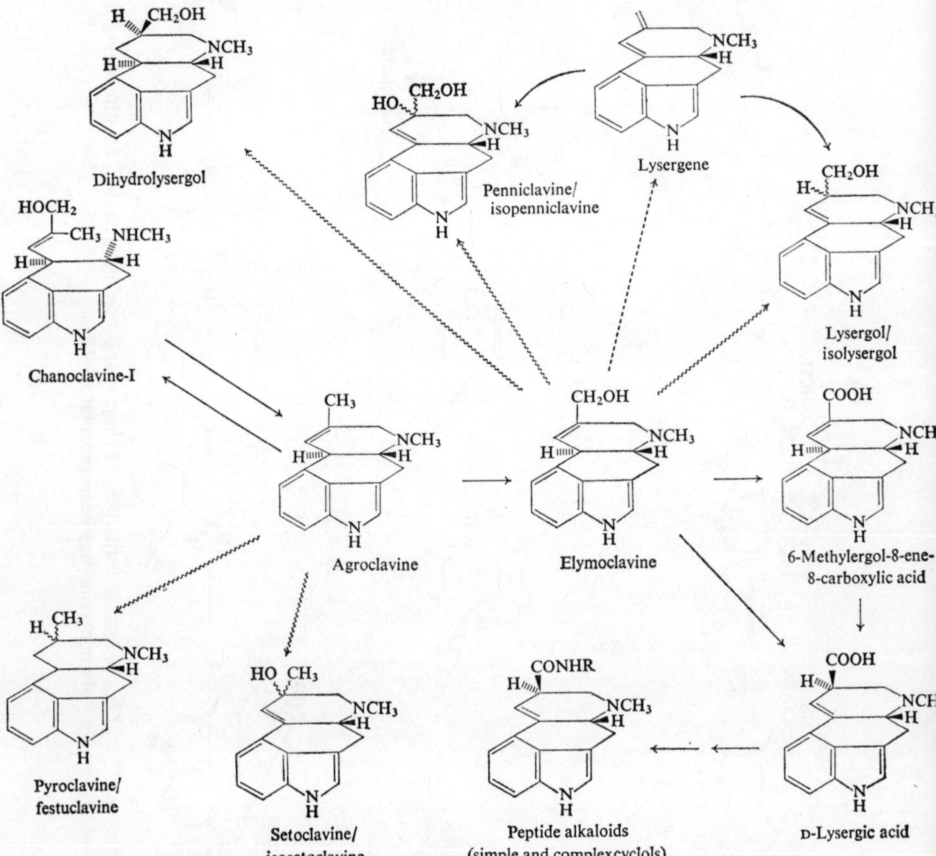

Figure 19. The main interrelationships of the peptide and clavine alkaloids.

2. Oxidative pathways operating within the clavine alkaloid series.
3. The main oxidative route from agroclavine to lysergic acid, which is at some stage NADPH dependent.
4. Double bond shifts from $\Delta^{8,9}$ to $\Delta^{9,10}$ positions.

The main interrelationships have been reviewed (see Agurell[15] and Ramstad[16]) and are shown in Fig. 19.

1. *Double bond reduction*

Until recently the dihydroclavine alkaloids were the only known true saturated ergoline derivatives, for example pyroclavine, festuclavine, costaclavine and fumigaclavines A and B (Table 2). Recently the first dihydrolysergic acid alkaloid, dihydroergosine[164] has been reported.

2. *Hydroxylation and epoxidation of ring D of the clavine alkaloids*

Studies of the mechanism of hydroxylation of ring D of agroclavine and elymoclavine have been particularly fruitful (see Ramstad[16]). However, unlike the oxidation of agroclavine to elymoclavine which is quite restricted

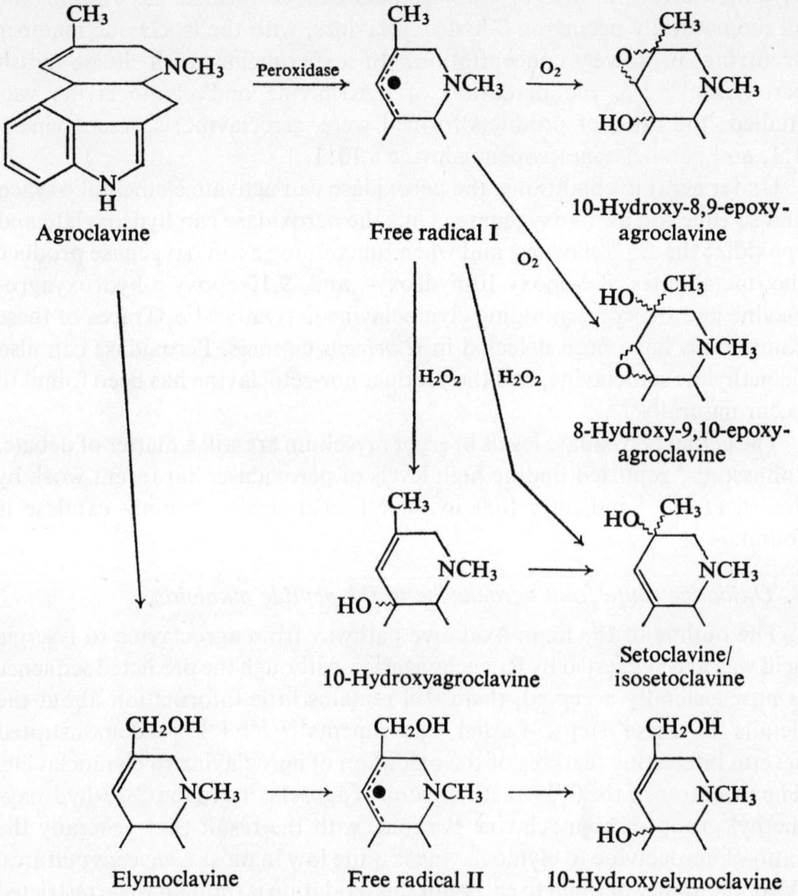

Agroclavine

Free radical I

10-Hydroxy-8,9-epoxy-
agroclavine

8-Hydroxy-9,10-epoxy-
agroclavine

10-Hydroxyagroclavine

Setoclavine/
isosetoclavine

Elymoclavine

Free radical II

10-Hydroxyelymoclavine

Figure 20.

4

to *Claviceps* and which is NADPH dependent[72], the ability to effect 8-hydroxylation of agroclavine and elymoclavine is in no way restricted, and is not uncommon in fungi[15,165-167] and plant homogenates[168,169]. Due to the work of Agurell, Ramstad, Taylor and their coworkers[168,170,171], some insight into the mechanism of this hydroxylation has been achieved; recent reports described enzymatic conversion *in vitro*[168,171,172]. It was deduced from their studies that the enzyme peroxidase attacks the labile C-10 hydrogen, abstracting one electron, and generating the free radicals I and II (see Fig. 20). These radicals were considered to react with hydroxyl radicals from hydrogen peroxide, forming 8- and 10-hydroxyclavines.

Isomerization of the double bond from $\Delta^{8,9}$ to $\Delta^{9,10}$ results in a new asymmetric centre at C-8. Two isomers can be formed, accounting for all the naturally occurring 8-hydroxyclavines, with the isoclavine member occurring in lower concentration. In experiments with horse-radish peroxidase[168,172], the oxidation of agroclavine and elymoclavine was studied; the ratio of products formed were, setoclavine:isosetoclavine = 3:1, and penniclavine:isopenniclavine $\leqslant$ 10:1.

Under aerobic conditions, the peroxidase can activate elemental oxygen and so function as an oxygenase. Thus the peroxidase can hydroxylate and epoxidize the $\Delta^{8,9}$-clavines, and when functioning as an oxygenase produce the metabolites 8,9-epoxy-10-hydroxy- and 9,10-epoxy-8-hydroxyagro-clavine and the corresponding elymoclavine derivatives[172]. Traces of these compounds have been detected in *Claviceps* cultures. Peroxidase can also demethylate setoclavine, and the product nor-setoclavine has been found to occur naturally[168].

The actual peroxidase levels in ergot mycelium are still a matter of debate. Johnsson[173] reported finding high levels of peroxidase, but recent work by Jindra *et al.*[174] indicates that in their fungal strains, mainly catalase is found.

3. *Oxidative route from agroclavine to the peptide alkaloids*

The outline of the main oxidative pathway from agroclavine to lysergic acid was first suggested by Rochelmeyer[175]. Although the predicted sequence is now generally accepted, there still remains little information about the details of these steps. Earlier experiments[145,146,165,176] demonstrated several interesting features of the oxidation of agroclavine to elymoclavine. The oxidation of the C-17-methyl group of agroclavine to the C-17-hydroxy-methyl group of elymoclavine is rapid, with the result that generally the ratio of agroclavine to elymoclavine is quite low in most *Claviceps* cultures. Also, the ability of fungi to carry out this oxidation is thought to be restricted to *Claviceps* species and perhaps to some other fungi producing agroclavine and elymoclavine. This is in contrast to the ability of organisms to perform 8-hydroxylation of the clavine alkaloids, which is more widespread.

Agurell[15] using a cell homogenate of *Claviceps* 47A, demonstrated the oxidation of ^{14}C-agroclavine to elymoclavine, obtaining a 60% conversion after 14 hours incubation. This high level of conversion was considered to support the direct intermediate role of agroclavine and to preclude prior degradation. Experiments with submerged cultures using doubly labelled intermediates[15] have confirmed these findings. Further, Floss *et al.*[150] obtained evidence that the C-17-hydroxymethyl-oxygen of elymoclavine originates from molecular oxygen. Jindra *et al.*[174] considered that this oxidation differed from the 8-hydroxylation in that it did not involve a substrate-specific peroxidase, although Johnsson[173] claimed that he found a correlation between the ratio of elymoclavine and agroclavine and the amount of peroxidase in the fungal mycelium, which was thought to indicate an active role by peroxidase in this oxidation.

Ogunlana *et al.*[153] recently reported the conversion of chanoclavine-I to elymoclavine in crude supernatant fractions from *Claviceps* strain 231. No labelled agroclavine was found after the incubation and the extract was incapable of converting agroclavine into elymoclavine. As a result, these workers considered the possibility that at least in this strain, agroclavine may not be an intermediate in the biosynthetic sequence leading from chanoclavine-I to elymoclavine. Cavender and Anderson[119] prepared a partially purified cell-free supernatant fraction from *C. purpurea* PRL-1980 which catalysed the conversion of tryptophan to chanoclavine-I, chanoclavine-II, agroclavine and elymoclavine, but some purification of the supernatant was necessary in order to bring about these reactions. Aerobic incubation of this enzyme system with [14]C- or ^{3}H-agroclavine and liver concentrate or an NADPH-generating system (NADP, glucose-6-phosphate and glucose-6-phosphate dehydrogenase), resulted in the conversion of agroclavine to elymoclavine with efficiencies of 15% for an 18 hour incubation. By comparing the rates of synthesis of elymoclavine from agroclavine *in vivo* and *in vitro*, it was concluded that in this strain agroclavine was indeed an intermediate leading to elymoclavine, catalysed by an NADPH-dependent agroclavine hydroxylase.

Similarly, ^{14}C-elymoclavine and ^{3}H-elymoclavine have been found[136, 176, 177] to incorporate into lysergic acid and the corresponding moiety of the peptide alkaloids. Other studies by Agurell and Johnsson[178] tested both agroclavine and elymoclavine as precursors of ergometrine and ergotamine, and the results were interpreted as supporting the biosynthetic sequence agroclavine to elymoclavine to peptide alkaloids. High specific incorporations of the order of 29–55% were obtained. Mothes and

Gröger[163,177] have unequivocably established the role of the clavine alkaloids in the biosynthesis of lysergic acid derivatives. 17-^{14}C-Elymoclavine was fed to strains producing peptide alkaloids, and degradation of the derived lysergic acid resulted in the finding that 60% of the activity was present in the carboxyl group. These observations have been confirmed by Floss *et al.*[179] Agurell[106,107,180] followed the complete sequence from tryptophan to lysergylmethylcarbinolamide in *C. paspali* by competitive feeding experiments, and found that all intermediates between tryptophan and the lysergic acid alkaloid, except agroclavine, showed an incorporation supporting this sequence (Fig. 21). Although no details of the oxidation of

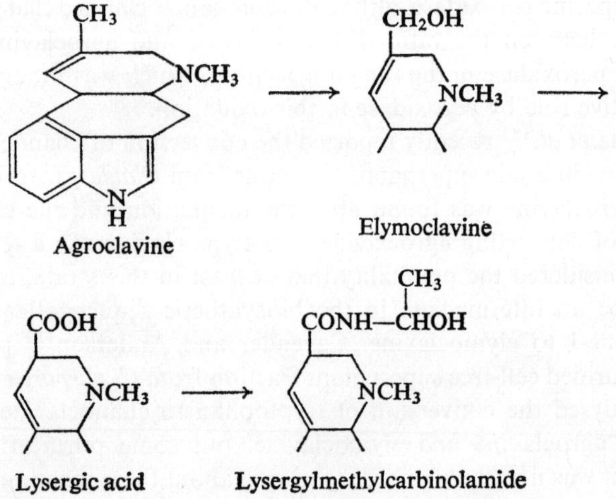

Agroclavine

Elymoclavine

Lysergic acid

Lysergylmethylcarbinolamide

Figure 21.

elymoclavine are known, Floss *et al.*[179] have argued that it may take place via an aldehyde intermediate leading to lysergic acid.

The role of 6-methylergol-8-ene-8-carboxylic acid is not understood. That it may play a role as an intermediate in the biosynthesis of the lysergic acid alkaloids was first shown by Agurell[180]. However, its position in the main oxidative scheme is uncertain as yet. Similarly, the biosynthetic status of the recently discovered clavicipitic acid (3) has not been ascertained.

Lysergic acid appears to function as a precursor of the lysergyl nucleus of the simple and complex peptide alkaloids. Agurell[107,180] found that lysergic acid functioned as a precursor of ergometrine and lysergylmethylcarbinolamide; Minghetti *et al.*[181] also reported that ^{3}H-lysergic acid was efficiently incorporated into ergometrine.

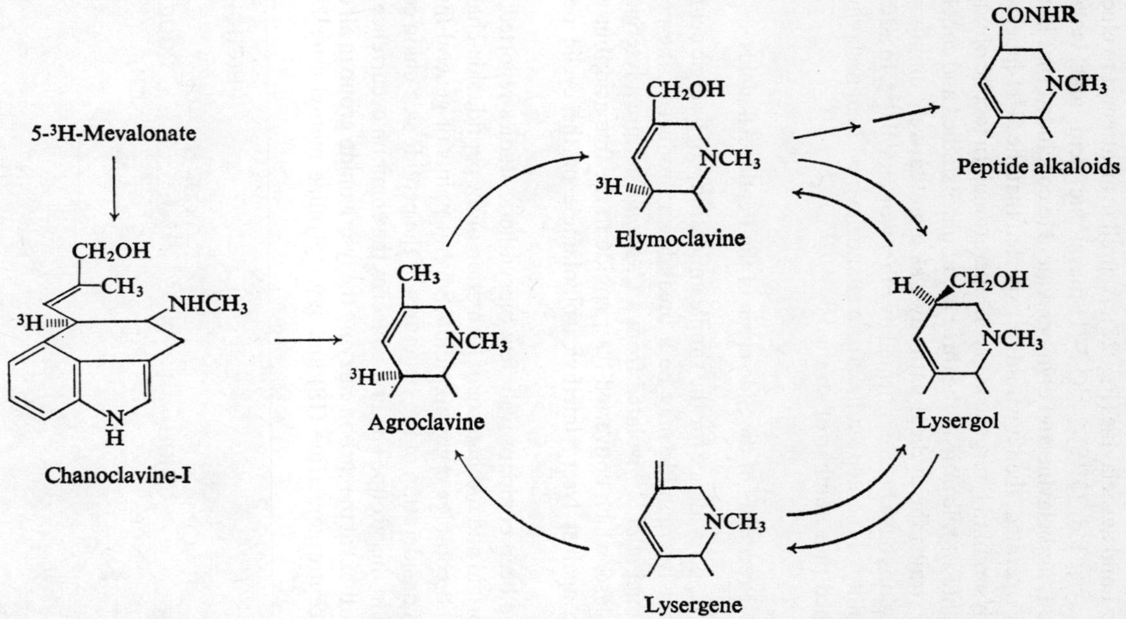

Figure 22. The cyclical pathway of Abe. This scheme would predict the loss of label from 10-³H-chanoclavine-I during a cycle passing through $\Delta^{8,9}$- to $\Delta^{9,10}$-clavines.

Abe *et al.*[30, 32, 182, 183] have examined a range of *Claviceps* strains and found yet another type of pathway which appears to operate in a cyclical manner, and which shows a novel step including the interconvertibility of chanoclavine-I and agroclavine (Fig. 22). Agurell in an attempt to elucidate these pathways, fed (3RS)-2-^{14}C-5-^{3}H-mevalonic acid and isolated[15] chanoclavine-I, agroclavine and elymoclavine. He concluded that if a cyclical pathway operated, then it would be expected that the 5-^{3}H-label would be lost if the lysergol to lysergene step was operational. In fact he obtained a nearly constant isotopic ratio in the three metabolites studied and concluded that in these strains the cyclical pathway was quantitatively unimportant. These conclusions have been confirmed by other workers[151, 184]. In addition, the irreversibility of the step leading from chanoclavine-I to agroclavine has been confirmed in a number of strains[131, 145, 178].

D.   Precursors of the Side-chain of the Peptide Alkaloids

There is, as yet, no evidence for the simple amidation[15] of lysergic acid, and it appears that lysergamide may be a breakdown product of the peptide alkaloids. Amici *et al.*[184a] isolated from a *C. paspali* culture a lysergic acid amide hydrolase, which converted lysergamide to the free acid, but which was inactive against lysergylmethylcarbinolamide or the cyclic peptide alkaloids.

There have been comparatively few biosynthetic studies reported, concerning the origin and mechanism of formation of the peptide side-chains of the simple and complex ergot alkaloids. Several points of general interest arise from a consideration of the structures of the peptide side-chain constituents found in this group. The first concerns the common occurrence of an oxygen substituent at the carbon carrying the lysergamido group in all of the alkaloids except ergometrine (**18**) and lysergylvaline methyl ester[196] (**19**) (Fig. 23).

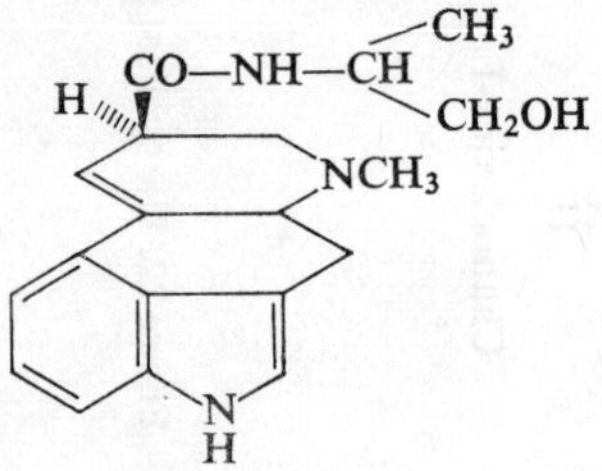

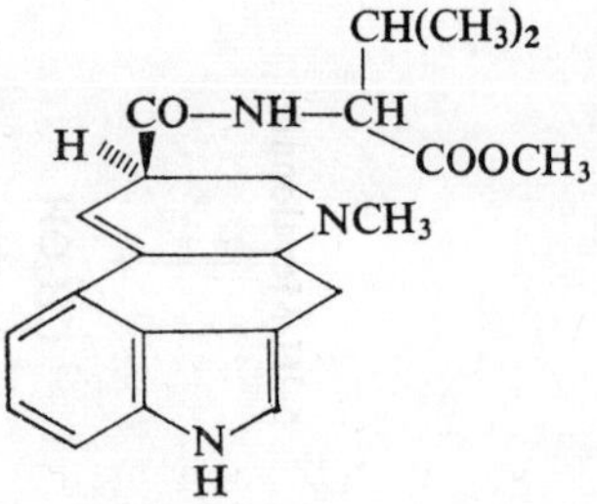

Figure 23.

Agurell[15] pointed out that it is unlikely that a hydroxylated amino acid is a direct precursor, but rather that hydroxylation is probably the result of a secondary process. This conclusion is consistent with precursor feeding experiments (see later). Ramstad[16] pointed out that whatever the origin and mechanism of introduction of the oxygen substituent at the α-position of the first amino acid, it also serves the important function of stabilizing the cyclic side-chain moiety of the di- and tripeptide ergot alkaloids[185]. This involves the formation of cyclols, as in the tripeptide group or a six-membered lactone as in the case of ergosecaline (**12**).

Proline is always the terminal amino acid in the tripeptide alkaloids and as suggested by Ramstad[16], its incorporation into the diketopiperazine ring (see Fig. 25) characteristic of these alkaloids, may represent an early cyclization step in the formation of a cyclol tripeptide alkaloid. Proline is also a constituent of the diketopiperazines L-prolyl-L-phenylalanine from

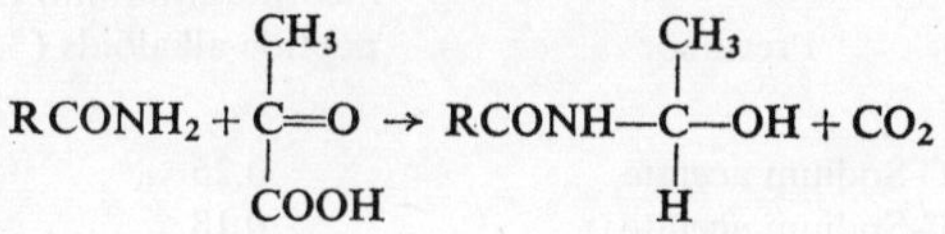

$$
\begin{array}{ccc}
\text{CH}_3 & & \text{CH}_3 \\
| & & | \\
\text{RCONH}_2 + \text{C}{=}\text{O} & \rightarrow & \text{RCONH}{-}\text{C}{-}\text{OH} + \text{CO}_2 \\
| & & | \\
\text{COOH} & & \text{H}
\end{array}
$$

Figure 24. A scheme requiring the derivation of the carbinolamide oxygen atom directly from pyruvate. Condensation of pyruvate with a derivative of lysergic acid and ammonia is proposed, although lysergamide itself does not function as a precursor[15].

Rosellinia necatrix and L-prolyl-L-valine anhydride which is a product of *Aspergillus ochraceus* as well as *R. necatrix*[187].

An alternative non-oxidative derivation of the cyclol oxygen would involve the combination of lysergic acid and ammonia with pyruvyl derivatives, in which the carbonyl-oxygen becomes in ergotamine (**11**) the heterocyclic cyclol oxygen atom, or analogously in lysergylmethylcarbinol-amide (**9**, R = CONHCHOHCH$_3$) the hydroxyl substituent as indicated in Fig. 24. This mechanism would be consistent with the results of double-labelling studies by Castagnoli *et al.*[108]

From a recent compilation (1964) of the known fungal metabolites[186], it is evident that the same amino acids which are components of the ergot peptide alkaloids, also predominate among the amino acid constituents of other nitrogenous products of fungi. The following list indicates the frequency of appearance of individual amino acids (number in parentheses) among the products included in the above list of fungal products, exclusive of the ergot alkaloids: valine (8), leucine (7), proline (6), alanine (5), isoleucine (4), phenylalanine (4), serine (4), tryptophan (3), cysteine (3), aspartate (3) and glycine (3); lysine, glutamate, histidine and threonine only appear once and methionine and arginine not at all. It should be noted that this list does not include any streptomyces metabolites.

1. N-(α-Hydroxyethyl)lysergamide (lysergylmethylcarbinolamide)

Various side-chain precursors have been suggested for this simple peptide alkaloid. Agurell[15] considered that ethylamine may be a precursor, followed by hydroxylation to give the methylcarbinolamide side-chain, but only negative results were obtained with ^{14}C-ethylamine. Other precursors were tried by Agurell using *C. paspali* 458/3 (Table 7).

It appeared from the initial studies of Castagnoli *et al.*[188] that alanine could function as a precursor of this side-chain.

Table 7. Precursor feeding experiments with *C. paspali* 458/3, producing *N*-(α-hydroxyethyl)lysergamide[15]

Precursor	Incorporation into the peptide alkaloids (%)
1-^{14}C-Sodium acetate	0.26
2-^{14}C-Sodium acetate	0.18
1-^{14}C-Acetamide	0.18
L-(U-^{14}C)-Alanine	0.40, 0.22

Gröger *et al.*[189] investigated side-chain precursors of *N*-(α-hydroxyethyl)-lysergamide using both ^{14}C- and ^{15}N-labelled substrates. In these experiments the isotopes were added after 2–3 days fermentation. With L-(U-^{14}C)-alanine feeding experiments, it was found that 57 to 71 % of the total label was located in the side-chain of *N*-(α-hydroxyethyl)lysergamide, determined as the acetaldehyde DNP derivative. ^{15}N-Labelling experiments were carried out using DL-^{15}N-alanine, DL-^{15}N-aspartic acid and L-(amide-^{15}N)-glutamine. The results showed that nearly equal incorporation of ^{15}N into the nucleus and side-chain had taken place, but in the case of ^{15}N-glutamine more of the label went into the nucleus than into the side-chain. Time-course experiments with DL-^{15}N-alanine were carried out over a period of 8 hours to 4 days, and the results indicated an active transamination of the alanine, even over short time intervals. On the basis of these results it was concluded that alanine was the precursor of the side-chain including the nitrogen atom. Castagnoli *et al.*[108] studied the incorporation of a series of precursors into the nucleus and side-chain of this alkaloid, but these experiments were carried out on the ninth day of the fermentation (see Table 8).

From these results it can be seen that relatively less alanine-carbon enters the side-chain (40%) than was observed by Floss (57–71%). Castagnoli *et al.*[108] also studied the incorporation of doubly labelled L-(U-^{14}C,^{15}N)-alanine into the side-chain of *N*-(α-hydroxyethyl)lysergamide with the results shown in Table 9.

From these values it can be seen that very little nitrogen enters the ergoline nucleus of lysergylmethylcarbinolamide, in contrast to the results of

Table 8. Incorporation of ^{14}C-labelled precursors into
N-(α-hydroxyethyl)lysergamide[108]

Precursor	Radioactivity (μCi/flask)	Efficiency of incorporation into total alkaloid (%)	Radioactivity of alkaloid (I) in side-chain (%)
L-(U-^{14}C)-Alanine	15	3.7	41.3
L-(U-^{14}C)-Alanine	25	1.7	39.1
DL-(2-^{14}C)-Alanine	25	1.4	39.7
DL-(1-^{14}C)-Alanine	15	0.31	
DL-(1-^{14}C)-Alanine	20	0.05	
2-^{14}C-Sodium pyruvate	15	1.6	38.9
2-^{14}C-Sodium pyruvate	25	1.3	20.1
2-^{14}C-Sodium pyruvate	20	3.2	37.0
1-^{14}C-Sodium pyruvate	15	0.20	
DL-(3-^{14}C)-Serine	25	2.7	0.7
1-^{14}C-Sodium acetate	25	10.1	< 0.1
DL-(2-^{14}C)-Mevalonic acid lactone	20	1.3	<0.1
DL-(3-^{14}C)-Tryptophan	15	24.0	<0.1
DL-(3-^{14}C)-Tryptophan	20	19.8	
2-^{14}C-Indole	20	5.1	<0.1
^{14}C-Sodium formate	15	6.6	0.3

Precursors were added to 9-day shake cultures and the alkaloid was recovered after incubation for a further 24 h. The methylcarbinolamide side-chain activity was determined by radioassay of the acetaldehyde dimedone derivative obtained by pyrolysis of the alkaloid.

Gröger *et al.*[189], which were obtained using younger cultures (2 to 3 days old). Unexpectedly, it was shown that nitrogen enters the lysergyl side-chain more efficiently than do carbon atoms 2 and 3 of DL-alanine. It was concluded from this work that (*a*) alanine may be a direct precursor of the alkaloid, but that its carbon skeleton and its amino group are incorporated after equilibration with their respective endogenous pools, which may be of different sizes, or (*b*) that alanine is not the direct precursor, but that its carbon skeleton and amino group are reutilized at different rates after cleavage of the C—N bond and conversion into an obligatory intermediate,

Table 9. Incorporation of L-(U-^{14}C,[15]N)-alanine into N-(α-hydroxyethyl)lysergamide[108]

Compound	Expt. no.	Specific radioactivity (^{14}C)(μCi/mmol)	Enrichment (^{15}N) (atoms % excess)	Ratio (^{15}N/^{14}C) (atoms % excess per μCi/mmol)	Side-chain ^{15}N/^{14}C ratio	
					Alanine ^{15}N/^{14}C ratio	
Alanine	1	59.4	97	1.6	Expt. 1	1.5
	2	59.9	97	1.6		
	3	20.1	97	4.8		
Methylcarbinolamide side-chain	1	2.13	5.16	2.4	Expt. 2	1.4
	2	0.95	2.19	2.3		
	3	0.16	1.96	12.0		
Lysergic acid	1	4.13	0.35	0.08	Expt. 3	2.5
	2	1.50	0.165	0.11		
	3	0.28	0.140	0.50		

The specific radioactivity of ^{14}C-alanine refers to C-2 and C-3 only and is equivalent to 66.6% of the total specific radioactivity. All results relating to the methyl-carbinolamide side-chain and the lysergic acid moiety have been corrected for dilution of the labelled material by unlabelled alkaloid.

for example pyruvate (see Fig. 24). Minghetti *et al.*[181] reported studies of the biosynthesis of the side-chain of this alkaloid using both L-(U-^{14}C)-alanine and L-^{14}C-alaninol as precursors. Both appeared to function with similar efficiencies of incorporation into the alkaloid (2.13 and 2.91 % respectively), with nearly equal specific incorporations into the carbinolamide side- chain. They concluded that L-alaninol may function as a precursor following its conversion to pyruvic acid or to a 3-carbon compound which was related to pyruvic acid.

2. Ergometrine

Biosynthetic studies of the side-chain of ergometrine have been carried out by two groups[181,190] (Table 10). Feeding experiments[190] with L-alaninol

Table 10. Biosynthesis of ergometrine[190]

| | | Specific incorporation into ergometrine (%) | |
Precursor	Incorporation into peptide alkaloid (%)	Nucleus	Alaninol side-chain
DL-(3-^{14}C)-Tryptophan	4.7	93	
D-^{14}C-Lysergic acid	6.9		
D-(8-^{3}H)-Lysergic acid	2.2		
L-(G-^{3}H)-Alanine	0.03		45
L-(U-^{14}C)-Alanine	0.13		37
1-^{14}C-Sodium acetate	0.28	92	3

Table 10 shows the results of precursor feeding experiments with a *C. paspali* ergometrine-producing strain. Degradation of the parent alkaloid for the side-chain amino acid was carried out under acidic conditions.

gave no incorporation into the alkaloid in a total of 8 experiments, and ^{3}H- and ^{14}C-DL-α-methylserines were not incorporated into ergometrine in 6 additional experiments. Interestingly, α-methylserine exists as a natural product of streptomyces species, in the antibiotic amicetin[191] and can be decarboxylated by some organisms to alaninol[192].

Minghetti *et al.*[181] studied the incorporation of L-(U-^{14}C)-alanine and L-^{14}C-alaninol into ergometrine. They found that alanine and alaninol were both efficiently incorporated (2 %), but that L-alaninol had a much greater ability to function as a source of the side-chain atoms of ergometrine relative to alanine, and concluded that L-alaninol is probably the direct precursor of the same residue in ergometrine.

3. Tripeptide alkaloids

With the more complex alkaloids, even fewer definitive studies have been reported. Paul[193] prepared homogenates of 2–3 week old sclerotia which he incubated for 24 hours at 30° with ^{14}C-labelled phenylalanine, alanine, serine and indole. Autoradiography of the purified alkaloids indicated that phenylalanine and alanine were incorporated into ergotamine and ergotoxine, but not into ergometrine. Serine and indole were not incorporated into any of the alkaloids. Vining and Taber[113] using low yielding peptide alkaloid strains of *C. purpurea* showed that tryptophan and phenylalanine labelled ergosine and ergotamine respectively, but they did not carry out

Table 11. Biosynthesis of ergotamine[194]

Side-chain constituent	Distribution of label in the side-chain constituents expressed as a percentage of the parent alkaloid	
	Ergotamine	Ergotoxine
Pyruvate-DNP derivative	12.7	
Proline	18.6	16.8
Phenylalanine	20.4	
Valine		19.7
Leucine		17.0

Table 11 shows the incorporation of the primary precursor L-(U-^{14}C)-alanine into the side-chain constituents of ergotamine and members of the ergotoxine group of alkaloids. The α-oxyamino acid acid was isolated as the dinitrophenylhydrazine (DNP) derivative.

rigorous degradation of the alkaloid to establish the exact labelling pattern. Recently, strains of *C. purpurea* producing high yields of ergotamine and members of the ergotoxine family of alkaloids have become available, and provide ideal systems for such investigations.

Majer *et al.*[194] reported studies of the biosynthesis of ergotamine using L-(U-^{14}C)-alanine, L-(U-^{14}C)-alaninol and [14]CH$_3$-methionine. It was found that L-(U-^{14}C)-alaninol did *not* incorporate into ergotamine. With L-(U-^{14}C)-alanine feeding experiments, the following results were obtained (see Table 11). From the table it can be readily seen that alanine incorporates more efficiently into phenylalanine and proline (and by difference lysergic acid) than into the α-oxyalanine substituent of ergotamine. [14]CH$_3$-Methionine was found to incorporate only into the nucleus of ergotamine.

Majer concluded from these results, that alanine was in rapid equilibrium with pyruvate and the precursors of phenylalanine and proline.

Minghetti *et al.*[181] in contrast, found that both L-(U-^{14}C)-alanine and L-^{14}C-alaninol were incorporated into ergotamine.

Bassett *et al.*[195] have carried out a detailed study of the precursors of the ergoline system and also the side-chain of the tripeptide alkaloid ergotamine (Tables 12, 13, 14). Both L-phenylalanine and L-proline were efficiently incorporated into the peptide alkaloid; L-phenylalanine specifically labelled

Table 12. Biosynthesis of ergotamine; precursors of the ergoline system[195]

			Degradation products of ergotamine			
Expt.	Precursor	Incorporation into peptide alkaloids (%)	Lysergic acid (%)	α-Oxy-alanine (%)	Proline (%)	Phenyl-alanine (%)
1	L-(3-^{14}C)-Tryptophan	3.57	95.3	0.12	0.09	0.92
2	L-(3-^{14}C)-Tryptophan	10.93				
3	DL-(2-^{14}C)-Mevalonic acid	5.3	101.4	0.04	0.07	0.29
4	DL-(2-^{14}C)-Mevalonic acid	6.24				
5	^{14}CH$_3$-Methionine	3.26	93.1	1.12	0.4	0.6
6	^{14}CH$_3$-Methionine	1.21				
7	^{14}C-Sodium formate	2.74		1.29	0.58	1.01
8	2-^{14}C-Sodium acetate	2.43				

In all cases except for experiment 6, the precursor was added at 10 days; in experiment 6 the methionine was added at 8 days. Experiments 1, 3, 5, 7 and 8 were incubated for 24 hours with the added precursor, the other experiments were harvested after 48 hours, only the 24 hour incubation samples were degraded. All experiments were carried out with a submerged culture of *C. purpurea*, grown on a fully defined medium containing 30% sucrose.

the phenylalanyl moiety of ergotamine (93–96% of total alkaloid activity), while proline labelled the prolyl unit slightly less specifically (73–78% of total activity). An examination of the amino acid metabolism of *C. purpurea* (Fr.) Tul. at 10 days, clearly showed that proline was much more readily metabolized than was phenylalanine at this stage of the fermentation.

The results of feeding experiments with L-(U-^{14}C)-alanine and DL-(1-^{14}C)-alanine as primary precursors, showed that both were incorporated with relatively low efficiency, into the peptide-alkaloid fraction. The labelling from L-(U-^{14}C)-alanine was found to be distributed throughout the ergotamine molecule but was primarily located in the α-oxyalanyl, prolyl

Table 13. Incorporation of ^{14}C-labelled phenylalanine and proline into ergotamine[a]

Precursor	Incubation time and activity added per flask		Incorporation into peptide alkaloids (%)[b]	Degradation products expressed as % parent alkaloid[c]			
	(hours)	(μCi)		Pyruvyl-pBP	Proline	Phenylalanine	Lysergic acid
L-(3-U-^{14}C)-Phenylalanine	48	55	2.5	0.2	4.6	96.1	2.87
L-(3-U-^{14}C)-Phenylalanine	48	25	5.03		1.03	93.64	
L-(3-U-^{14}C)-Phenylalanine	24	25	2.86	0.2	1.42	95.6	2
L-(U-^{14}C)-Proline	48	30	1.15	2.2	77.75	8.1	14.4
L-(U-^{14}C)-Proline	48	25	2.75		78.46	2.8	
L-(U-^{14}C)-Proline	24	25	2.27	1.8	74.39	1.34	21.48

[a] Precursors were added to 10-day shake cultures and the alkaloid recovered after a further 6.5, 24 or 48 hours. The ergotamine side-chain activity was determined by radioassay of phenylalanine, proline and the p-bromophenylhydrazone (pBP) derivative of pyruvic acid. The nucleus was radio-assayed as lysergic acid.
[b] Incorporation into the alkaloid is expressed as a percentage of the labelled precursor that was incorporated into the peptide alkaloid fraction.
[c] Incorporation into the degradation products of ergotamine, is expressed as a percentage of radioactivity in each fragment compared to the parent molecule, ergotamine.

Table 14. Incorporation of ^{14}C-labelled alanine into ergotamine[a]

Precursor	Incubation time and activity added per flask		Incorporation into peptide alkaloids (%)[b]	Degradation products expressed as a % of parent alkaloid[c]			
	(hours)	(μCi)		Pyruvyl-pBP	Proline	Phenylalanine	Lysergic acid
L-(U-^{14}C)-Alanine	24	30	0.26	52.5	25.13	6.6	20.3
L-(U-^{14}C)-Alanine	6.5	40	0.34	54.75	21.8	5.2	23.1
DL-(1-^{14}C)-Alanine	24	40	0.17	89.58	4.7	1.63	12.9
DL-(1-^{14}C)-Alanine	6.5	40	0.065	84.0	9.4	1.76	

[a,b,c] See footnotes for Table 13.

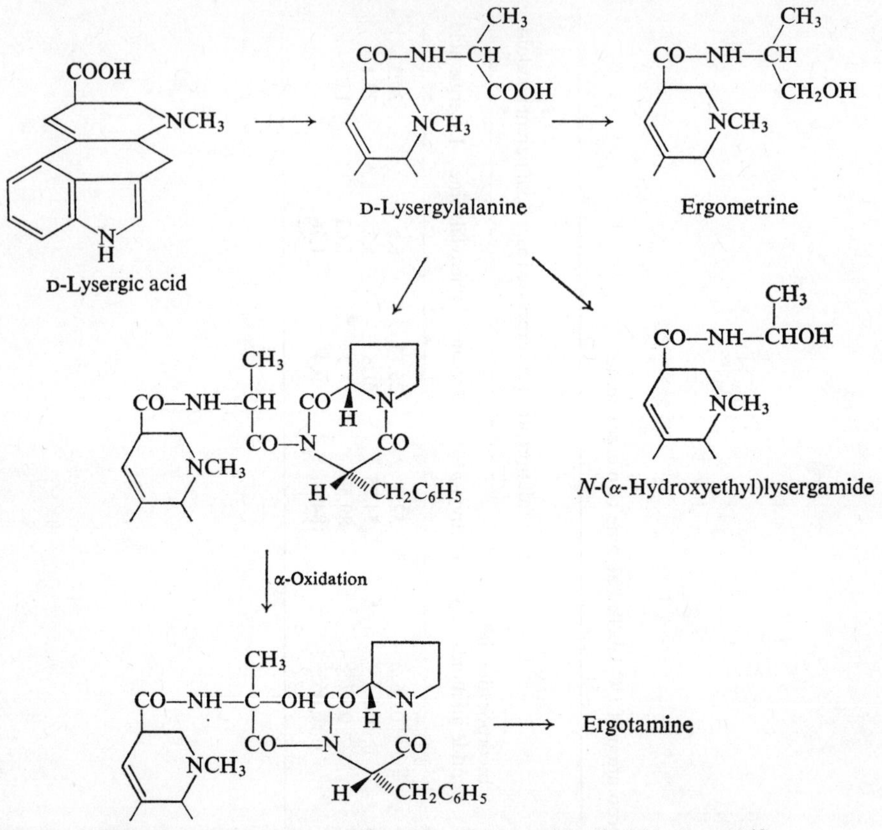

D-Lysergic acid

D-Lysergylalanine

Ergometrine

N-(α-Hydroxyethyl)lysergamide

α-Oxidation

Ergotamine

Figure 25. Proposed interrelationships of the peptide alkaloids (Agurell[15]).

and lysergyl groups. In comparison, DL-(1-^{14}C)-alanine labelled the α-oxyamino acid carbons of ergotamine, exclusively. Furthermore, an examination of the turnover of the amino acid pools of the mycelium at this stage of the fermentations, showed a pattern consistent with this distribution of labelling and with that observed for proline and phenylalanine.

These results, then, demonstrate that while all three carbon atoms of alanine are incorporated into the α-oxyalanine unit, carbon atoms 2 and 3 are also incorporated into the proline and lysergic acid residues.

A recent study[213] describes the incorporation of L-proline into ergotoxine in a manner closely paralleling the above ergotamine pathway.

4. Interconversions of the peptide alkaloids

In 1966 Agurell[15] postulated that lysergylalanine may be a precursor both of the simple peptide alkaloids ergometrine and N-(α-hydroxyethyl)-

lysergamide, and of the more complex peptide alkaloids (for details see Fig. 25). The interpretation of the results available at that time was based on structural similarities of the alkaloids, and on the biosynthetic implications of the structure of lysergylvaline methyl ester[196] **(19)**.

Bashadrjian *et al.*[197] reported experiments which involved the incorporation of the previously synthesized lysergylalanine (side-chain-2-^{14}C). This was fed to a 4-day submerged culture of *C. paspali* producing ergometrine and *N*-(α-hydroxyethyl)lysergamide. Degradation studies of the alkaloid indicated that the activity from lysergylalanine had entered ergometrine but not *N*-(α-hydroxyethyl)lysergamide. It appeared that nearly all of the labelling resided in the expected side-chain position of ergometrine. During the course of this study it was noted that addition of lysergylalanine to the strain of *Claviceps* used, which produced mainly ergometrine, caused a shift in alkaloid production, in favour of *N*-(α-hydroxyethyl)lysergamide formation.

The implication of results derived from studies of the relationship of ergometrine to the complex peptide ergot alkaloids seem to be less clear. Voigt *et al.*[198, 199] have reported the incorporation of ergometrine into ergotamine both by ripening sclerotia and by extracts from sclerotia. Minghetti *et al.*[181] were unable to repeat these results with their strains of *C. purpurea* grown in submerged culture. They found that 4,5-^{3}H-ergometrine did not incorporate into ergotamine although a measurable amount (3.5%) of ^{3}H-ergometrine appeared to be present in the mycelium at the end of the fermentation.

V. OTHER INDOLIC METABOLITES

A. Fungal Products

A recently described constituent of the mycelium of *C. paspali* Stevens and Hall is the metabolite paspalin[160], $C_{28}H_{39}O_2N$, which has not yet been completely characterized but which appears to be an example of the poly-isopentenylindole type related to the *Aspergillus echinulatus* product echinulin **(20)**[200].

Fungal products containing indolic C-4 substituents include, in addition to the ergot alkaloids, the hallucinogenic metabolites of *Psilocybe mexicana* Heim, namely psilocine **(21, R = H)** and psilocybine[186] **(21, R = —PO$_3$H$_2$)** and also the interesting 4-isopentenyl derivatives bissecodehydrocyclopiazonic acid **(22)**, cyclopiazonic acid **(23)** and its imine which are constituents of *Penicillium cyclopium*[156] (see Fig. 26). The latter product is derived from the former by a cyclization step, which links different carbons

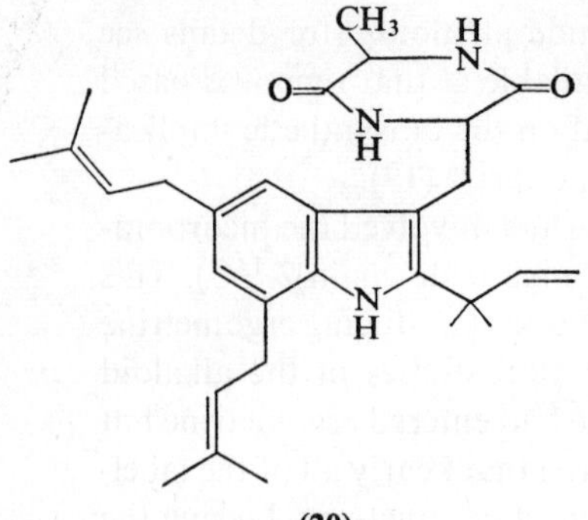

Figure 26.

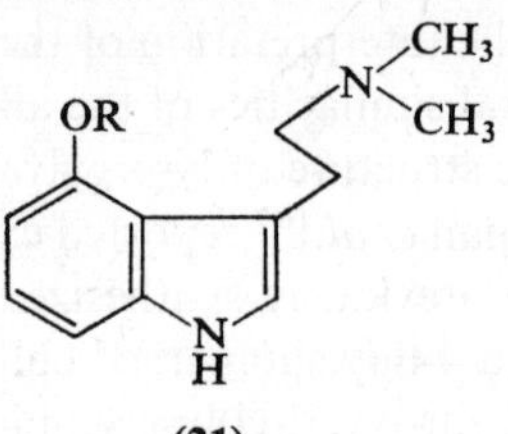

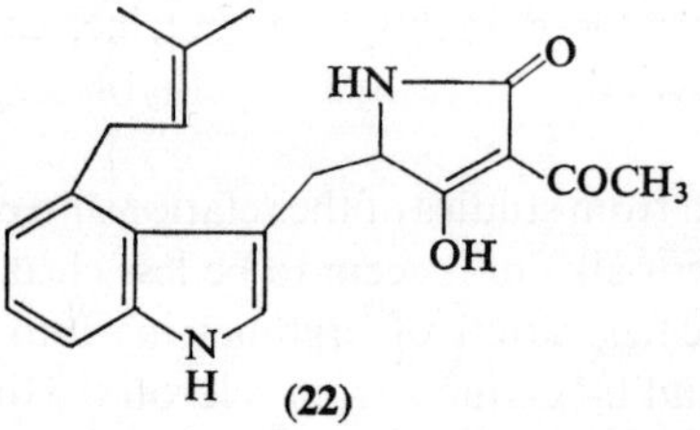

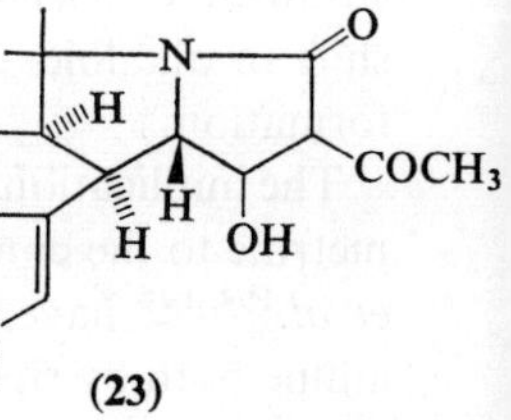

of the isopentenyl substituent and tryptophan side-chain to those involved in the formation of ring C of the ergot alkaloids.

B. Plant Products

The predominant group of plant indole alkaloids are typified by the penta-cyclic compound yohimbine (**8**) and the tetracyclic corynantheine (**24**). More than 800 of these alkaloids have been found, the structures of which contain in addition to a tryptamine nucleus, a common C_{10}- or C_9-moiety, depending upon the presence or absence of a carboxyl substituent. Two major variants of these structures are exemplified by the Aspidosperma alkaloid vindoline (**25**) and the Iboga alkaloid catharanthine (**26**); the structural relationships of the C_{10}-substituents are shown in Fig. 27 by the carbon skeletons associated with each of the four named alkaloids (heavy lines).

The early attractive biosynthetic hypotheses of Barger[203] and Hahn[204] attributed the origin of the C_{10}-unit to phenylalanine, plus two formalde-hyde or C_1-equivalents. An ingenious modification[205] involving 3,4-di-hydroxyphenylalanine which underwent cleavage between C-3 and C-4 (Woodward fission), was suggested to account for the seco alkaloids typified

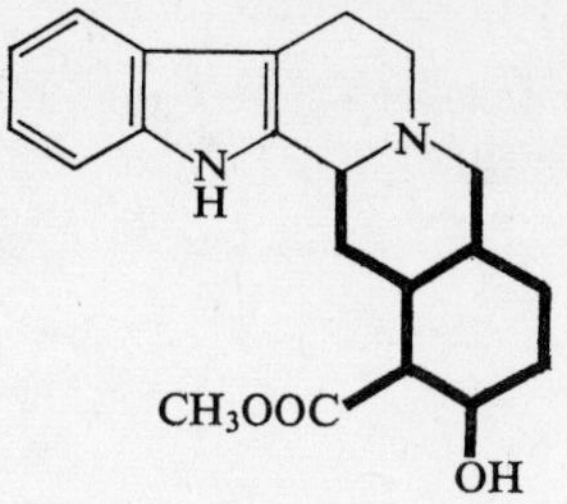

Figure 27.

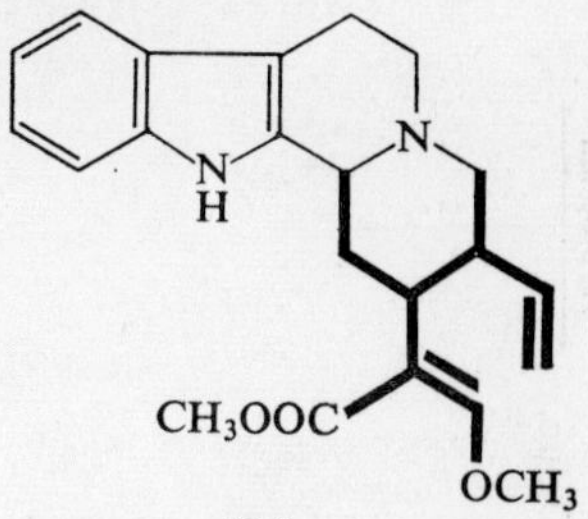

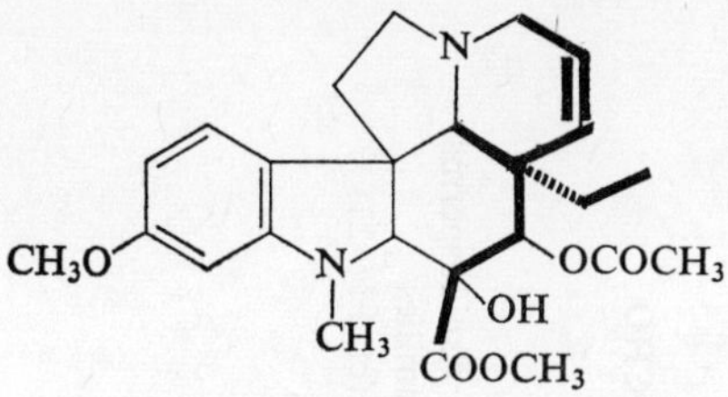

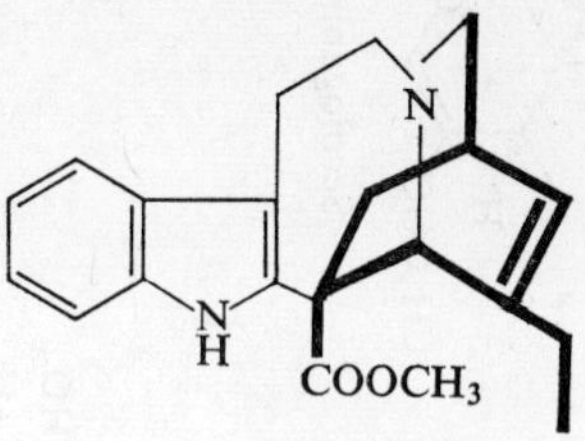

by strychnine and the subsequently characterized corynantheine (**24**). A later scheme of Wenkert and Bringi, invoked the direct utilization of the aromatic amino acid precursors shikimic acid and prephenic acid[206].

A monoterpenoid origin of the C_{10}-unit was proposed by Thomas[207] and by Wenkert[208]. The experimental confirmation of this hypothesis and the clarification of many details of intermediate steps have been achieved over the past six years in a parallel series of extremely elegant studies by the three independent groups of Arigoni, Battersby and Scott, the results of which were recently reviewed by Scott[209].

The hypothesis was based on the recognition of structurally common features of the C_9–C_{10} unit with an increasing number of cyclopentanoid monoterpenes. The subsequent characterization of loganin (**27**) as a member of this group of monoterpenes was of particular interest, as it coexists with strychnine alkaloids in the Strychnos fruit, and this relationship[210] appeared to provide additional indirect support for the hypothesis.

Recent studies[211, 212] have established the efficient utilization of geraniol, nerol, 10-hydroxygeraniol and hydroxynerol as precursors of loganin, which is itself readily incorporated into the indole alkaloids via secologanin and the key tryptamine-secologanin intermediate vincoside (**28**). This latter early indole intermediate acts as a precursor of the three main classes of

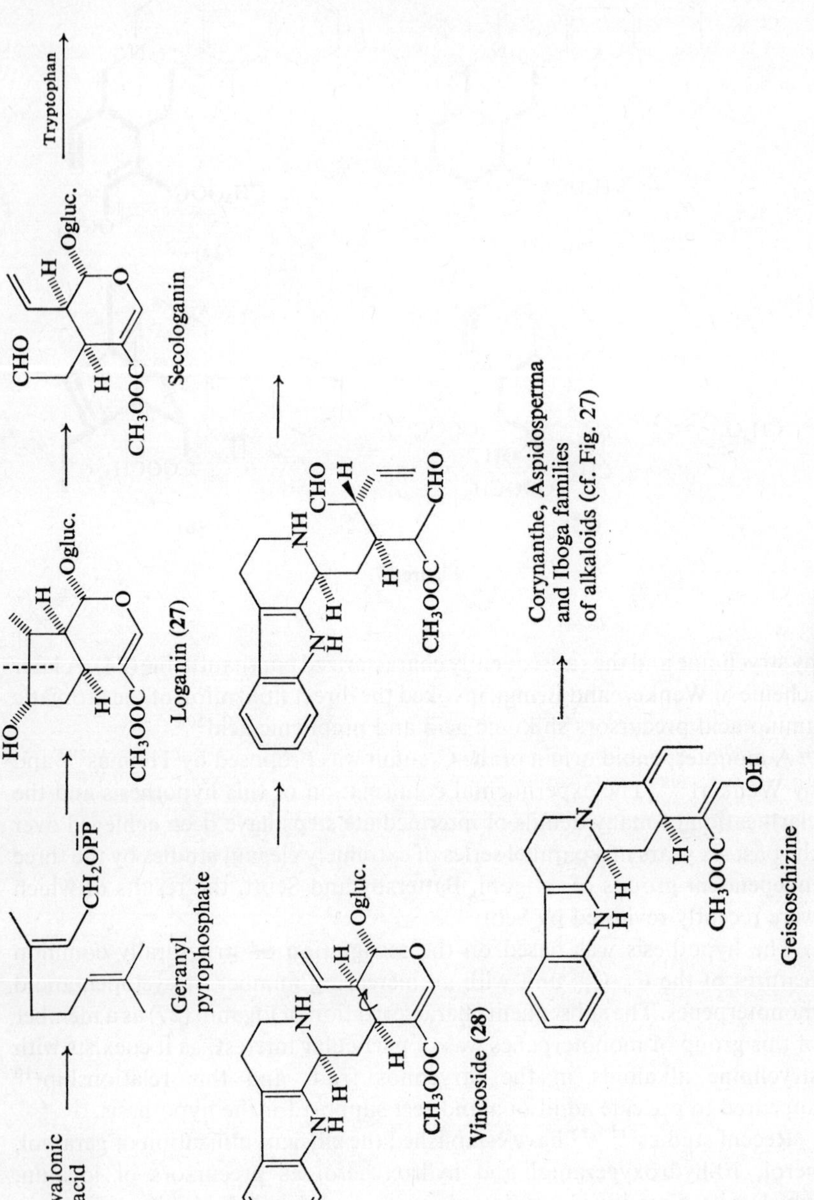

Mevalonic acid → Geranyl pyrophosphate → Loganin (27) → Secologanin → Tryptophan

Vincoside (28) → Geissoschizine → Corynanthe, Aspidosperma and Iboga families of alkaloids (cf. Fig. 27)

Figure 28.

alkaloid, of which the Aspidosperma and Iboga group require a rearrangement of the type proposed by Wenkert[208].

The crucial intermediates in the conversion of mevalonate and tryptophan to the plant indole alkaloids are indicated in Fig. 28.

These studies of fungal and plant indole alkaloids, demonstrate their mutual derivation from common primary precursors, tryptophan and mevalonate. However, as a direct isopentenylation of tryptophan is not involved in the biosynthesis of the above plant indole alkaloids, any chemotaxonomic relationship between these two groups must be extremely remote.

Addendum

Two new types of clavine have recently been described, namely cycloclavine (29)[214], which is a pentacyclic clavine constituent of the plant *Ipomea hildebrantii*, and the rugulovasins A and B, which are stereoisomers of structure 30[215] produced by the fungus *Penicillium concavorugulosum*.

The novel structural features of these alkaloids are of obvious biosynthetic interest. While cycloclavine can be envisaged as a rearrangement product of agroclavine, the reverse pathway would be biosynthetically inconsistent with the absence of a hydrogen atom at C-10 of cycloclavine, which in agroclavine has been shown to be derived from the 5-pro-S-hydrogen of mevalonate[130]. Rugulovasins A and B also lack hydrogen at C-10, and could conceivably arise in an analogous manner to 10-hydroxy-agroclavine (Fig. 20), by oxidation of the carboxylic acid equivalent of the tricyclic clavine chanoclavine-I (2) (chanoclavine-I aldehyde has already been shown to be an efficient precursor of elymoclavine[157]).

VI. REFERENCES

1. J. H. Birkinshaw and C. E. Stickings, 'Nitrogen containing metabolites of fungi' in *Progress in the Chemistry of Organic Natural Products*, Springer-Verlag, 1962.

2. G. Barger, *Ergot and Ergotism*, Gurney and Jackson, London, 1931.

3. F. J. Bové, *The Story of Ergot*, S. Karger, Basel, Switzerland–New York, 1970.

4. A. Hofmann, *Die Mutterkornalkaloide*, Ferdinand Enke–Verlag, Stuttgart, 1964.

5. D. X. Freedman and G. K. Aghajanian, *Lloydia*, **29**, 309 (1966).

6. P. G. Mantle, *J. Reprod. Fert.*, **18**, 81 (1969).

7. S. Agurell, *Experientia*, **20**, 25 (1964).

8. W. A. Taber and L. C. Vining, *Can. J. Microbiol.*, **4**, 611 (1958).

9. J. F. Spilsbury and S. Wilson, *J. Chem. Soc.*, 2085–2194 (1961).

10. M. Abe, S. Yamatodani, T. Yamono, Y. Zozu and S. Yamada, *Nippon Nogei Kagaku Kaishi*, **42**, 68 (1967).

11. M. Abe, 1st International Symposium, *Genetics of Industrial Micro-organisms*, Prague, Aug. 23, 1970.

12. A. Hofmann and H. Tscherter, *Experientia*, **16**, 414 (1960).

13. D. Stauffacher, H. Tscherter and A. Hofmann, *Helv. Chim. Acta*, **48**, 1379 (1965).

14. A. Der Marderosian, *Lloydia*, **30**, 23 (1967).

15. S. Agurell, *Acta Pharm. Suecica*, **3**, 71 (1966).

16. E. Ramstad, *Lloydia*, **31**, 327 (1968).

17. W. J. Kelleher, *Advan. Appl. Microbiol.*, **11**, 211 (1969).

18. R. Voigt, *Die Pharmazie*, **23**, 285, 353, 419 (1968).

19. A. Hofmann, H. Ott, R. Griot, P. A. Stadler and A. J. Frey, *Helv. Chim. Acta*, **46**, 2306 (1963).

20. H. Kobel, E. Schreier and J. Rutschmann, *Helv. Chim. Acta*, **47**, 1052 (1964).

21. N. Castagnoli Jr. and P. G. Mantle, *Nature*, **211**, 859 (1966).

22. P. G. Mantle and E. S. Waight, *Nature*, **218**, 581 (1968).

23. A. T. McPhail, G. A. Sim, A. J. Frey and H. Ott, *J. Chem. Soc. B*, 377–395 (1966).

24. A. Stoll and W. Schlientz, *Helv. Chim. Acta*, **38**, 585 (1955).

25. H. G. Floss, H. Guenther, U. Mothes and I. Becker, *Z. Naturforsch.*, **22B**, 399 (1967).

26. D. Stauffacher and H. Tscherter, *Helv. Chim. Acta*, **47**, 2186 (1964).

27. J. E. Robbers and H. G. Floss, *Tetrahedron Letters*, No. 23, 1857 (1969).

28. M. Abe, *Agr. Chem. Soc.*, (*Japan*), **22**, 2 (1948).

29. M. Abe, *Ann. Rep. Takeda Res. Labs.*, **10**, 73 (1951).

30. M. Abe and S. Yamatodani, *Progr. Indust. Microbiol.*, **5**, 205 (1964).

31. M. Abe, S. Yamatodani, T. Yamono and M. Kusumoto, *Agr. Biol. Chem. (Tokyo)*, **25**, 594 (1961).

32. M. Abe, S. Yamatodani, T. Yamono, Y. Zozu and S. Yamada *Nippon Nogei Kagaku Kaishi*, **42**, 68 (1967).

33. A. Hofmann, R. Brunner, H. Kobel and A. Brack, *Helv. Chim. Acta*, **40**, 1358 (1957).

34. A. Stoll and A. Hofmann, 'The ergot alkaloids', in *The Alkaloids* (Ed. R. H. F. Manske), Vol. 8, Academic Press, New York and London, 1965.

35. P. G. Mantle and C. Szyrczbek, in press.

36. W. W. Bonns, *Am. J. Bot.*, **9**, 339 (1922).

37. F. Arcamone, C. Bonino, E. B. Chain, A. Ferretti, P. Penella, A. Tonolo and L. Vera, *Nature*, **187**, 238 (1960).

38. F. Arcamone, E. B. Chain, A. Ferretti, A. Minghetti, P. Penella, A. Tonolo and L. Vera, *Proc. Roy. Soc.*, **155B**, 26 (1961).

39. L. R. Pacifici, W. J. Kelleher and A. E. Schwarting, *Lloydia*, **25**, 37 (1962).

40. D. Gröger and V. E. Tyler, *Lloydia*, **26**, 174 (1963).

41. H. Rossback, K. G. Buckel and H. Rochelmeyer, *Arzneimittel-Forsch.*, **6**, 690 (1956).

42. M. Abe, T. Yamono, Y. Zozu and M. Kusumoto, *U.S. Pat.*, 2,835 675 (1958).

43. A. Stoll, A. Brack, A. Hofmann and H. Kobel, *U.S. Pat.*, 2,809,920 (1957).
44. W. A. Taber and L. C. Vining, *Can. J. Microbiol.*, 3, 55 (1957).
45. A. Tonolo, *Nature*, 209, 1134 (1966).
46. A. M. Amici, A. Minghetti, T. Scotti, C. Spalla and L. Tognoli, *Experientia*, 22, 415 (1966).
47. A. M. Amici, A. Minghetti, T. Scotti, C. Spalla and L. Tognoli, *Appl. Microbiol.*, 15, 597 (1967).
48. A. M. Amici, A. Minghetti, T. Scotti, C. Spalla and L. Tognoli, *Appl. Microbiol.*, 17, 464 (1969).
49. W. A. Taber, *Developments in Industrial Microbiology, Am. Inst. Biol. Sci.*, 4, 295 (1963).
50. W. A. Taber, *Appl. Microbiol.*, 12, 321 (1964).
51. W. A. Taber, *Lloydia*, 30, 39 (1967).
52. S. Windisch and W. Bronn, *U.S. Pat.*, 2,936,266 (1960).
53. J. D. Bu'Lock and J. G. Barr, *Lloydia*, 31, 342 (1968).
54. J. D. Bu'Lock, 'Aspects of secondary metabolism of fungi', in *Biogenesis of Antibiotic Substances* (Eds. Z. Vanek and Z. Hostalek), Academic Press, New York and Academy of Sciences, Prague, 1965, pp. 61–71.
55. J. D. Bu'Lock, D. Hamilton, M. A. Hulme, A. J. Powell, D. Shepherd, H. M. Smalley and G. N. Smith, *Can. J. Microbiol.*, 11, 765 (1965).
56. J. D. Bu'Lock and D. Shepherd, *Experientia*, 21, 55 (1965).
57. R. A. Bassett, E. B. Chain, K. Corbett, A. G. F. Dickerson and P. G. Mantle. *Biochem. J.*, 127, 3P (1972).
58. A. Borrow, E. G. Jefferys, R. H. J. Kessell, E. C. Lloyd, P. B. Lloyd and I. S. Nixon, *Can. J. Microbiol.*, 7, 227 (1961).
59. W. A. Taber and L. C. Vining, *Can. J. Microbiol.*, 4, 611 (1958).
60. F. Arcamone, G. Cassinelli, G. Ferni, S. Penco, P. Penella and C. Pol, *Ergotamine Production and Metabolism of* C. purpurea *Strains in Stirred Fermenters*, Third International Fermentation Symposium, New Brunswick, New Jersey, 1968.
61. J. P. Rosazza, W. J. Kelleher and A. E. Schwarting, *Appl. Microbiol.*, 15, 1270 (1967).
62. O. P. Nizkovskaya and A. N. Shivrina, *Mikol. Fitopatol.*, 3, 66 (1969).
63. A. M. Amici, T. Scotti, C. Spalla and L. Tognoli, *Appl. Microbiol.*, 15, 611 (1967).
64. J. K. McDonald, V. H. Cheldelin and Tsoo E. King, *J. Microbiol.*, 80, 61 (1960).
65. J. K. McDonald, J. A. Anderson, V. H. Cheldelin and Tsoo E. King, *Biochem. Biophys. Acta*, 73, 533 (1963).
66. T. P. Singer, V. Massey and E. B. Kearney, *Arch. Biochem. Biophys.*, 69, 405 (1957).
67. B. K. Kim, W. J. Kelleher and A. E. Schwarting, *Lloydia*, 31, 422 (1968).
68. N. D. Zahid and R. M. Baxter, *Biogenesis of Ergot Alkaloids; Regulatory Factors in Induction*, presented at A.Ph.A. Academy of Pharmaceutical Sciences, Miama, May, 1968.
69. A. M. Conney, *Pharmacol. Rev.*, 19, 317 (1967).
70. S. H. Ambike, R. M. Baxter and N. D. Zahid, *Phytochemistry*, 9, 1953 (1970).
71. D. V. Parke, *The Biochemistry of Foreign Compounds*, Pergamon Press Ltd., London, 1968, p. 38.
72. J. Chen. Hsu and J. A. Anderson, *Chem. Commun.*, No. 20, 1318 (1970).
73. L. R. Pacifici, W. J. Kelleher and A. E. Schwarting, *Lloydia*, 26, 161 (1963).
74. F. Arcamone, E. B. Chain, A. Ferretti, A. Minghetti, P. Pannella and A. Tonolo, *Biochem. Biophys. Acta*, 57, 174 (1962).
75. H. G. Floss and U. Mothes, *Arch. Mikrobiol.*, 48, 213 (1964).

76. P. G. Mantle, unpublished.

77. K. Corbett, unpublished.

78. F. Lingens, W. Goebel and H. Uesselar, *European J. Biochem.*, **2**, 442 (1967).

79. F. Lingens, W. Goebel and H. Uesseler, *Naturwissenschaften*, **54**, 141 (1967).

80. B. Sprossler and F. Lingens., *Z. Physiol. Chem.*, **351**, 448 (1970).

81. B. Sprossler and F. Lingens., *Z. Physiol. Chem.*, **351**, 967 (1970).

82. E. Teuscher, *Pharmazie*, **20**, 778 (1965).

83. E. Teuscher, *Flora*, **155**, 80 (1964).

84. P. G. Mantle and A. Tonolo, *Trans. Br. Mycol. Soc.*, **51**, 499 (1968).

85. N. Castagnoli and A. Tonolo, International Congress Microbiology Symposium, Moscow 31, 1966.

86. A. Tonolo, *Eva-Udvardy-Nagy*, *Acta Microbiol. Akad. Sci. Hung,,* **15**, 29 (1968).

87. H. Mathes and O. H. Kürscher, *Arch. Pharm.*, **269**, 88 (1931).

88. K. E. Bharucha and F. D. Gunstone, *J. Chem. Soc.*, **123**, 610 (1957).

89. L. J. Morris and S. W. Hall, *Lipids*, **1**, 188 (1966).

90. L. J. Morris, S. W. Hall and A. T. James, *Biochem. J.*, **199**, 29C-30C (1966).

91. P. G. Mantle, L. J. Morris and S. W. Hall, *Trans. Br. Mycol. Soc.*, **53**, 441 (1969).

92. A. S. Perlin and W. A. Taber, *Can. J. Chem.*, **41**, 2278 (1963).

93a. K. W. Buck, A. W. Chen, A. G. Dickerson and E. B. Chain, *J. Gen. Microbiol.*, **51**, 337 (1968).

93b. A. G. Dickerson, P. G. Mantle and C. A. Szezyrbek, *J. Gen. Microbiol.*, **60**, 403 (1970).

94. T. Perenyi, E. Udvardy-Nagy, G. Waek and E. K. Novak, *Acta Microbiol. Acad. Sci. Hung.*, **13**, 205 (1966).

95. T. Perenyi, E. Udvardy-Nagy and E. K. Novak, *Acta Microbiol. Acad. Sci. Hung.*, **15**, 17 (1968).

96. F. Arcamone, W. Barbieri, G. Casinelli and C. Pol, *Carbohydrate Res.*, **14**, 65 (1970).

97. R. A. Bassett and A. G. Dickerson, unpublished.

98. A. Grein, *J. Microbiol.*, **15**, 817 (1967).

99. Sir R. Robinson, *Structural Relations of Natural Products*, Oxford University Press, Oxford, 1955.

100. R. J. Suhadolnik, L. H. Henderson, I. B. Hanson and Y. H. Loo, *J. Am. Chem. Soc.*, **80**, 3153 (1958).

101. K. Mothes, F. Weygand, D. Gröger and H. Grisebach, *Z. Naturforsch.*, **13B**, 41 (1958).

102. W. A. Taber and L. C. Vining, *Chem. Ind. (London)*, 1218 (1959).

103. D. Gröger, H. J. Wendt, K. Mothes and F. Weygand, *Z. Naturforsch.*, **14B**, 355 (1959).

104. R. M. Baxter, S. I. Kandel and A. Okany, *Chem. Ind. (London)*, 266 (1960).

105. D. Gröger and D. Erge, *Pharmazie*, **19**, 775 (1964).

106. S. Agurell, *Acta Pharm. Suecica*, **3**, 11 (1966).

107. S. Agurell, *Acta Pharm. Suecica*, **3**, 23 (1966).

108. N. Castagnoli, Jr., K. Corbett, E. B. Chain and R. Thomas, *Biochem. J.*, **117**, 451 (1970).

109. H. Plieninger, R. Fischer, G. Keilich and H. D. Orth, *Ann. Chem.*, **642**, 214 (1961).

110. D. Gröger, D. Erge and H. G. Floss, *Z. Naturforsch.*, **20B**, 856 (1965).

111. H. G. Floss, U. Mothes and H. Günther, *Z. Naturforsch.*, **19B**, 784 (1964).

112. D. Gröger, H. J. Wendt, K. Mothes and F. Weygand, *Z. Naturforsch.*, **14B**, 335 (1959).

113. W. A. Vining and L. C. Taber, *Can. J. Microbiol.*, **9**, 291 (1963).
114. H. G. Floss and U. Mothes, *Z. Naturforsch.*, **18B**, 866 (1963).
115. H. G. Floss, U. Mothes, D. Onderka and U. Hornemann, *Z. Naturforsch.*, **20B**, 133 (1965).
116. R. M. Baxter, S. I. Kandel and A. Okany, *Chem. Ind. (London)*, 1453 (1961).
117. H. G. Floss and D. Gröger, *Z. Naturforsch.*, **19B**, 393 (1964).
118. H. G. Floss and D. Gröger, *Z. Naturforsch.*, **18B**, 519 (1963).
119. F. L. Cavender and J. A. Anderson, *Biochem. Biophys. Acta*, **208**, 345 (1970).
120. R. M. Baxter, S. I. Kandel, A. Okany and R. G. Pyke, *Can. J. Chem.*, 2936 (1964).
121. D. Gröger, K. Mothes, H. Simon, H. G. Floss and F. Weygand, *Z. Naturforsch.*, **15B**, 141 (1960).
122. E. H. Taylor and E. Ramstad, *Nature*, **188**, 494 (1960).
123. A. J. Birch, B. J. McLaughlin and H. Smith, *Tetrahedron Letters*, No. 7, 1 (1960).
124. D. Gröger, *Arch. Pharm.*, **292**, 389 (1959).
125. E. H. Taylor and E. Ramstad, *J. Pharm. Sci.*, **50**, 681 (1961).
126. H. Plieninger, H. Immel and A. Völkl, *Ann. Chem.*, **706**, 223 (1967).
127. S. Bhattachargi, A. J. Birch, A. Brack, A. Hofmann, H. Kobel, D. C. C. Smith, H. Smith and J. Winter, *J. Chem. Soc.*, 421 (1962).
128. R. M. Baxter, S. I. Kandel and A. Okany, *J. Am. Chem. Soc.*, **84**, 2997 (1962).
129. H. G. Floss, U. Hornemann, N. Schilling, K. Kelly, D. Gröger and D. Erge, *J. Am. Chem. Soc.*, **90**, 6500 (1968).
130. M. Seiler, W. Acklin and D. Arigoni, *Chem. Commun.*, 1394 (1970).
131. R. M. Baxter, S. I. Kandel, A. Okany and K. L. Tam, *J. Am. Chem. Soc.*, **84**, 4350 (1962).
132. R. M. Baxter, S. I. Kandel and A. Okany, *Tetrahedron Letters*, No. 17, 596 (1961).
133. H. G. Floss, personal communication to Seiler *et al.*, ref. 130.
134. H. G. Floss, *Chem. Commun.*, 804 (1967).
135. H. Plieninger, R. Fischer and V. Liede, *Angew. Chem.*, **74**, 430 (1962).
136. S. Agurell, *Dissertation*, Purdue University, Lafayette, Jan. 1962.
137. H. Plieninger, M. Hoebel and V. Liede, *Chem. Ber.*, **96**, 1618 (1963).
138. F. Weygand, H. G. Floss and U. Mothes, *Tetrahedron Letters*, 873 (1962).
139. F. Weygand, H. G. Floss, U. Mothes, D. Gröger and K. Mothes, *Z. Naturforsch.*, **19B**, 202 (1964).
140. H. Plieninger, R. Fischer and V. Liede, *Ann. Chem.*, **672**, 223 (1964).
141. J. E. Robbers and H. G. Floss, *Arch. Biochem. Biophys.*, **126**, 967 (1968).
142. S. Agurell and Jan-Erik Lindgren, *Tetrahedron Letters*, 5127 (1968).
143. G. Popjak and J. W. Cornforth, *Biochem. J.*, **101**, 553 (1966).
144. A. Hofmann, R. Brunner, H. Kobel and A. Brack, *Helv. Chim. Acta*, **40**, 1358 (1957).
145. S. Agurell and E. Ramstad, *Arch. Biochem. Biophys.*, **98**, 457 (1962).
146. U. Mothes, *Dissertation*, München, 1964.
147. W. Acklin, T. Fehr and D. Arigoni, *Chem. Commun.*, 799 (1966).
148. T. Fehr, W. Acklin and D. Arigoni, *Chem. Commun.*, 801 (1966).
149. D. Arigoni, *Some Aspects of Mevalonoid Biosynthesis*, in Symposium on Organic Chemical Approaches to Biosynthesis, London, 1965.
150. H. G. Floss, H. Guenther, D. Gröger and D. Erge, *J. Pharm. Sci.*, **56**, 1675 (1965).
151. D. Gröger, D. Erge and H. G. Floss, *Z. Naturforsch.*, **21B**, 827 (1966).
152. R. Voigt, M. Bornshein and G. Rabitzsch, *Pharmazie*, **22**, 326 (1967).
153. E. O. Ogunlana, B. J. Wilson, V. E. Tyler and E. Ramstad, *Chem. Commun.*, 775 (1970).

154. H. G. Floss, U. Hornemann, N. Schilling, D. Gröger and D. Erge, *Chem. Commun.*, 105 (1967).

155. H. G. Floss, *Chem. Commun.*, 804 (1967).

156. C. W. Holzapfel, R. D. Hutchinson and D. C. Wilkins, *Tetrahedron*, **26**, 5239 (1970).

157. B. Naidoo, J. M. Cassady, G. E. Blair and H. G. Floss, *Chem. Commun.*, 471 (1970).

158. H. G. Floss, D. Gröger and D. Erge, Fifth International Symposium on the Chemistry of Natural Products, London, 1968, Abstracts, p. 72.

159. H. G. Floss, personal communication.

160. T. Fehr, *Ph.D. Dissertation*, ETH Zürich, 1967.

161. R. Thomas, unpublished.

162. S. Agurell, *Acta Pharm. Suecica*, **2**, 357 (1965).

163. D. Gröger, H. R. Schutte and K. Stoll, *Z. Naturforsch.*, **18B**, 850 (1963).

164. P. G. Mantle and E. S. Waight, *Nature*, **218**, 581 (1968).

165. J. Belveau and E. Ramstad, *Lloydia*, **29**, 234 (1966).

166. A. Brack, A. Hofmann, F. Kalberger, H. Kobel and J. Rutachmann, *Arch. Pharm.*, **294**, 230 (1961).

167. V. E. Tyler, D. Erge and D. Gröger, *Planta Medica*, **13**, 315 (1965).

168. E. H. Taylor, K. J. Goldner, S. F. Pong and H. R. Shough, *Lloydia*, **29**, 239 (1966).

169. D. Gröger, *Planta Medica*, **11**, 444 (1963).

170. E. H. Taylor and E. Ramstad, Eleventh Pacific Science Congress, Tokyo, August, 1966.

171. W.-n Lin, E. Ramstad and E. H. Taylor, *Lloydia*, **30**, 202 (1967).

172. W. Chen, *Thesis*, Purdue University, Lafayette, Indiana, 1967.

173. M. Johnsson, *Physiol. Plantarium*, **17**, 547 (1964).

174. A. Jindra, E. Ramstad and H. G. Floss, *Lloydia*, **31**, 190 (1968).

175. H. Rochelmeyer, *Pharm. Ztg.*, **103**, 1269 (1958).

176. S. Agurell and E. Ramstad, *Tetrahedron Letters*, 501 (1961).

177. K. Mothes, K. Winkler, D. Gröger, H. G. Floss, U. Mothes and F. Weygand, *Tetrahedron Letters*, 933 (1962).

178. S. Agurell and M. Johnsson, *Acta Chem. Scand.*, **18**, 2285 (1964).

179. H. G. Floss, H. Guenther, D. Gröger and D. Erge, *Z. Naturforsch.*, **21B**, 128 (1966).

180. S. Agurell, *Acta Pharm. Suecica*, **3**, 65 (1966).

181. A. Minghetti and F. Arcamone, *Experientia*, **25**, 926 (1969).

182. M. Abe, S. Yamatodani, T. Yamono, Y. Zozu and S. Yamada, *Beitr. Biochem. Physiol. Naturstoffen. Festschr.*, 19 (1965).

183. M. Abe, Third Symposium on Biochemistry and Physiology of the Alkaloids, Halle-Saale, June 1965.

184. R. M. Baxter, S. I. Kandel, A. Okany and K. L. Tam, *Acta Chem. Scand.*, **18**, 2285 (1964).

184a. A. M. Amici, A. Minghetti and C. Spalla, *Biochim. Appl.* **12**, 50 (1966).

185. R. G. Griot and A. J. Frey, *Tetrahedron*, **19**, 1661 (1963).

186. S. Shibata, S. Natori and S. Udogawa, *List of Fungal Products*, University of Tokyo Press, Tokyo, 1964.

187. Y.-S. Chen, *Bull. Agric. Chem. Soc. (Japan)*, **24**, 372 (1960).

188. N. Castagnoli, E. B. Chain and R. Thomas, Fourth Internat. Symp. Chem. Nat. Prod., Stockholm, 1966, Abstract p. 165.

189. D. Gröger D. Erge and H. G. Floss, *Z. Naturforsch.*, **23B**, 177 (1968).

190. U. Nelson and S. Agurell, *Acta Chem. Scand.*, **23**, 3393 (1969).

191. E. H. Flynn, J. W. Hinman, E. L. Caron and D. O. Wolf, Jr., *J. Am. Chem. Soc.*, **75**, 5867 (1953).
192. H. G. Aaslestad and A. D. Larson, *J. Bacteriol.*, **88**, 1296 (1964).
193. A. G. Paul, *Ph.D. Thesis*, Connecticut, 1957.
194. J. Majer, J. Keybal and I. Komersova, *Folia Microbiologica*, **12**, 489 (1967).
195. R. A. Bassett, E. B. Chain and K. Corbett, *Biochem. J.*, **122**, 58 (1971).
196. W. Schlientz and R. Brunner, *Experientia*, **19**, 397 (1963).
197. G. Bashadrjian, H. G. Floss, D. Gröger and D. Erge, *Chem. Commun.*, 418 (1969).
198. R. Voigt and M. Bornshein, *Pharmazie*, **20**, 521 (1965).
199. R. Voigt and M. Bornshein, *Pharmazie*, **21**, 380 (1966).
200. G. Casnati, A. Quilico and A. Ricca, *Gazz. Chim. Ital.*, **92**, 129 (1962).
201. C. W. Holzapfel, *Tetrahedron*, **24**, 2101 (1968).
202. C. W. Holzapfel and D. C. Wilkins, Fifth Internat. Symp. Chem. Nat. Prod., London, July 1968, Abstract C65.
203. G. Barger and C. Scholz, *Helv. Chim. Acta*, **16**, 1343 (1933).
204. G. Hahn and H. Werner, *Ann. Chem.*, **520**, 123 (1935).
205. R. B. Woodward, *Nature*, **162**, 155 (1948).
206. E. Wenkert and N. V. Bringi, *J. Am. Chem. Soc.*, **81**, 1474 (1959).
207. R. Thomas, *Tetrahedron Letters*, 544 (1961).
208. E. Wenkert, *J. Am. Chem. Soc.*, **84**, 98 (1962).
209. A. I. Scott, *Curr. Accounts Chem. Res.*, **3**, 158 (1970).
210. R. Thomas in *Biogenesis of Antibiotic Substances* (Eds. Z. Vanek and Z. Hostalek), Czech. Acad. Sci. Prague, 1965, p. 155.
211. S. Escher, P. Loew and D. Arigoni, *Chem. Commun.*, 823 (1970).
212. A. R. Battersby, S. H. Brown and T. G. Payne, *Chem. Commun.*, 827 (1970).
213. D. Gröger and D. Erge, *Z. Naturforsch.*, **25B**, 196 (1970).
214. D. Stauffacher, P. Niklaus, H. Tscherter, H. P. Weber and A. Hofmann, *Tetrahedron*, **25**, 5879 (1969).
215. S. Yamatodani, Y. Asahi, A. Matsukura, S. Ohmomo and M. Abe, *Agr. Biol. Chem. (Tokyo)*, **34**, 485 (1970).

The Biosynthesis of Plant Sterols

L. J. GOAD   DEPARTMENT OF BIOCHEMISTRY,
THE UNIVERSITY,

and

T. W. GOODWIN   P.O. BOX 147, LIVERPOOL, U.K.

I. INTRODUCTION

The plant kingdom is endowed with a considerable number of diverse sterols whose structural elucidations have provided many challenging problems to the chemist[1-3]. However while studies on cholesterol biosynthesis in animals have continued unabated for more than two decades[4-8] it is only in the past 6 years that intensive efforts have been made to resolve the various biosynthetic problems unique to plant sterols[8-12]. In this period, no doubt stimulated by the extensive work upon animal sterols, rapid progress has been made which has revealed challenging and hitherto unconsidered facets of sterol production in plants. In particular a

113

picture is emerging which shows that striking differences in sterol bio-synthetic mechanisms occur in different classes of organisms and which may perhaps be shown to have some evolutionary implications.

II. STRUCTURE AND DISTRIBUTION

The present review is restricted principally to a consideration of the 3β-monohydroxy sterols found in fungi, algae and higher plants. It is convenient on both structural and biosynthetic grounds to divide the sterols into three groups which may be referred to as (*a*) triterpenes or 4,4-dimethyl sterols, (*b*) 4-methyl sterols and (*c*) 4-demethyl sterols or simply sterols. The naturally occurring plant sterols are listed in Table 1 together with their original or principal plant sources.

Representatives of most plant classes have been examined and with few exceptions found to contain sterols. The sterols produced by different species will not be discussed in detail here but a few general comments concerning the types of 4-demethyl sterols present in different classes of plants are relevant to the problems of phytosterol biosynthesis.

It was originally reported[104, 133] that blue–green algae (Cyanophyceae) lacked sterols but recent work has shown that *Anacystis nidulans*[134], *Fremyella diplosiphon*[134] and *Phormidium luridum*[135] contain cholesterol (**42**) and 24-ethyl cholesterol mixtures although these sterols are difficult to extract from the cells[134].

In the Phaeophyceae fucosterol (**67**) predominates but small amounts of 24-methylene cholesterol (**51**) and cholesterol (**42**) have been observed in several species[68, 136, 137]. The possibility that saringasterol (**69**) and 24-keto-cholesterol (**47**) found in some brown algae arise by autoxidation during the drying process prior to extraction of the seaweed has been discussed[68].

The predominant sterol of the Rhodophyceae is cholesterol (**42**)[59–61] with 22-dehydrocholesterol (**43**)[61, 62] and desmosterol (**44**)[61, 138] present in some species. Considerable variation in the sterol composition of different samples of *Rhodymenia palmata* has been observed[138, 139]. Some samples contained principally desmosterol and others mainly cholesterol. The variations did not appear to be seasonal or environmental and indicate that it is advisable, if possible, to analyse several samples of a plant material to ascertain the 'normal' sterols of the species in question. While C_{27}-sterols are dominant in most red algae small quantities of C_{28}- and C_{29}-sterols are present in some species[138–140] and indeed a C_{28}-sterol is the major component in *Rytiphlea tinctoria*[141].

The sterol composition of the Chlorophyceae and Chrysophyceae is variable and a wide range of C_{27}-, C_{28}- and C_{29}-sterols have been

Table 1. Naturally occurring plant sterols and related compounds

(*a*) *Triterpenes and 4,4-dimethyl sterols*		
Lanosterol; 5α-lanosta-8,24-dien-3β-ol	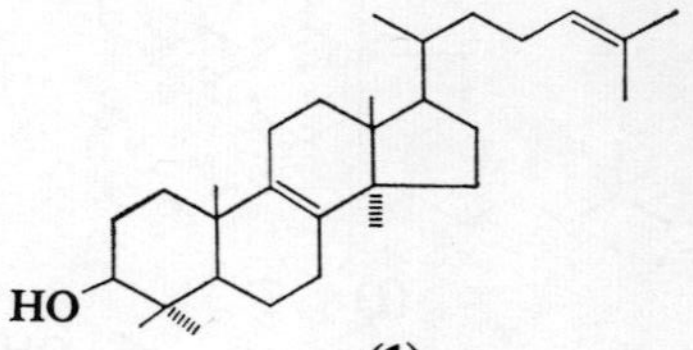 (1)	*Saccharomyces cerevisiae*[13,14], *Inonotus obliquus*[15], *Euphorbia* species[16–18]
23-Hydroxylanosterol; (23S)-23-hydroxy-5α-lanosta-8,24-dien-3β-ol	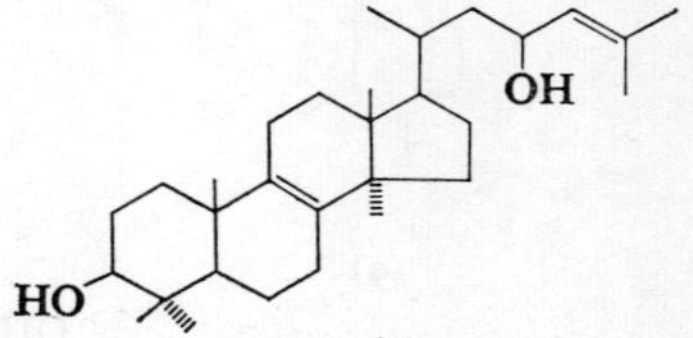 (2)	*Scleroderma aurantium*[98,99]
24-Methylene dihydrolanosterol; 4,4,14α-trimethyl-5α-ergosta-8,24(28)-dien-3β-ol; eburicol	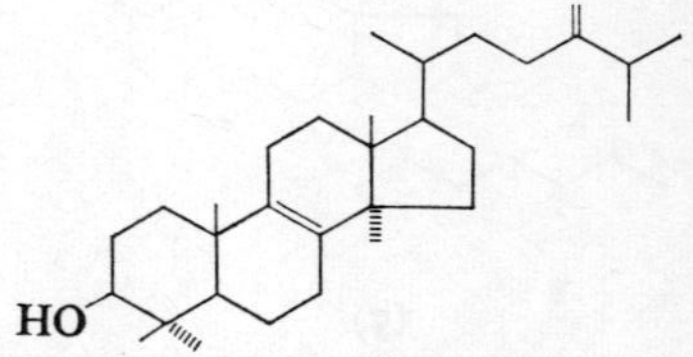 (3)	*Phycomyces blakesleeanus*[19,20] *Fomes officinalis*[21] *Physarum polycephalum*[22]
Parkeol; 5α-lanosta-9(11),24-dien-3β-ol	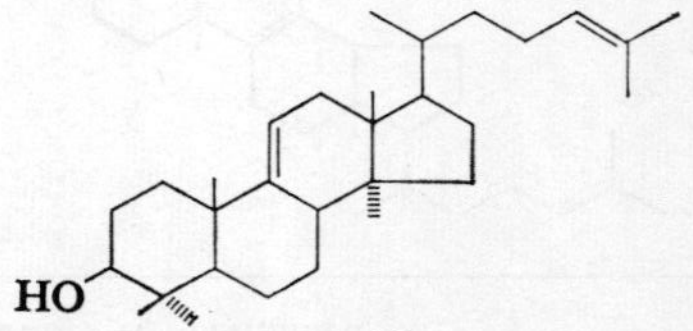 (4)	*Butyrospermum parkii*[23]

Table 1.—*continued*

4,4-Dimethylzymosterol; 4,4-dimethyl-5α-cholesta-8,24-dien-3β-ol	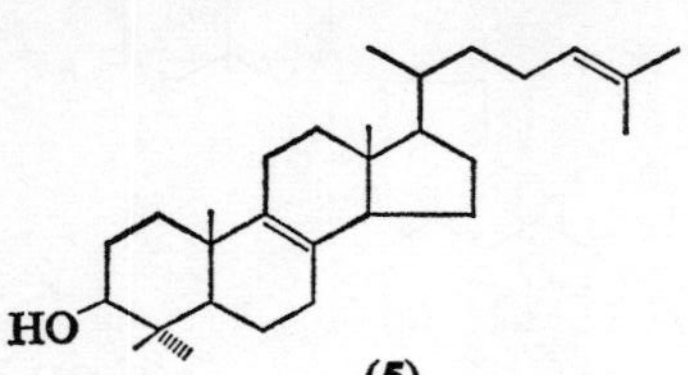 **(5)**	*Saccharomyces cerevisiae*[14]
4,4-Dimethyl-5α-ergosta-8,24(28)-dien-3β-ol	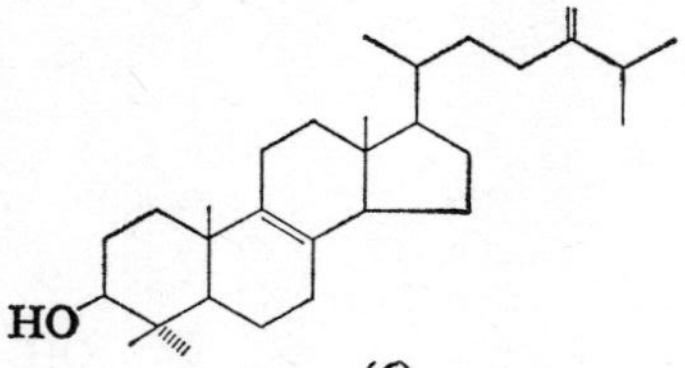 **(6)**	*Phycomyces blakesleeanus*[19, 20]
Cycloartenol; 9β,19-cyclo-5α-lanost-24-en-3β-ol	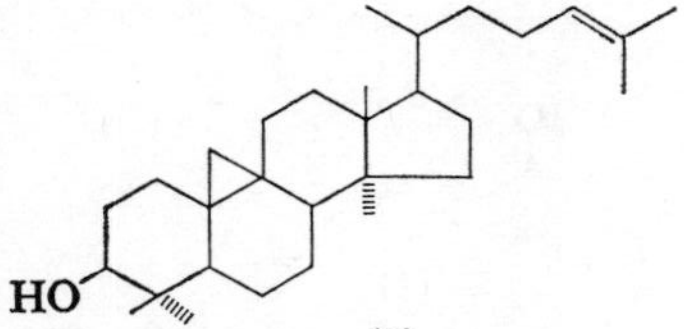 **(7)**	*Strychnos nux-vomica*[24] *Euphorbia* species[18]
Cycloartanol; 9β,19-cyclo-5α-lanostan-3β-ol	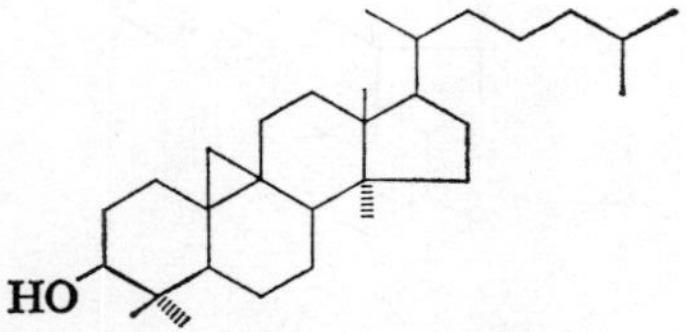 **(8)**	*Oryza sativa*[25] *Polypodium vulgare*[41]

5 β-Cycloorystenol; 9β,19-cyclo-5α-lanost-25-en-3β-ol	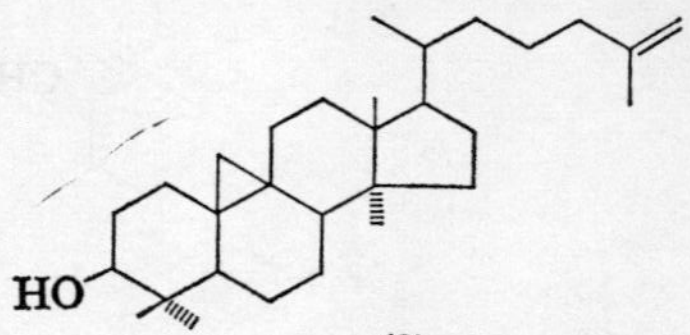 (9)	*Oryza sativa*[155]
24-Methylene cycloartanol; 4,4,14α-trimethyl-9β,19-cyclo-5α-ergost-24(28)-en-3β-ol	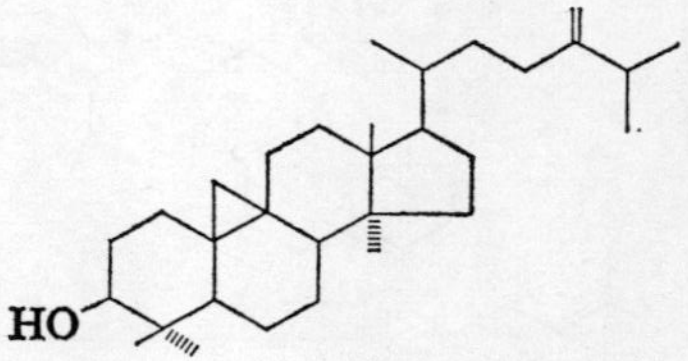 (10)	*Oryza sativa*[26]
Cyclolaudenol; 4,4,14α-trimethyl-9β,19-cyclo-5α-ergost-25-en-3β-ol	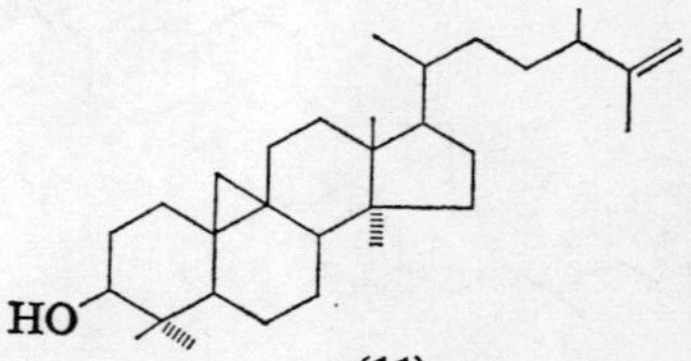 (11)	*Papaver somniferum*[27] *Polypodium vulgare*[28]
Cyclosadol; 4,4,14α-trimethyl-9β,19-cyclo-5α-ergost-23-en-3β-ol	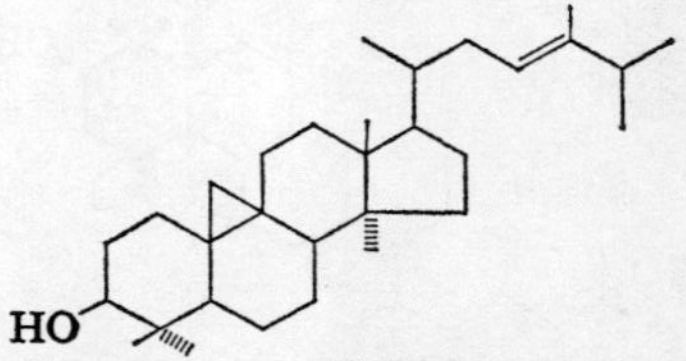 (12)	*Zea mays*[29]

Table 1.—*continued*

Compound	Source
(13) Cycloart-23-en-3β,25-diol; 9β,19-cyclo-5α-lanost-23-en-3β,25-diol	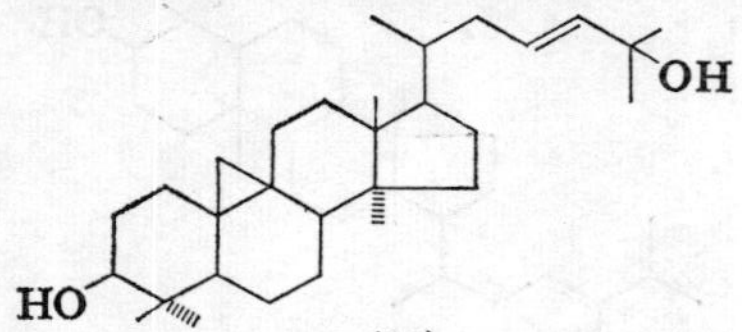 *Euphorbia cyparissias*[100] *Tillandsia usneoides*[101]
(14) Cycloneolitsin; 3β-methoxy-24,24-dimethyl-9β,19-cyclo-5α-lanost-25-ene	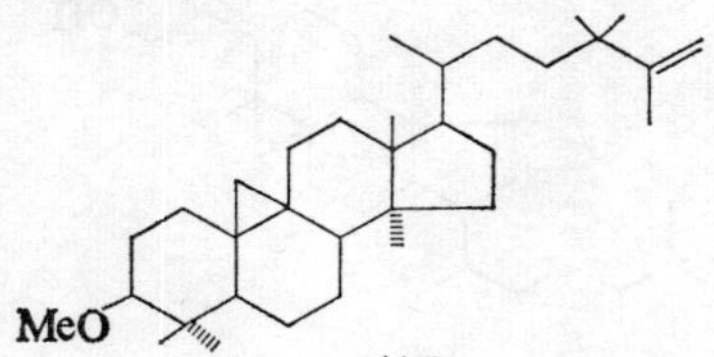 *Neolitsea dealbata*[39]
(15) Euphol; 5α,13α,14β,17α-lanosta-8,24-dien-3β-ol	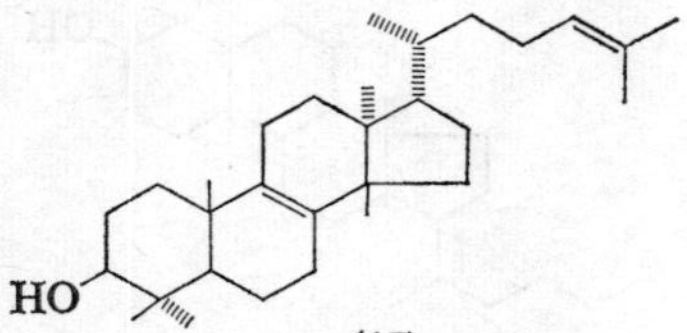 *Euphorbia* spp.[3,30]
(16) Tirucallol; 5α,13α,14β,17α,20S-lanosta-8,24-dien-3β-ol	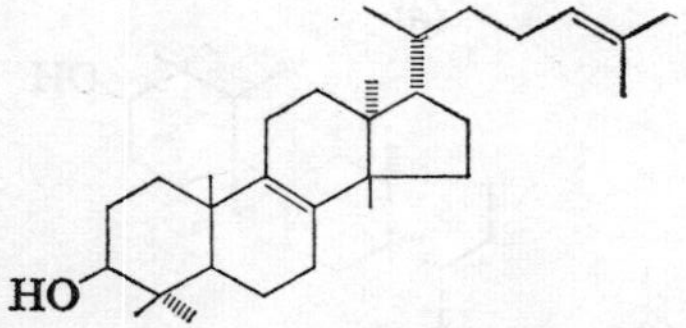 *Euphorbia tirucalli*[31] *E. triangularis*[32]

Butyrospermol; 5α,13α,14β,17α-lanosta-7,24-dien-3β-ol	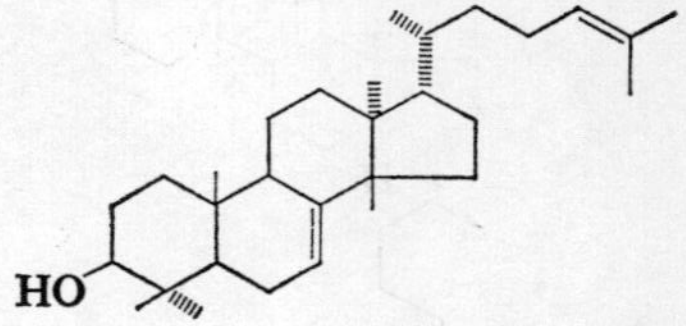 (17)	*Butyrospermum parkii*[33,34] *Maclura pomifera*[35]
Euphorbol; 4,4,14α-trimethyl-5α,13α,14β,17α,20S-ergosta-8,24(28)-dien-3β-ol	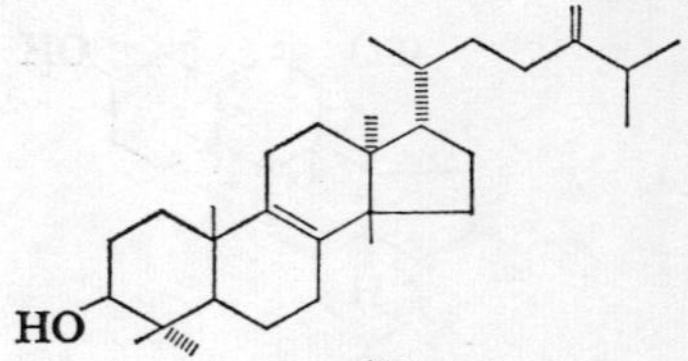 (18)	*Euphorbia triangularis*[32] *E. resinifera*[36]
Dammaradienol; 8-methyl-18-nor-5α-lanosta-20,24-dien-3β-ol	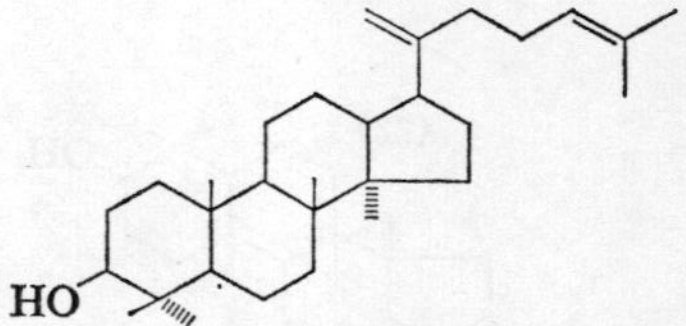 (19)	Dipterocarpaceae resins[37,37a]
Dammarenediol; 8-methyl-18-nor-5α-lanosta-24-en-3β,20ξ-diol	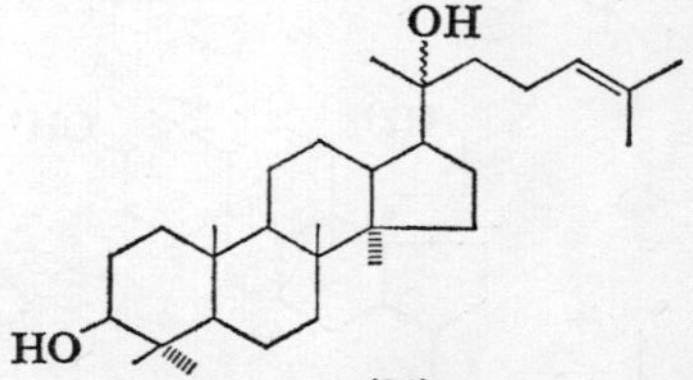 (20)	Dipterocarpaceae resins[37,37a]

Table 1.—*continued*

Carnaubadiol; 8-methyl-18-nor-5α-ergost-25-en-3β,20ξ-diol	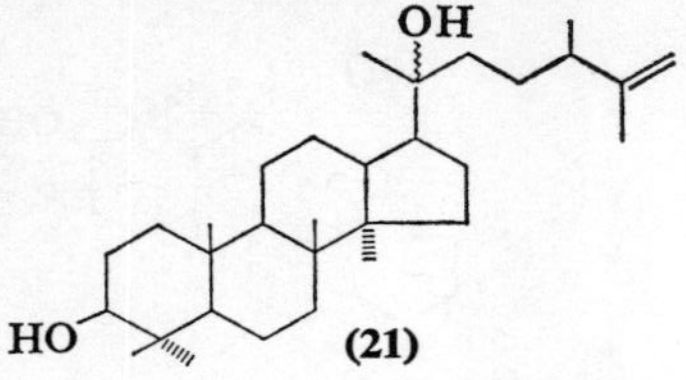 (21)	*Copernicia prunifera*[156]
Aglaiol; 24ξ,25-epoxy-8-methyl-18-nor-5α-lanost-20-en-3β-ol	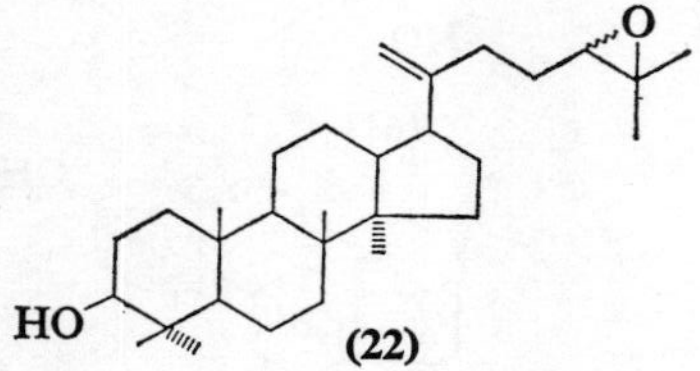 (22)	*Aglaia odorata*[38]
3β-Hydroxyprotosta-17(20),24-diene; 8α-methyl-18-nor-5α,13α,14β-lanosta-E-17(20),24-dien-3β-ol	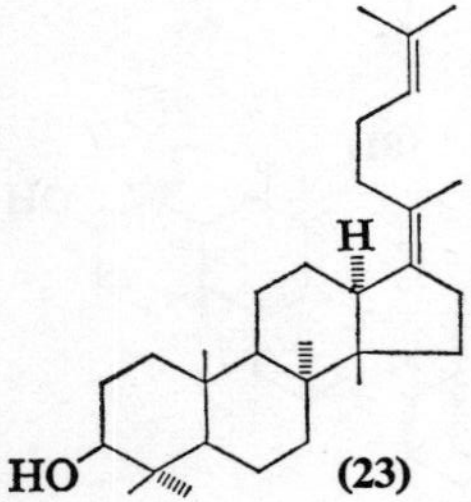 (23)	*Cephalosporium caerulens*[258]
3β,31-Dihydroxyprotosta-17(20),24-diene; 8α-methyl-18-nor-5α,13α,14β-lanosta-E-17(20),24-dien-3β,31-diol	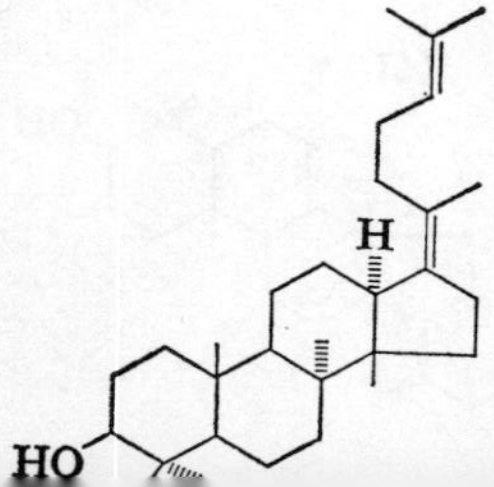	*Cephalosporium caerulens*[259]

(b) 4-Methyl sterols

31-Nor-cycloartenol; 4α,14α-dimethyl-9β,19-
cyclo-5α-cholest-24-en-3β-ol

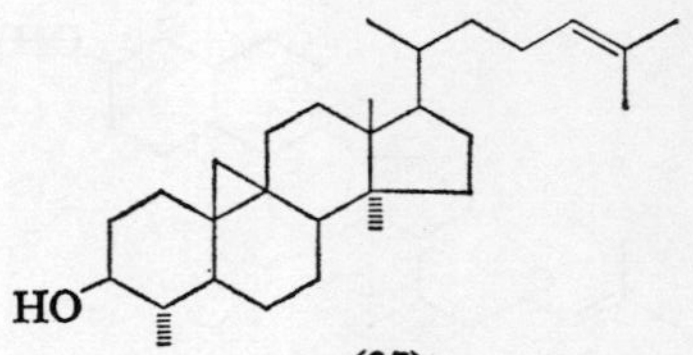

(25)

Carnegiea gigantea[40]

31-Nor-cycloartanol; 4α,14α-dimethyl-9β,19-
cyclo-5α-cholestan-3β-ol

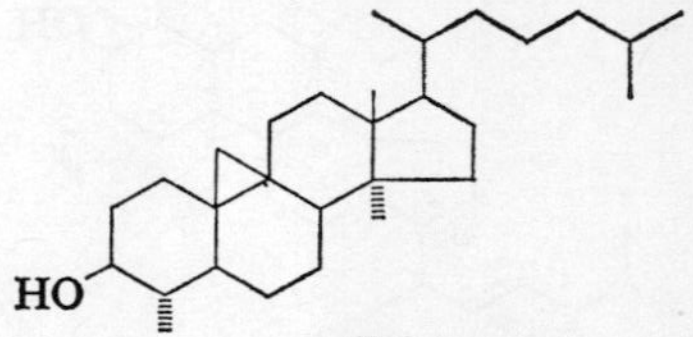

(26)

Polypodium vulgare[41]
Smilex aspera[42]

Cycloeucalenol; 4α,14α-dimethyl-9β,19-cyclo-
5α-ergost-24(28)-en-3β-ol

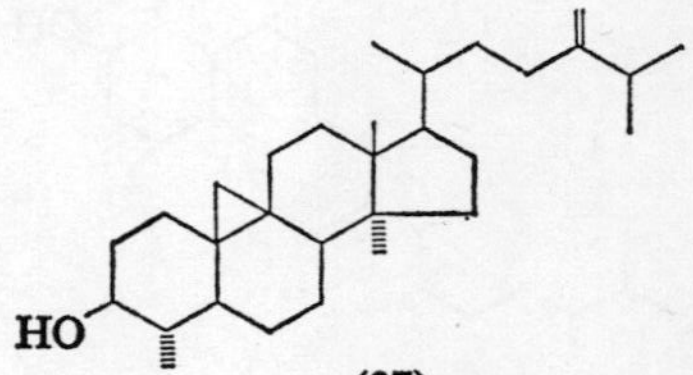

(27)

Eucalyptus microcorys[43]
Swietenia mahagoni[44]

31-Nor-cyclolaudenol; 4α,14α-dimethyl-9β,19-
cyclo-5α-ergost-25-en-3β-ol

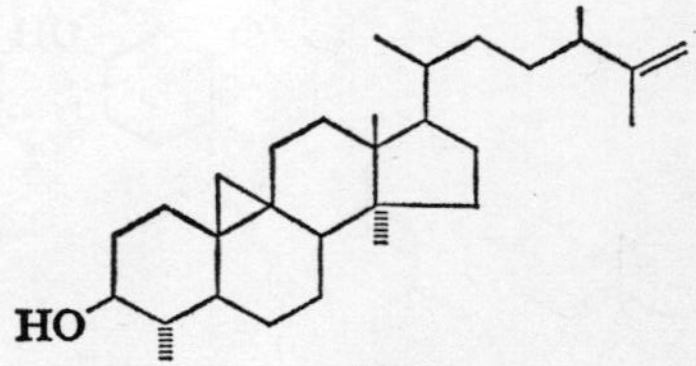

(28)

Polypodium vulgare[28]

Table 1.—*continued*

31-Nor-lanosterol; 4α,14α-dimethyl-5α-cholesta-8,24-dien-3β-ol	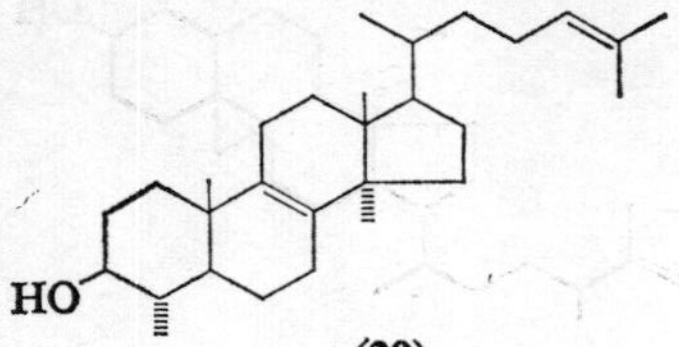 (29)	*Funtumia latifolia*[46] *Solanum tuberosum*[47]
Obtusifoliol; 4α,14α-dimethyl-5α-ergosta-8, 24(28)-dien-3β-ol	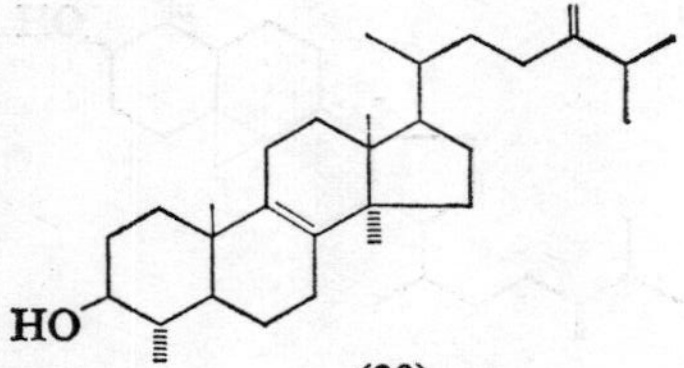 (30)	*Euphorbia obtusifolia*[48]
Lophenol; 4α-methyl-5α-cholest-7-en-3β-ol	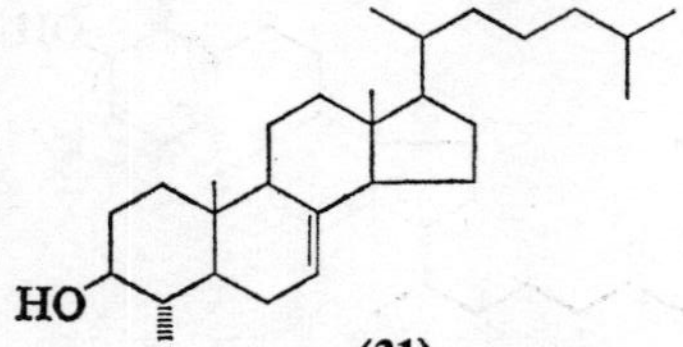 (31)	*Lophocereus shottii*[49]
24,25-Dehydrolophenol; 4α-methyl-5α-cholesta-7,24-dien-3β-ol	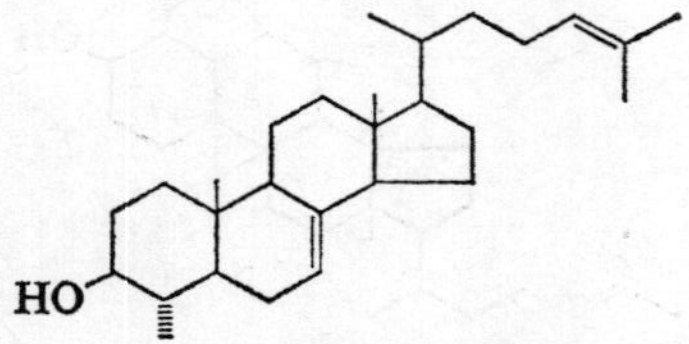 (32)	*Funtumia latifolia*[46]

4α-Methylzymosterol; 4α-methyl-5α-cholesta-8,24-dien-3β-ol	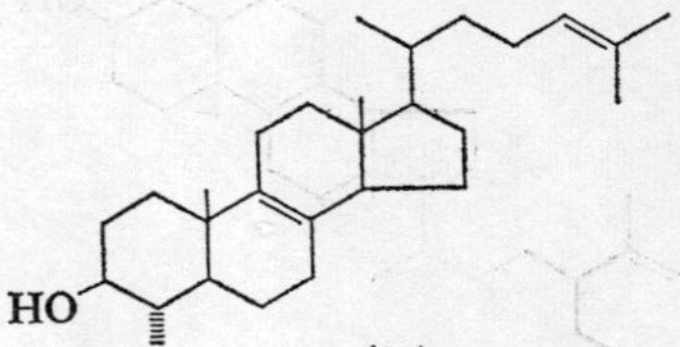 (33)	*Saccharomyces cerevisiae*[14]
4α-Methyl-5α-cholesta-8(14),24-dien-3β-ol	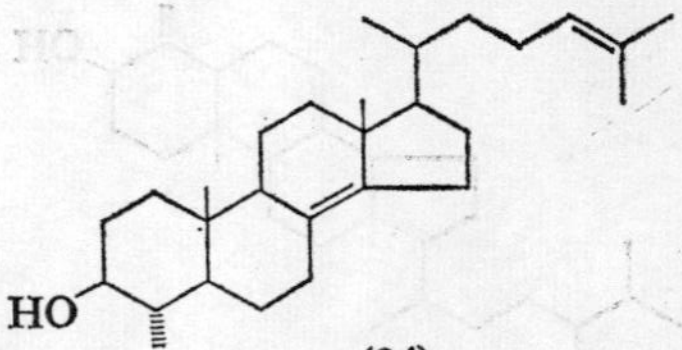 (34)	*Saccharomyces cerevisiae*[50]
4α-Methyl-5α-ergosta-8,24(28)-dien-3β-ol	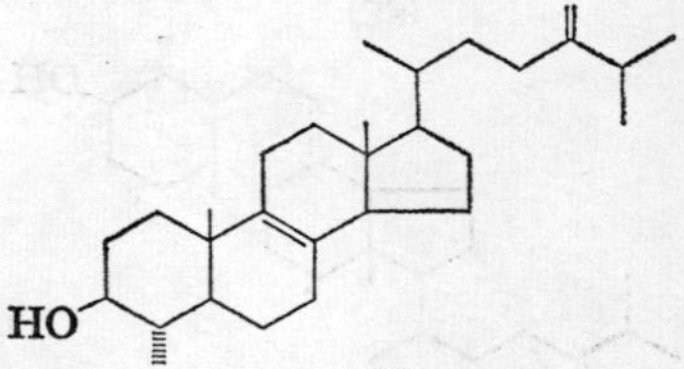 (35)	*Saccharomyces cerevisiae*[50]
24-Methylene lophenol; 4α-methyl-5α-ergosta-7,24(28)-dien-3β-ol	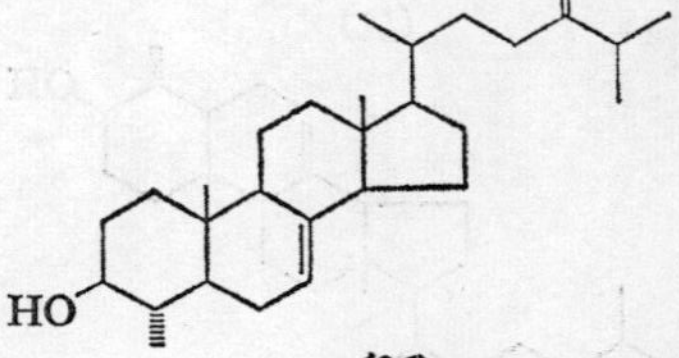 (36)	*Solanum tuberosum*[51]

Table 1.—*continued*

24-Methyl lophenol; 4α,24ξ-dimethyl-5α-cholest-7-en-3β-ol	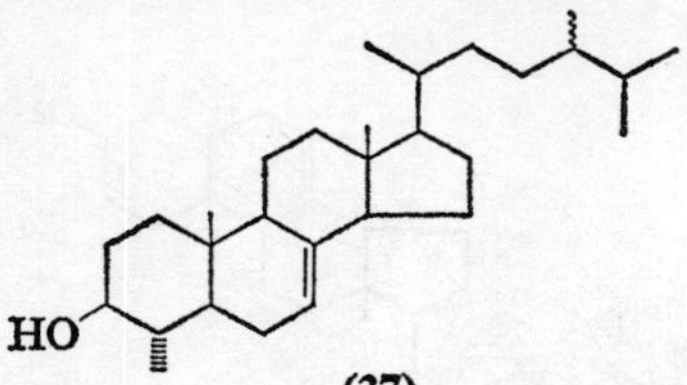 (37)	*Solanum tuberosum*[51] *Caffea arabica*[52] *Citrus paradisi*[45]
24-Ethylidene lophenol, citrostadienol; 4α-methyl-5α-stigmasta-7,Z-24(28)-dien-3β-ol	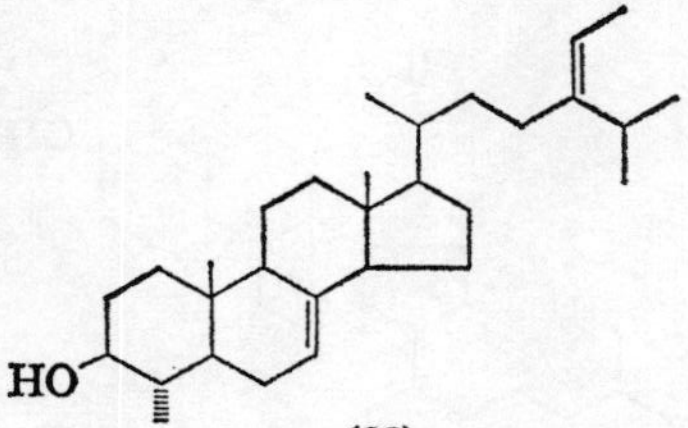 (38)	*Citrus* spp.[53–56] *Solanum tuberosum*[57] *Betula verrucosa*[58]
24-Ethyl lophenol; 4α-methyl-24ξ-ethyl-5α-cholest-7-en-3β-ol	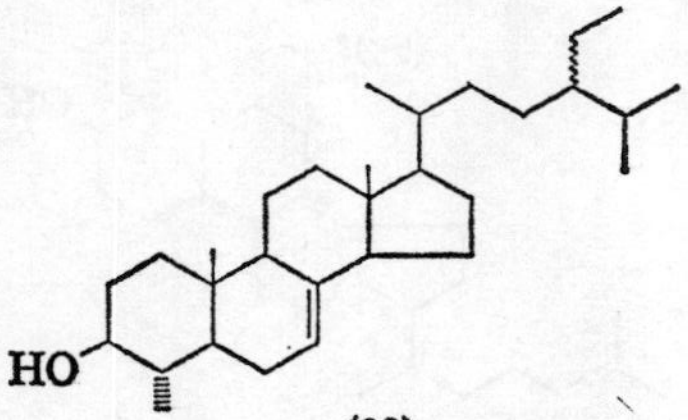 (39)	*Caffea arabica*[52] *Citrus paradisi*[45]
4α-Methyl-(24S)-24-ethyl-5α-cholesta-7,25-dien-3β-ol	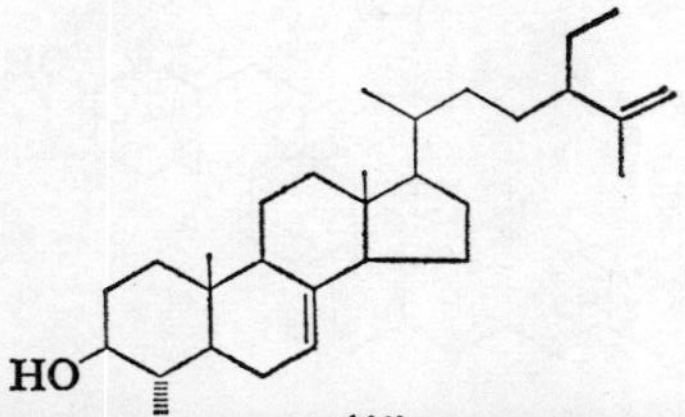	*Clerodendrum campbellii*[128]

4β-Methyl stigmasta-7,24(28)-dienol;
   4β-methyl-5α-stigmasta-7,Z-24(28)-dien-3β-ol

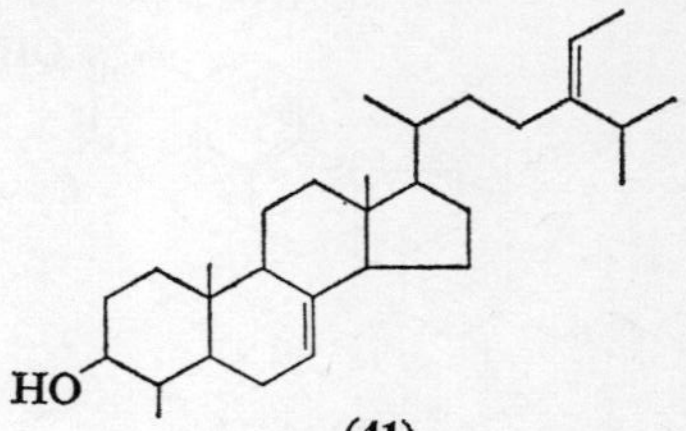

Calendula officinalis[132]

(41)

(c) *4-Demethyl sterols*

Cholesterol; cholest-5-en-3β-ol

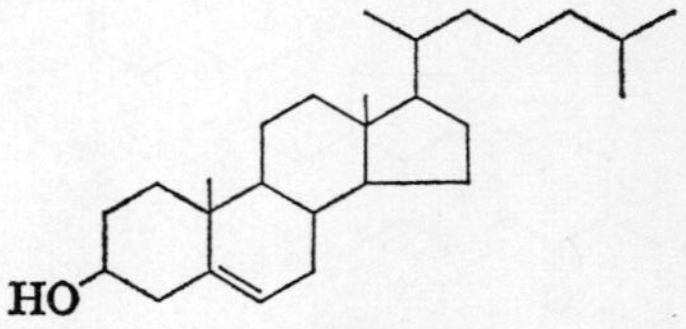

Rhodophyceae[59–61]
Small amounts in many plants

(42)

22-Dehydrocholesterol; cholesta-5,22-dien-3β-ol

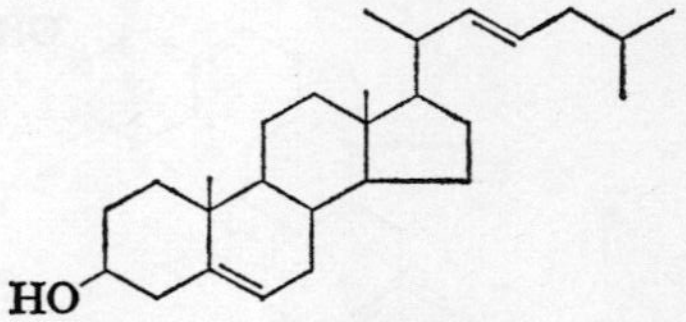

Hypnea japonica[62]

(43)

Desmosterol; cholesta-5,24-dien-3β-ol

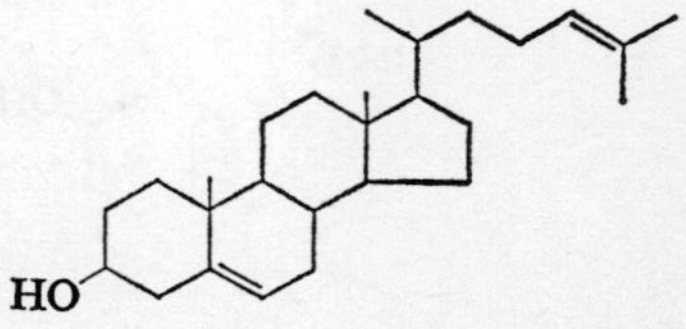

Pollen[63]
Rhodymenia palmata[61]
Funtumia latifolia[46]

(44)

Table 1.—*continued*

Zymosterol; 5α-cholesta-8,24-dien-3β-ol	(45)	*Saccharomyces cerevisiae*[64]
22-Hydroxycholesterol; (22S)-22-hydroxycholest-5-en-3β-ol	(46)	*Narthecium ossifragum*[65,66]
24-Ketocholesterol; 24-oxo-cholest-5-en-3β-ol	(47)	*Pelvetia canaliculata*[67] *Ascophyllum nodosum*[68]
Peniocerol; 5α-cholest-8-en-3β,6α-diol	(48)	*Peniocereus fosterianus*[69] *Wilcoxia viperina*[70]

Macdougallin; 14α-methyl-5α-cholest-8-en-3β,6α-diol

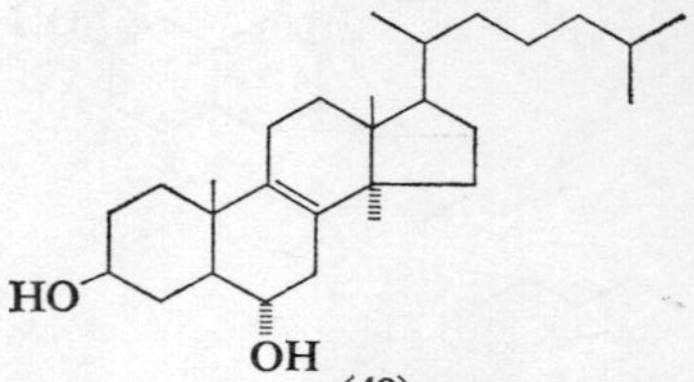

Peniocereus macdougallii[71]

(49)

Pollinastanol; 14α-methyl-9β,19-cyclo-5α-
cholestan-3β-ol

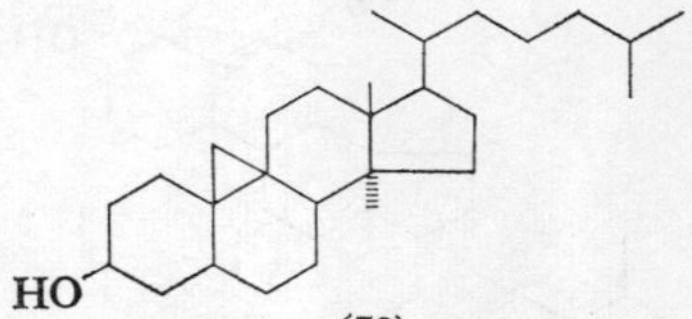

Pollen[72,73,73a]
Polypodium vulgare[74]

(50)

24-Methylene cholesterol; ergosta-5,24(28)-dien-3β-ol

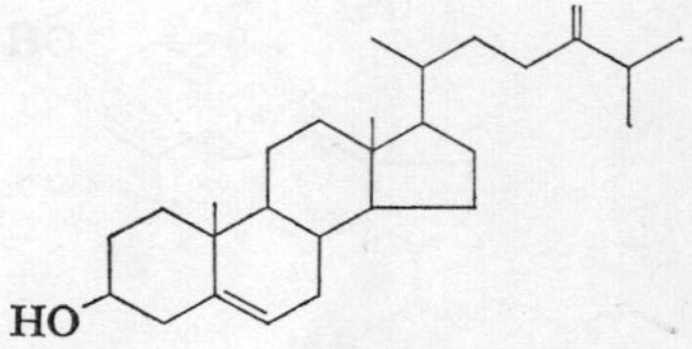

Pollen[73a,75,76]
Rhodophyceae[59,60]
Avena sativa[77]

(51)

Episterol; 5α-ergosta-7,24(28)-dien-3β-ol

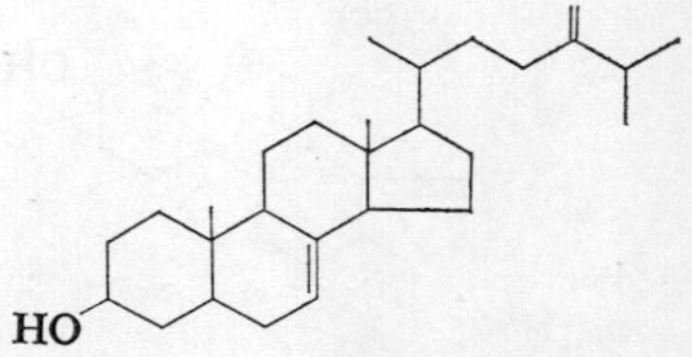

Saccharomyces cerevisiae[78−80]
Phycomyces blakesleeanus[19,20]

(52)

Table 1.—*continued*

Compound		Source
Fecosterol; 5α-ergosta-8,24(28)-dien-3β-ol	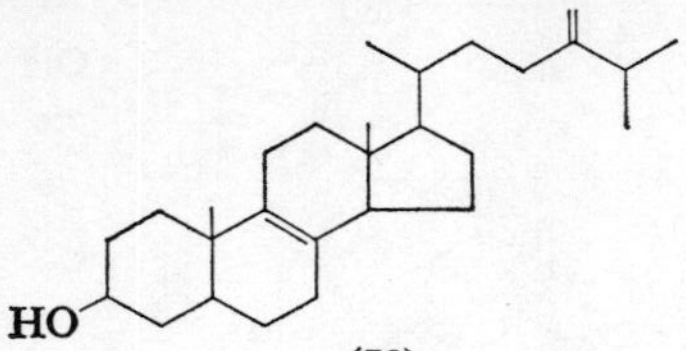 (53)	*Saccharomyces cerevisiae*[81]
5-Dehydroepisterol; ergosta-5,7,24(28)-trien-3β-ol	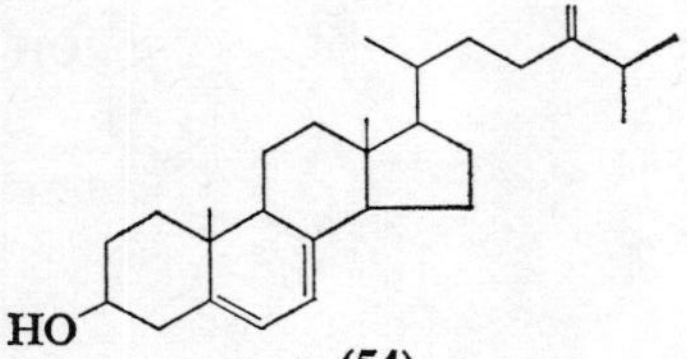 (54)	*Phycomyces blakesleeanus*[83]
Ergostatetraeneol, ergosta-5,7,22,24(28)-tetraen-3β-ol	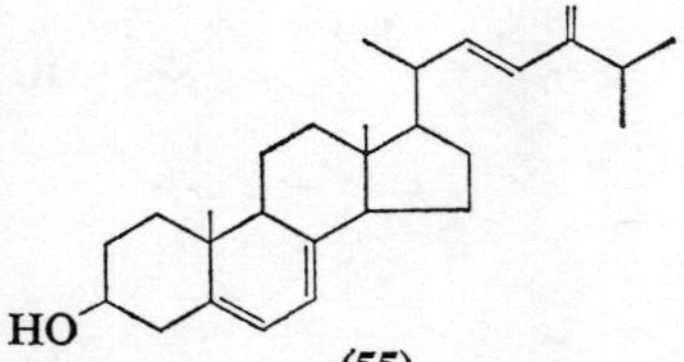 (55)	*Saccharomyces cerevisiae*[84,85]
Ergost-5-en-3β-ol; (24S)-24-methylcholest-5-en-3β-ol	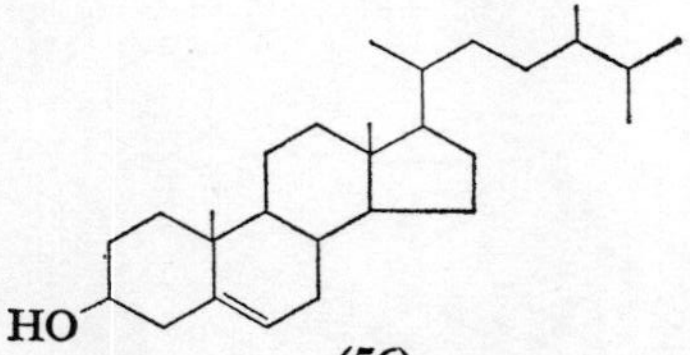 (56)	*Chlorella ellipsoidea*[86] *Ochromonas danica*[87]

Brassicasterol; ergosta-5,22-dien-3β-ol;
(24R)-24-methylcholesta-5,22-dien-3β-ol

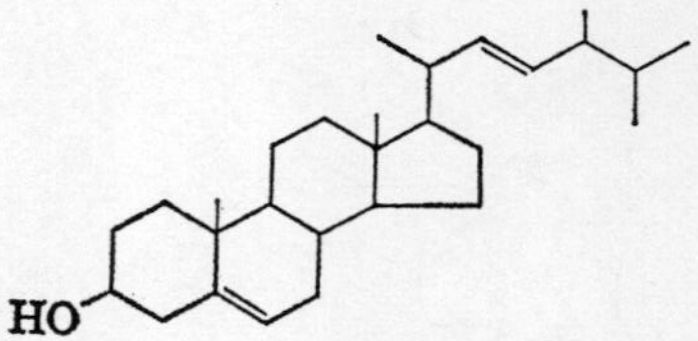

Brassica rapa[88]
Chlorella spp.[86]
Ochromonas danica[87]

(57)

Fungisterol; 5α-ergost-7-en-3β-ol

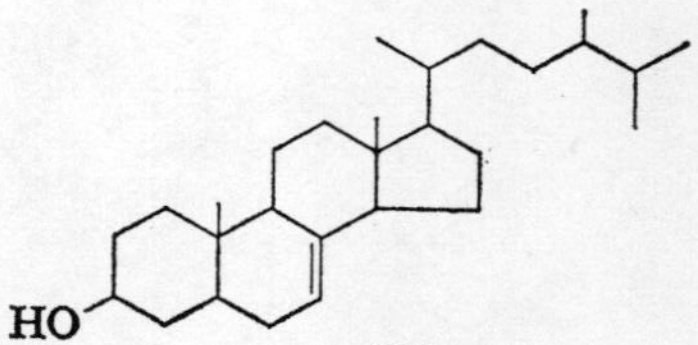

Amanita phalloides[89]
Candida utilis[82]
Chlorella vulgaris[116]

(58)

5-Dihydroergosterol; 5α-ergosta-7,22-dien-3β-ol

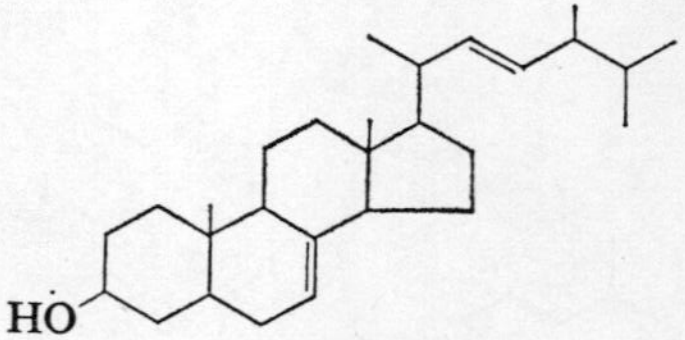

Saccharomyces cerevisiae[90]

(59)

Ergosterol; ergosta-5,7,22-trien-3β-ol

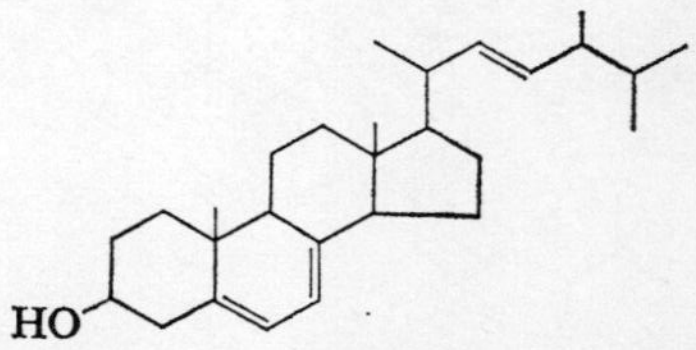

Many fungi[91–94]

(60)

Table 1.—*continued*

Ergosta-5,7,14,22-tetraen-3β-ol	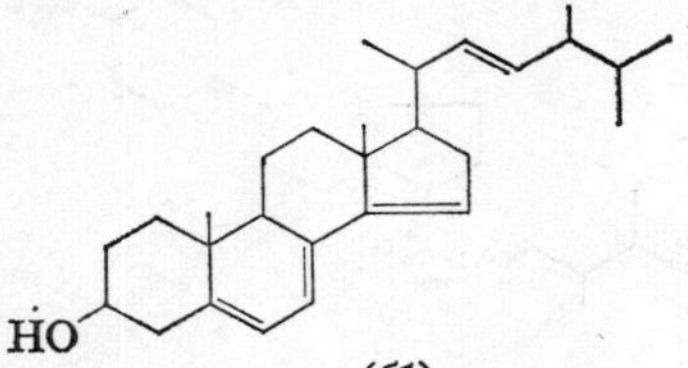 (61)	*Aspergillus niger*[95]
22-Dihydroergosterol; ergosta-5,7-dien-3β-ol	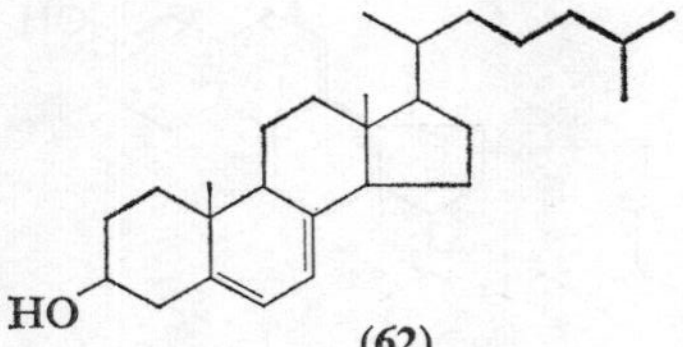 (62)	*Polyporus pargamenus*[96]
Ascosterol; 5α-ergosta-8,23-dien-3β-ol	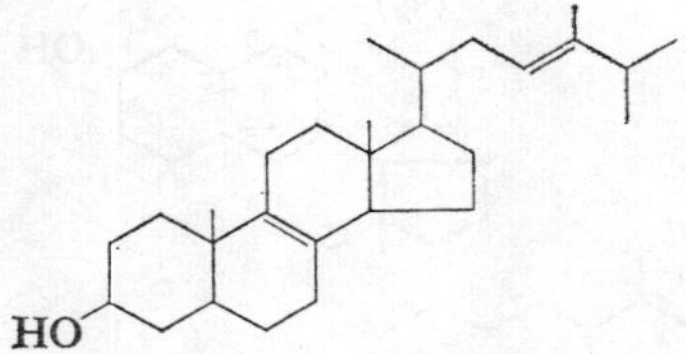 (63)	*Saccharomyces cerevisiae*[97]
Ergosta-5,8,17(20)-trien-3β-ol	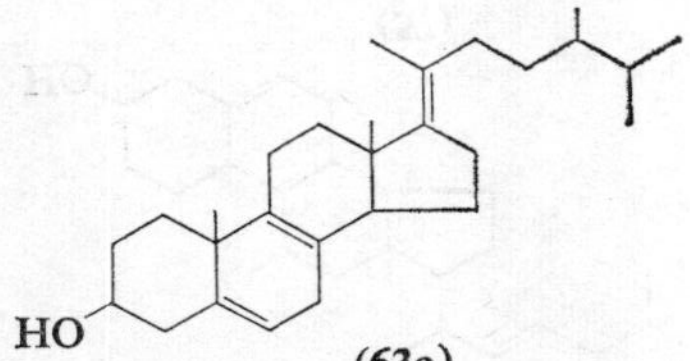 (63a)	*Edgeworthia papyrifera*[436]

Campestanol; (24R)-24-methyl-5α-cholestan-3β-ol	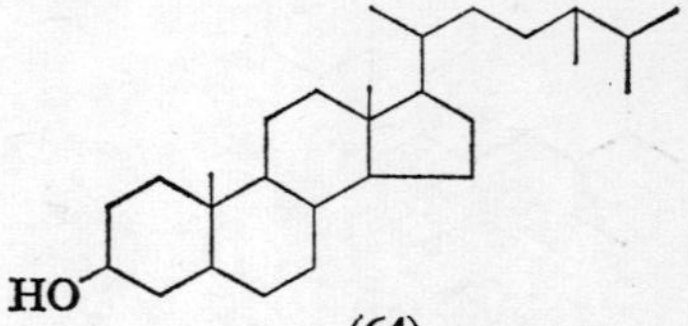 (64)	*Zea mays*[102] *Physarum polycephalum*[22]
Campesterol; (24R)-24-methylcholest-5-en-3β-ol	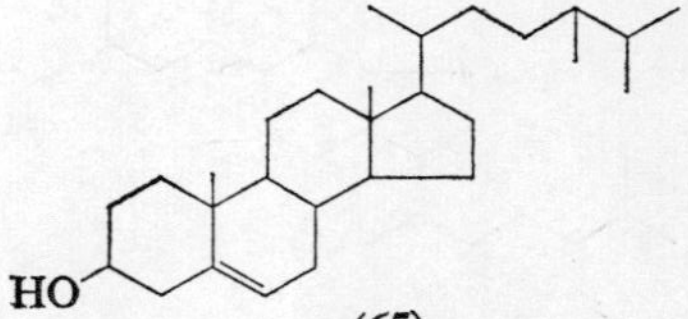 (65)	Many plants[93,94]
Haliclonasterol; (24R)-24-methyl-(20S)-cholest-5-en-3β-ol	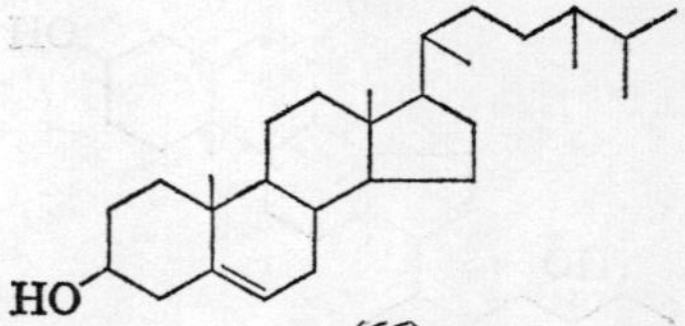 (66)	*Monostroma nitidum*[103]
Fucosterol; stigmasta-5,E-24(28)-dien-3β-ol	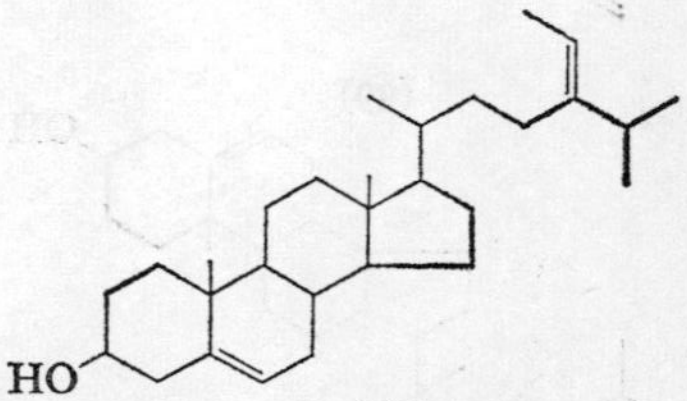 (67)	Phaeophyceae[60,104,105]

Table 1.—*continued*

Sargasterol; (20S)-stigmasta-5,24(28)-dien-3β-ol	(68)	*Sargassum ringoldianum*[106]
Saringasterol; stigmasta-5,28-dien-3β,24ξ-diol	(69)	*Sargassum ringoldianum*[131] *Ascophyllum nodosum*[68]
28-Isofucosterol, avenasterol; stigmasta-5,Z-24(28)-dien-3β-ol	(70)	*Avena sativa*[107,108] *Enteromorpha linza*[103] *Enteromorpha intestinalis*[109] *Ulva lactuca*[109]
Δ^7-Avenasterol; 5α-stigmasta-7,Z-24(28)-dien-3β-ol	(71)	*Avena sativa*[77] *Vernonia anthelmintica*[110]

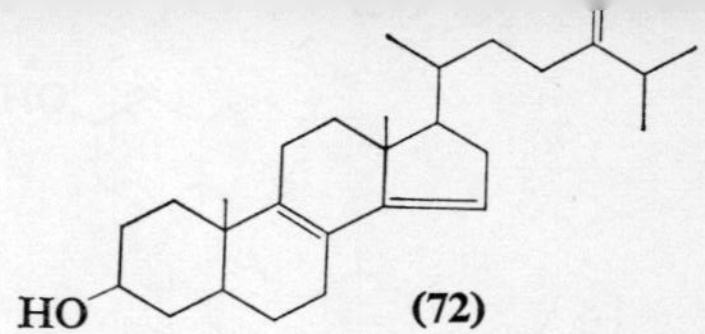

Vernosterol; 5α-stigmasta-8(14),15,Z-24(28)-
trien-3β-ol

Vernonia anthelmintica[112]
(cf. with ref. 111)

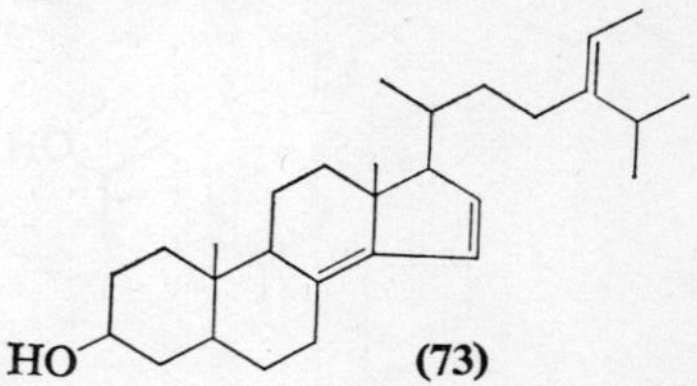

(72)

Clionasterol; (24S)-24-ethylcholest-5-en-3β-ol

Chlorella spp.[86]
Conopharyngia durissima[113]

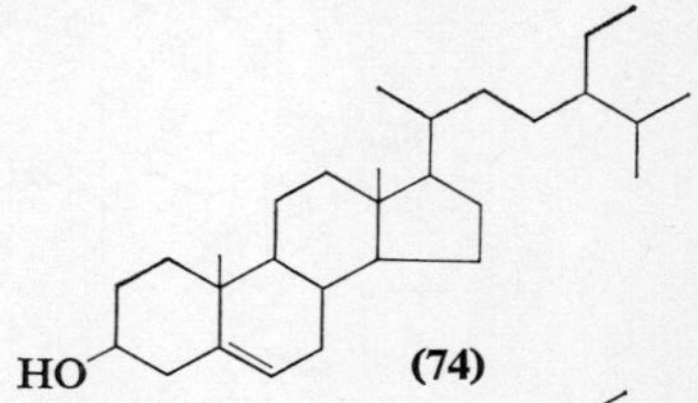

(73)

Poriferasterol; (24R)-24-ethylcholesta-5,22-dien-3β-ol

Ochromonas malhamensis[87]

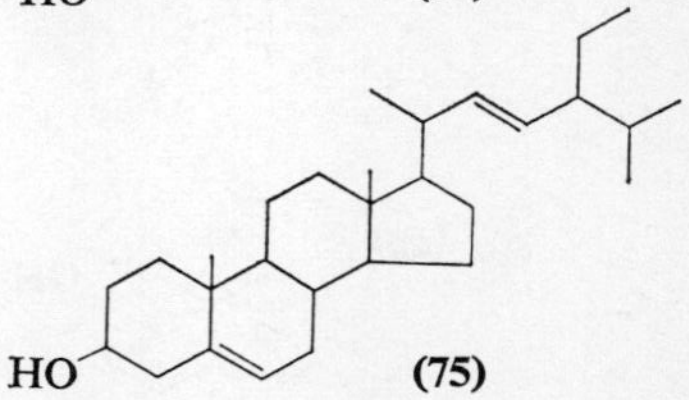

(74)

(75)

7-Dehydroporiferasterol; (24R)-24-ethylcholesta-
5,7,22-trien-3β-ol

Ochromonas danica[87]

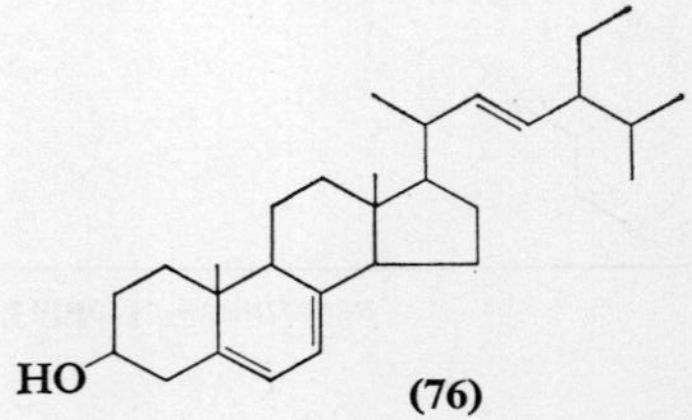

(76)

Table 1.—*continued*

Chondrillasterol; (24R)-24-ethyl-5α-cholesta-7,22-dien-3β-ol	(77)	*Chlorella vulgaris*[86,116] *Scenedesmus obliquus*[114] *Navicula pelliculosa*[115]
22-Dihydrochondrillasterol; (24S)-24-ethyl-5α-cholest-7-en-3β-ol	(78)	*Chlorella vulgaris*[116]
Stigmastanol; 5α-stigmastan-3β-ol	(79)	*Calycanthus floridas*[117] *Acacia* spp.[118] *Zea mays*[102] *Physarum polycephalum*[22]
Sitosterol; stigmast-5-en-3β-ol; (24R)-24-ethylcholest-5-en-3β-ol	(80)	Many plants[93,94]

Dictyostelium discoideum[119]

Many plants[93,94]

Lophocereus scottii[49]
Acacia spp.[118]
Bupleurum falcatum[120]

Acacia spp.[118]
Spinacea oleracea[121]
Medicago sativa[122]

(81)

(82)

(83)

(84)

22-Dehydrostigmastanol; 5α-stigmast-22-en-3β-ol

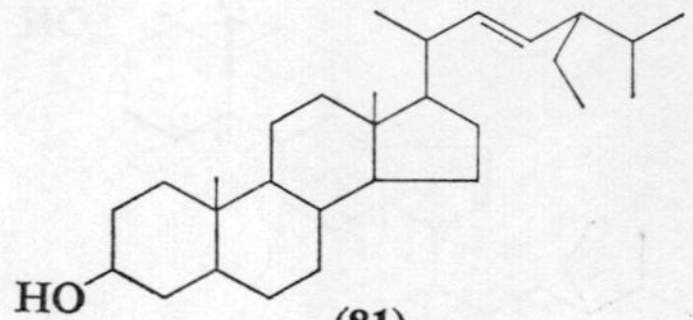

Stigmasterol; stigmasta-5,22-dien-3β-ol;
(24S)-24-ethylcholesta-5,22-dien-3β-ol

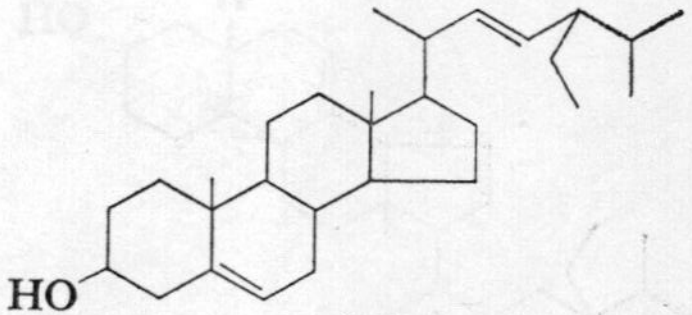

Stigmast-7-enol; 5α-stigmast-7-en-3β-ol

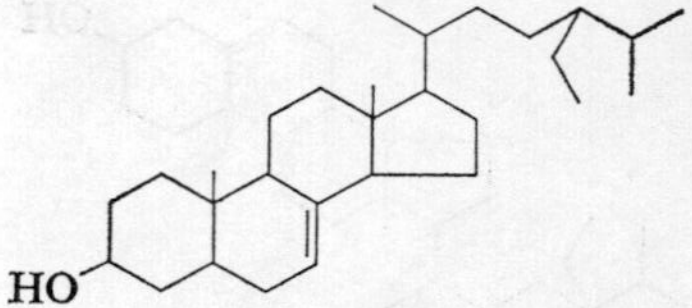

Spinasterol; 5α-stigmasta-7,22-dien-3β-ol

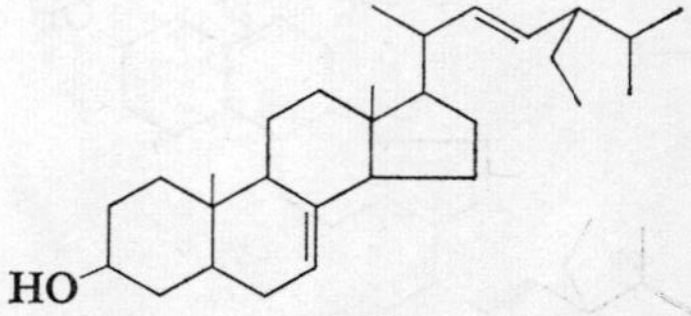

Table 1.—*continued*

7-Dehydrostigmasterol; stigmasta-5,7,22-trien-3β-ol	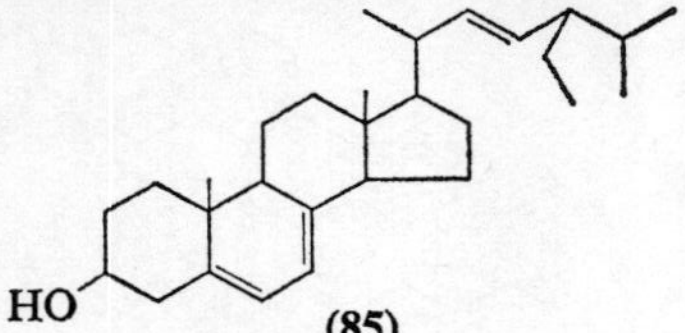 (85)	*Phellodendron amurense*[123]
Celsianol; stigmasta-5,9(11)-dien-3β-ol	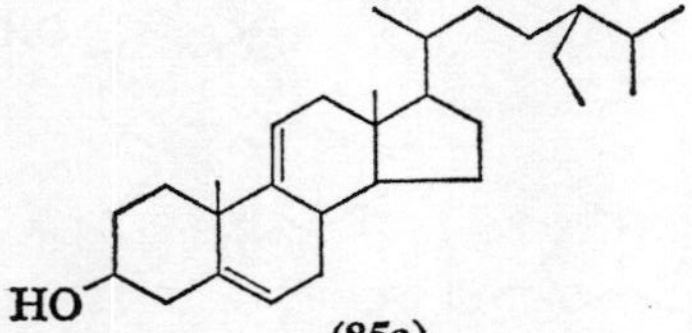 (85a)	*Celsia coromandeliana*[437]
5β-Stigmast-7-en-3α-ol	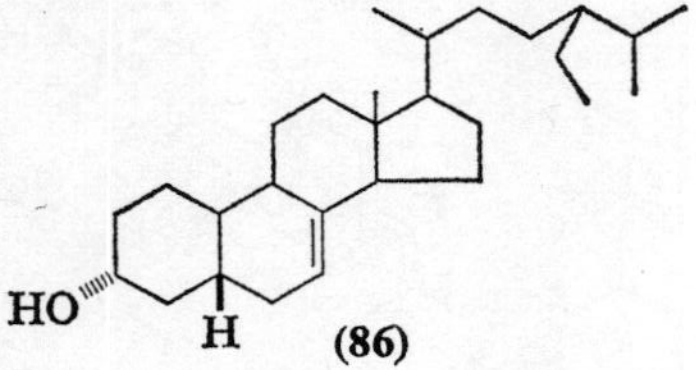 (86)	*Daemia extensa*[124]
5β-Stigmast-8(14)-en-3α-ol	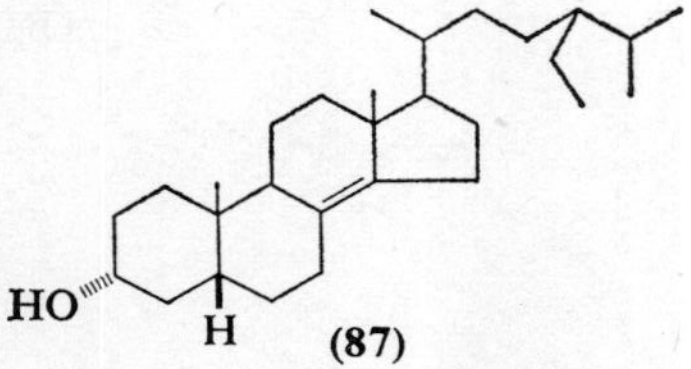 (87)	*Daemia extensa*[124]

5α-Stigmasta-8(14),22-dien-3β-ol	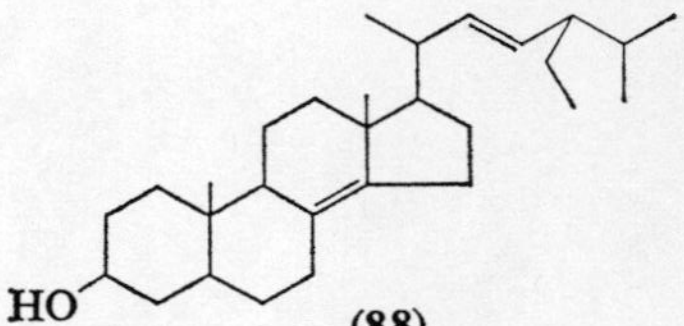 **(88)**	*Aplopappus heterophyllus*[125]
Clerosterol; stigmasta-5,25-dien-3β-ol	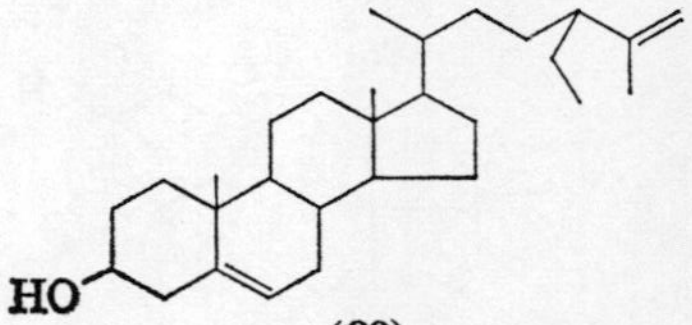 **(89)**	*Clerodendron infortunatum*[126] *Momordica charantia*[127]
(24S)-24-Ethylcholesta-5,22,25-trien-3β-ol	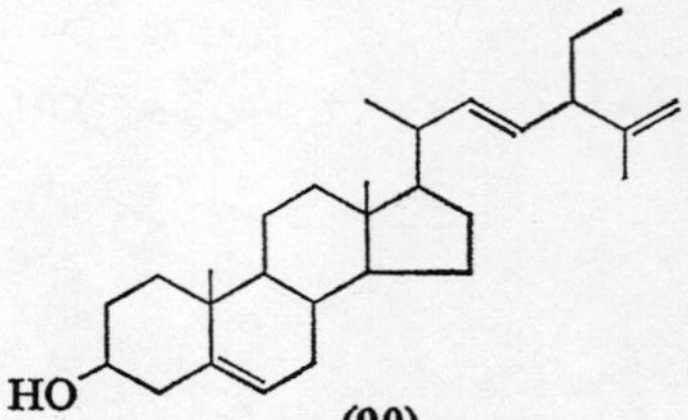 **(90)**	*Clerodendrum campbellii*[128]
5α-Stigmasta-7,25-dien-3β-ol	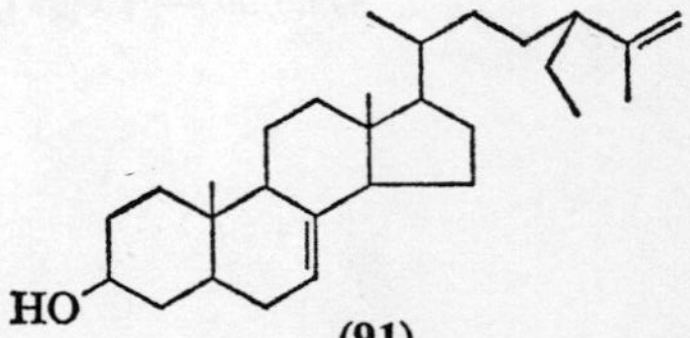 **(91)**	*Momordica charantia*[129]

Table 1.—*continued*

5α-Stigmasta-7,22,25-trien-3β-ol	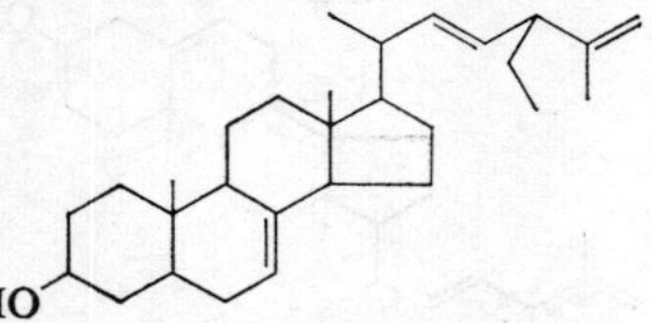 HO (92)	*Momordica charantia*[129]
Elasterol; 5α-stigmasta-7,16,25-trien-3β-ol	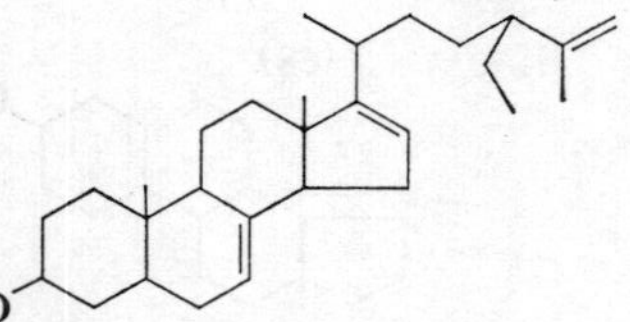 HO (93)	*Ecballium elaterium*[130]

reported[86, 87, 103, 109, 114–116, 136, 142–149]. In most cases where the C_{28}- and C_{29}-sterols have been rigorously characterized they have the 24S configuration[86, 87, 114–116, 147, 148] similar to fungal sterols but in contrast with the sterols of most higher plants which are 24R compounds*. However reports on some algal species indicate that they either contain 24R sterols or that the C-24 configuration has not been determined[142–146]. In view of the biogenetic implications a careful examination of many more species, with particular attention to C-24 configuration in the C_{28}- and C_{29}-sterols seems warranted to determine if 24S sterols are indeed characteristic of algae.

In the fungi C_{28}-sterols predominate; ergosterol (60) is the principal sterol found in the majority of the Ascomycetes and Basidiomycetes examined[93, 94, 150], the other sterols found are principally ergostanol derivatives and can be regarded as ergosterol precursors. Among the Phycomycetes considerable variation is apparent in the sterol content. Species in the order Mucorales contain mainly ergosterol (60)[19, 20, 151] while members of the Peronosporales contain negligible amounts of sterol[151, 152] and require exogenous sterol for maximum growth and reproduction[152, 153]. Fungi of the orders Saprolegniales and Leptomitales contain[151] mixtures of cholesterol (42), desmosterol (44), 24-methylene cholesterol (51) and fucosterol (67) and so differ markedly from the other orders of fungi in their sterol biosynthetic capabilities. The slime moulds also appear to be unique among fungi since the two species so far examined, *Dictyostelium discoideum*[119, 154] and *Physarum polycephalum*[22], contain sterols of the C_{27}, C_{28} and C_{29} series and in particular produce the fully saturated compounds campestanol (64) and stigmastanol (79).

The Bryophyta and Pteridophyta have not been extensively examined but campesterol (65) and sitosterol (80) have been identified as constituents of some mosses[157, 158], ferns[159] and a horsetail[160].

The sterol mixtures from a number of Gymnosperms[161, 162] contain principally sitosterol (80) with smaller amounts of campesterol (65) and cholesterol (42). Sitosterol (80) and campesterol (65) together with stigmasterol (82), which all have the same configuration at C-24, are also the most common sterols found in the Angiosperms[93, 94]. However some species belonging to the Verbenacaeae and Cucurbitaceae are proving interesting exceptions since they contain C_{29}-sterols possessing a 25-methylene group[126–130, 163–165] such as compounds 89–93†. The sterols of the majority

---

* These configurations apply to sterols with the fully saturated side-chain. Introduction of double bonds may change the allocation of configuration from R to S and *vice versa*.

† Sterols 89, and 91–93 are shown in Table 1 as stigmasterol derivatives. However recent synthetic work[166, 167, 428] indicates that they may have the opposite configuration at C-24 as shown for sterol 90[128].

of Angiosperms analysed have a Δ^5 bond[93,94]; however Δ^7 sterols appear to be more common among species belonging to the Cucurbitaceae[129,130,165] and Leguminoseae[118,121,122] families.

III. SQUALENE BIOSYNTHESIS

The initial stages of sterol biosynthesis are anaerobic and involve the elaboration of the C_{30}-hydrocarbon squalene (**101**) from the C_6-compound mevalonic acid (**94**) via a sequence of phosphorylated compounds (Scheme

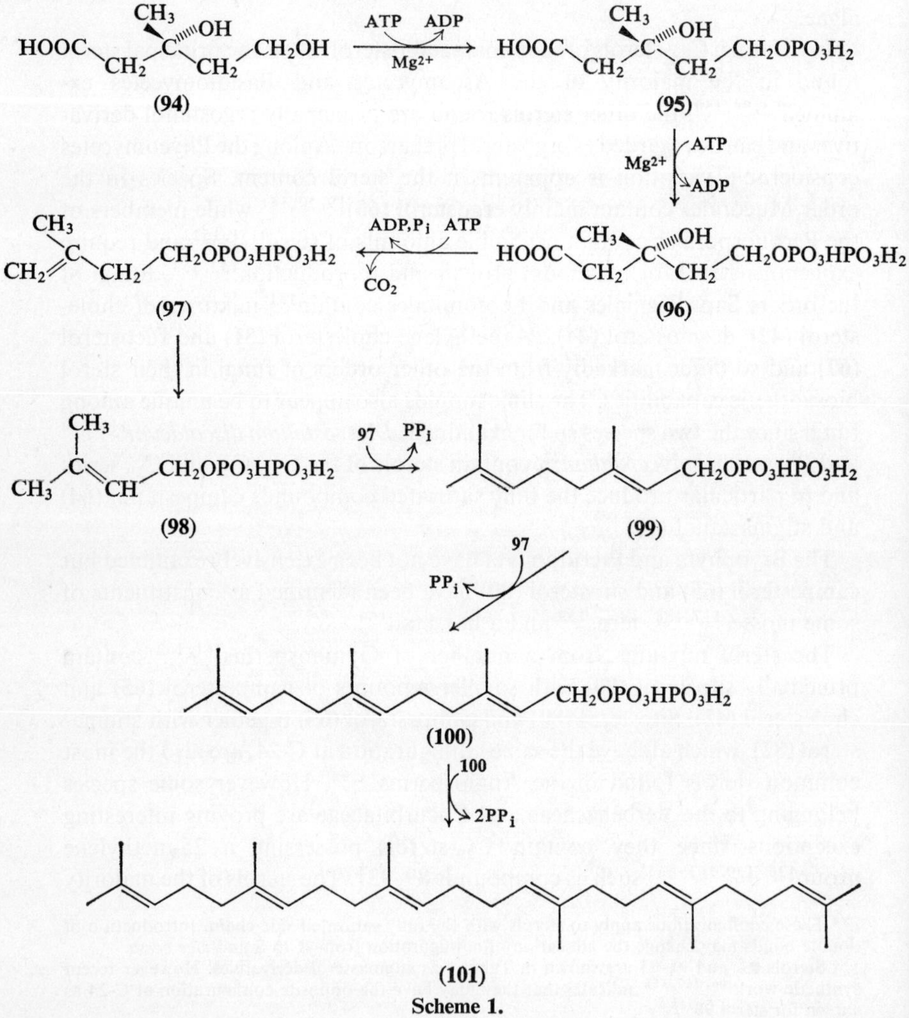

Scheme 1.

1). The route was initially elucidated using liver and yeast preparations and is well documented[168, 169]. The first step, catalysed by mevalonic kinase[170–173], produces mevalonic acid 5-phosphate (**95**) which is further phosphorylated by phosphomevalonate kinase[174–176] to give mevalonic acid 5-pyrophosphate (**96**). This is then followed by a dehydration and decarboxylation reaction mediated by 5-diphosphomevalonic anhydro-decarboxylase[177–179] to produce isopentenyl pyrophosphate (**97**). Isomerization of isopentenyl pyrophosphate by the enzyme isopentenyl pyrophosphate isomerase[180–183] leads to dimethylallyl pyrophosphate (**98**) which can then condense with further molecules of isopentenyl pyrophosphate in the presence of farnesyl synthetase to give first geranyl pyrophosphate (**99**) and then farnesyl pyrophosphate (**100**)[184, 185]. Finally two molecules of farnesyl pyrophosphate (**100**) are converted into one molecule of squalene (**101**) by a reaction requiring NADPH and a microsomal enzyme[177, 186–188].

The detailed stereochemistry of the reactions leading to squalene has been defined by the elegant work of Cornforth, Popják and their coworkers who developed synthetic methods for the preparation of mevalonic acid (**94**) labelled stereospecifically with deuterium or tritium at C-2, C-4 and C-5[189–191]. This work, although performed using enzymes of animal origin, will be briefly described since studies on some aspects of phytosterol biosynthesis have relied entirely upon the use of these stereospecifically labelled species of mevalonic acid.

The isomerization of isopentenyl pyrophosphate (**97**) to dimethylallyl pyrophosphate (**98**) and the subsequent condensations to give farnesyl pyrophosphate (**100**) all involve the loss of a C-2 proton from each of the isopentenyl pyrophosphate molecules involved (Scheme 2). In isopentenyl pyrophosphate (**97**) C-2 is derived from C-4 of the precursor mevalonic acid (**94**). It was thus shown[192, 193] using (4R)-4-^{2}H$_1$-mevalonic acid (**102a**) and (4S)-4-^{2}H$_1$-mevalonic acid (**102b**) that the protons lost in the above reactions are all derived from the 4-pro-S position of mevalonic acid (Scheme 2), the 4-pro-R hydrogen atom of mevalonic acid is retained and occupies the positions indicated in squalene (**103**).

The use of (2R)-2-^{2}H$_1$-mevalonic acid (**104**) and (2S)-2-^{2}H$_1$-mevalonic acid (**105**) enabled Cornforth *et al.*[194] to demonstrate that the conversion of mevalonic acid 5-pyrophosphate (**96**) into isopentenyl pyrophosphate (**97**) involves the *trans* elimination of the carboxyl and hydroxyl groups by a mechanism such as that depicted by **106** in Scheme 3. Moreover the availability of these species of stereospecifically labelled mevalonate permitted the investigation of the mechanism of condensation of isopentenyl pyrophosphate (**107**) with dimethylallyl pyrophosphate (**108**) and geranyl pyrophosphate. The results obtained[194] reveal that the allylic group (**108**) is

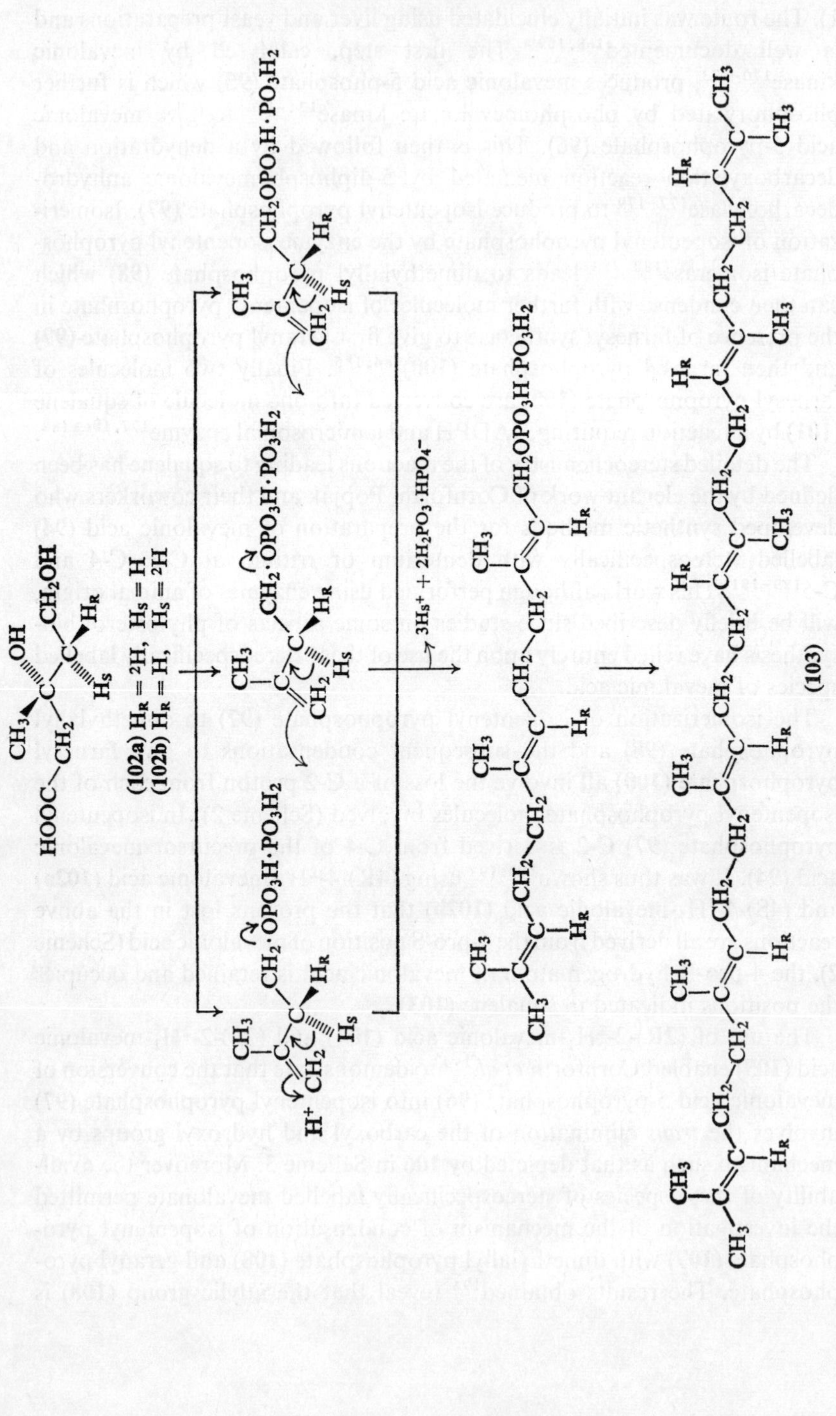

$(102a)\ H_R = {}^2H,\ H_S = H$
$(102b)\ H_R = H,\ H_S = {}^2H$

$3H_S^+ + 2H_2PO_3 \cdot HPO_4^-$

(103)

Scheme 2.

added to the vinylic carbon of isopentenyl pyrophosphate (**107**) on that side of the double bond from which the groups $-CH_2CH_2OP_2O_6{}^{3-}$, $-CH_3$, $-H_A$ and $-H_B$ are seen in a clockwise sequence (Scheme 4). A mechanism for this reaction has been formulated[192, 194] in which there is first a *trans* addition of the allylic group and a suitable electron-donating group X^-, possibly a constitutent group of the condensing enzyme, to give **109**. This is then followed by a *trans* elimination of X^- and the 2-pro-R proton from C-2 to leave geranyl pyrophosphate (**110**). In the condensation

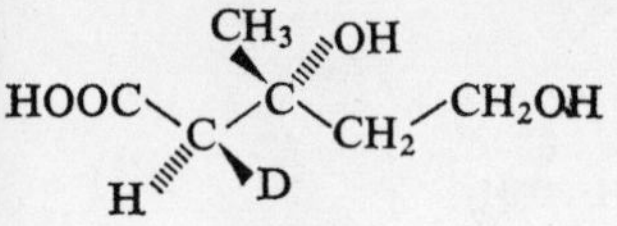

(104)

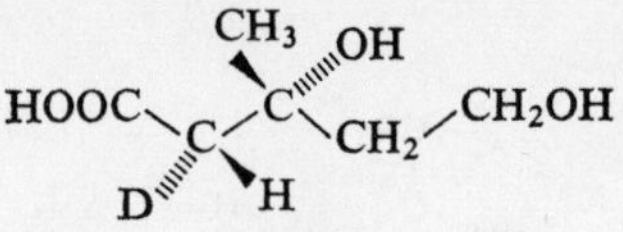

(105)

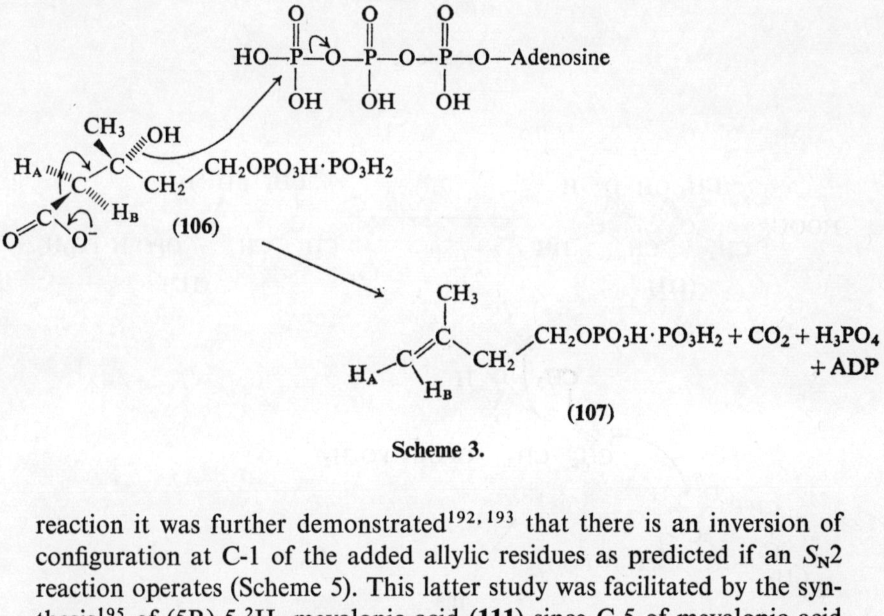

(106)

$$HO-P(=O)(OH)-O-P(=O)(OH)-O-P(=O)(OH)-O-Adenosine$$

$$CH_2OPO_3H\cdot PO_3H_2 + CO_2 + H_3PO_4 + ADP$$

(107)

Scheme 3.

reaction it was further demonstrated[192, 193] that there is an inversion of configuration at C-1 of the added allylic residues as predicted if an S_N2 reaction operates (Scheme 5). This latter study was facilitated by the synthesis[195] of (5R)-5-^{2}H$_1$-mevalonic acid (**111**) since C-5 of mevalonic acid becomes C-1 of the derived allylic pyrophosphate (**112**).

The coupling of two farnesyl pyrophosphate molecules to give squalene (Scheme 6) is similarly a stereospecific process[192] whereby there is an inversion of configuration at C-1 of one of the farnesyl pyrophosphate residues (**113**). In the other farnesyl residue (**114**) there is an exchange of the 1-pro-S hydrogen atom with the 4-pro-S hydrogen atom of NADPH whilst the farnesyl 1-pro-R hydrogen is retained in its original configuration[192, 193, 195]

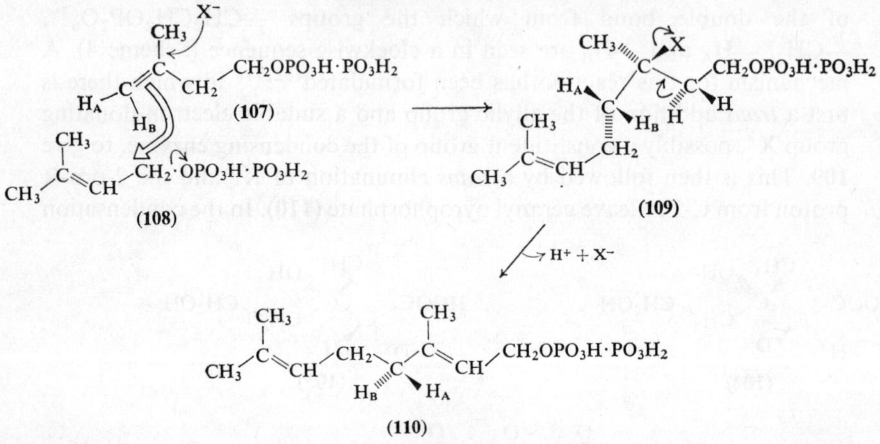

(107)

(108)

(109)

H⁺ + X⁻

(110)

Scheme 4.

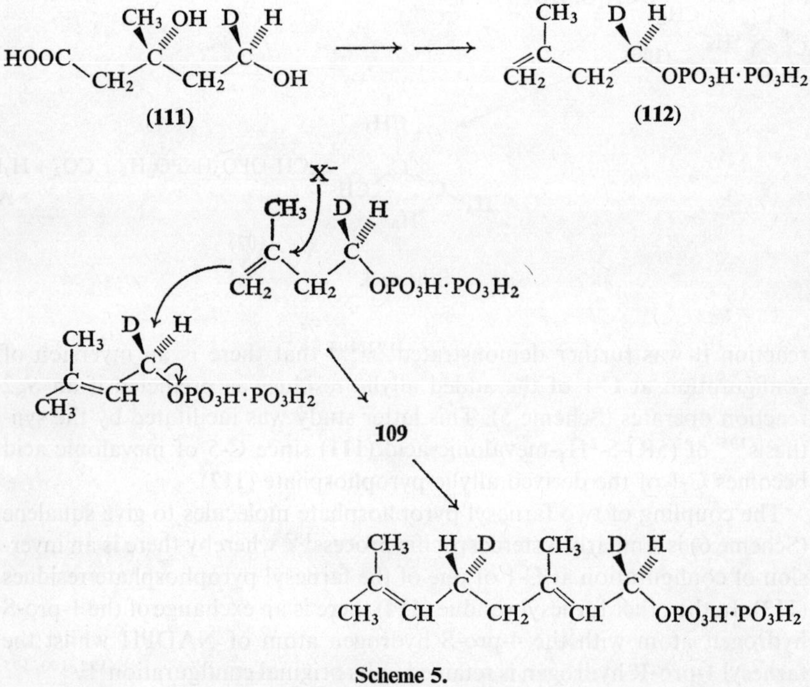

(111)

(112)

109

Scheme 5.

in the squalene produced **(115)**. The coupling of two farnesyl pyrophosphate molecules is now known to proceed with the intermediate formation of a presqualene pyrophosphate[446-448].

The biosynthesis of squalene in plant tissues has been shown to proceed by the same route as that operative in animal tissues. Germinating seeds of the pea (*Pisum sativum*) readily incorporated 2-^{14}C-mevalonic acid into squalene[196] which was subsequently shown to have the same labelling pattern as squalene biosynthesized in animal tissues[197]. Mevalonate kinase and phosphomevalonate kinase enzymes have been demonstrated in a number

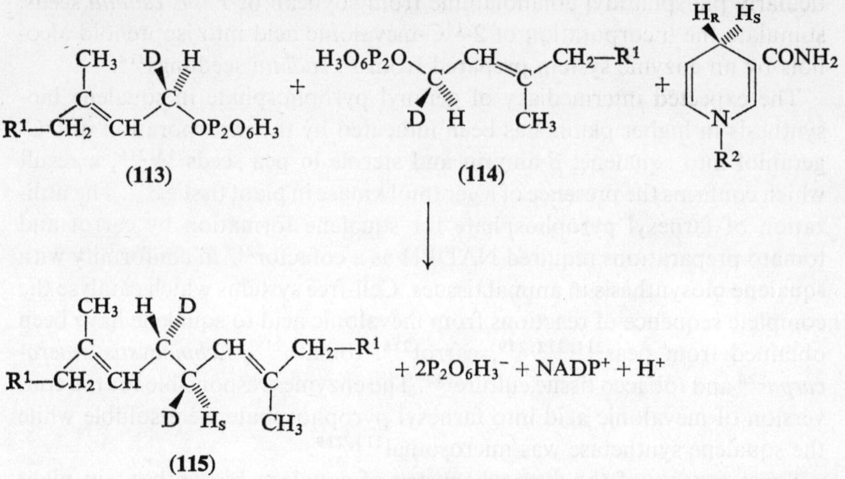

Scheme 6.

of plant tissues including *Cucurbita pepo*[198], sweet potato[199], *Pisum sativum*[200], Valencia orange juice vesicles[201], *Kalanchoë crenata* callus culture[202], *Phaseolus vulgaris*[203], *Hevea brasiliensis* latex[204,205] and *Pinus radiata*[206-208]. For maximum mevalonate kinase activity a divalent metal ion was required[198,204], Mn^{2+} being more efficient than Mg^{2+}. The mevalonate kinase of some plant tissues was inhibited by reagents which react with thiol groups[198,201,204]. In contrast to pig liver kinase[176] the mevalonate kinases from yeast[170], *Cucurbita pepo*[198] and *Hevea brasiliensis* latex[204] were not activated by the addition of thiol reagents although the kinase from the orange juice vesicles apparently does require reducing agents for full activity[201]. Optimum pH values in the range 5.5 to 7.5 have been cited for the mevalonate kinase isolated from various plant tissues[170,198,201,204,208], whilst two isoenzymes have been reported in the leaves of *Phaseolus vulgaris*[209], one in the chloroplast with an optimum pH of 7.5 and the second, an exochloroplastidic kinase, with an optimum at pH 5.5.

The further conversion of mevalonate 5-pyrophosphate into isopentenyl pyrophosphate has been observed with cell-free preparations from sweet potato[199], *Kalanchoë crenata* callus tissues[202], *Pisum sativum*[200] and *Hevea brasiliensis* latex[210] and two isopentenyl pyrophosphate isomerase fractions have been obtained from pumpkin fruit[210a]. With a pea system the synthesis of dimethylallyl, geranyl and farnesyl pyrophosphates was also obtained[200]. In some plant homogenates the accumulation of free geraniol and farnesol, rather than the pyrophosphates, has been noted[206, 210–213, 220] and results from phosphatase activity in the preparations[206, 210]. Phospholipids, particularly phosphatidyl ethanolamine from soybean or *Pinus radiata* seeds, stimulate the incorporation of 2-^{14}C-mevalonic acid into isoprenoid alcohols by an enzyme system prepared from *P. radiata* seedlings[214].

The expected intermediacy of geranyl pyrophosphate in squalene biosynthesis in higher plants has been indicated by the incorporation of ^{14}C-geraniol into squalene, β-amyrin and sterols in pea seeds[215, 216], a result which confirms the presence of a geraniol kinase in plant tissues[217]. The utilization of farnesyl pyrophosphate for squalene formation by carrot and tomato preparations required NADPH as a cofactor[218] in conformity with squalene biosynthesis in animal tissues. Cell-free systems which catalyse the complete sequence of reactions from mevalonic acid to squalene have been obtained from peas[211, 212, 219], carrot[218], tomato[218], *Echinocystis macrocarpa*[220] and tobacco tissue culture[213]. The enzymes responsible for the conversion of mevalonic acid into farnesyl pyrophosphate were soluble while the squalene synthetase was microsomal[213, 219].

Some aspects of the stereochemistry of squalene biosynthesis in plant tissues have been demonstrated using the various types of mevalonic acid labelled stereospecifically with tritium. In all the cases examined the reactions have been found to proceed by mechanisms analogous to those operative in animal tissues. The incubation of (4R)-4-^{3}H$_1$-mevalonic acid with carrot root[221], tomato[222], pea seedlings[223], *Solanum tuberosum*[224], *Fucus spiralis*[225, 226], *Phycomyces blakesleeanus*[227] and *Aspergillus fumigatus*[228, 229] gave squalene which retained six tritium atoms, showing that, as in animals, the 2-pro-R hydrogen atom of isopentenyl pyrophosphate is eliminated during squalene synthesis (Scheme 2). The incorporation of six tritium atoms into the squalene produced by various plant tissues[230, 231] when incubated with (5R)-5-^{3}H$_1$-mevalonic acid indicates that the coupling of the two farnesyl pyrophosphate molecules proceeds by a mechanism similar, if not identical, to that operative in animal tissues (Scheme 6). This conclusion is further substantiated by the observed incorporation of only eleven tritium atoms from a possible twelve when 5-^{3}H$_2$-mevalonic acid was incorporated into squalene by tomato slices[230]. The stereochemical fate of the C-2 hydrogen atoms of mevalonic acid during plant squalene biosynthesis is assumed to be the same as in animal systems (Schemes 3 and 4) and this has been substantiated in the case of squalene biosynthesized by yeast homogenates incubated with $(2R)$-2-$^3H_1$- and $(2S)$-2-$^3H_1$-mevalonic acids[232]. However the incorporation of $(2R)$-2-$^3H_1$,2-^{14}C- and $(2S)$-2-$^3H_1$, 2-^{14}C-mevalonic acid into the squalene and sterols of several plant

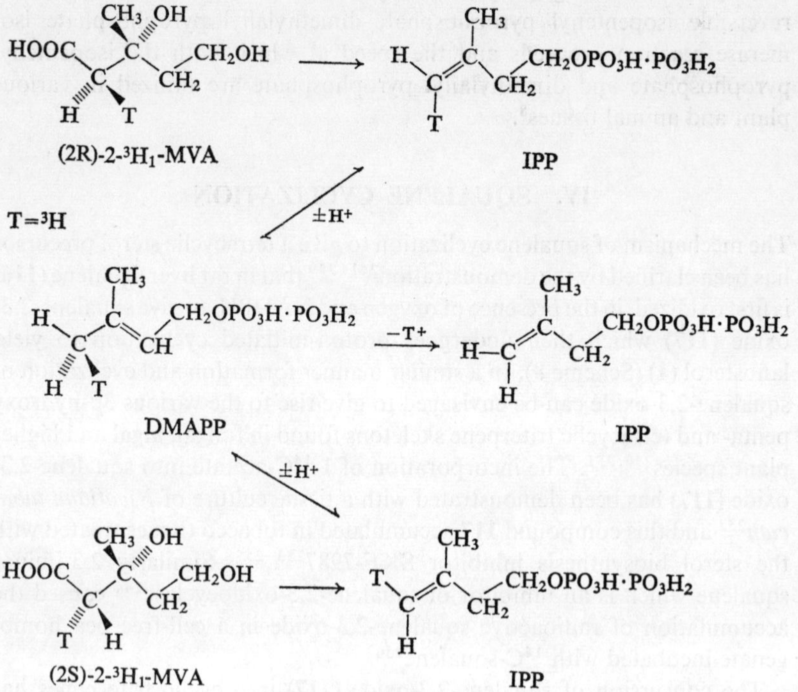

CH₃ OH   CH₃

HOOC—C—C—CH₂OH ⟶ H—C=C—CH₂OPO₃H·PO₃H₂ (CH₂)

(2R)-2-³H₁-MVA   IPP

T = ³H   ±H⁺

DMAPP   −T⁺   IPP

±H⁺

HOOC—C—C—CH₂OH ⟶ (2S)-2-³H₁-MVA   IPP

Scheme 7.

tissues[233–236] has revealed that the interpretation of results may be complicated by the reversible action of isopentenyl pyrophosphate isomerase[180, 182, 237]. The isolated squalene and sterols[233–236] had lower $^3H\!:\!^{14}C$ ratios than anticipated from the $^3H\!:\!^{14}C$ ratios of the starting mevalonates. This excessive elimination of tritium can best be explained[8] by the reversibility[180, 182, 237] of the isopentenyl pyrophosphate–dimethylallyl pyrophosphate isomerase reaction as shown in Scheme 7. It is also apparent that this reversible reaction could lead to an inversion of configuration in the derived squalene of a portion of the originally stereospecific tritium label of the parent mevalonic acid. The extent of the loss of tritium and scrambling of stereospecific label by this isomerase reaction apparently varies with different biological systems. Thus with liver[238-240] and yeast[232] homogenates the loss of tritium during the incorporation of (2R)- and (2S)-2-^{3}H$_1$-mevalonic acids into squalene was minimal, with *Larix decidua*, *Fucus spiralis* and *Aspergillus niger*[236] there was moderate tritium loss whilst with *Ochromonas malhamensis*[233] and *Aspergillus fumigatus*[234] the loss of tritium was high. Such findings are presumably a reflection of the rates at which the reversible isopentenyl pyrophosphate–dimethylallyl pyrophosphate isomerase reaction proceeds and the speed at which both the isopentenyl pyrophosphate and dimethylallyl pyrophosphate are utilized in various plant and animal tissues[8].

IV. SQUALENE CYCLIZATION

The mechanism of squalene cyclization to give a tetracyclic sterol precursor has been clarified by the demonstration[241-249] that in rat liver squalene (**116**) is first oxidized, in the presence of oxygen and NADPH, to give squalene-2,3-oxide (**117**) which then undergoes proton-initiated cyclization to yield lanosterol (**1**) (Scheme 8). In a similar manner formation and cyclization of squalene-2,3-oxide can be envisaged to give rise to the various 3β-hydroxy penta- and tetracyclic triterpene skeletons found in fungal, algal and higher plant species[250, 251]. The incorporation of 1-^{14}C-acetate into squalene-2,3-oxide (**117**) has been demonstrated with a tissue culture of *Nicotiana tabacum*[252] and this compound **117** accumulated in tobacco tissues treated with the sterol biosynthesis inhibitor SKF-7987[253-255]. Similarly 2,3-iminosqualene which is an inhibitor of squalene-2,3-oxidocyclase[244] caused the accumulation of radioactive squalene-2,3-oxide in a cell-free pea homogenate incubated with ^{14}C-squalene[256].

The conversion of squalene-2,3-oxide (**117**) into cyclic triterpenes has been demonstrated in several plant species. In the fungus *Fusidium coccineum* squalene-2,3-oxide (**117**) was incorporated[257] into fusidic acid (**118**) which may be derived by suitable oxidative changes from 3β-hydroxy-protosta-17(20),24-diene (**23**) which can arise by squalene-2,3-oxide (**117**) cyclization as indicated in Scheme 9. Indeed 3β-hydroxyprotosta-17(20),24-diene has been isolated from *Cephalosporium caerulens*[258] and its formation from 2-^{14}C-mevalonic acid by a cell-free extract of *Emericellopsis salmosynnemata* reported[260].

Lanosterol (**1**) has been identified in a number of fungi such as yeast[13, 14], *Inonotus obliquus*[15], *Aspergillus fumigatus*[20], *Phycomyces blakesleeanus*[20] and *Agaricus campestris*[261] and has been produced from squalene-2,3-oxide (**117**) by cell-free homogenates of yeast[262] and *Phycomyces blakesleeanus*[263].

In a higher plant the formation of the pentacyclic triterpene, β-amyrin
was observed when a cell-free homogenate obtained from germinated pea

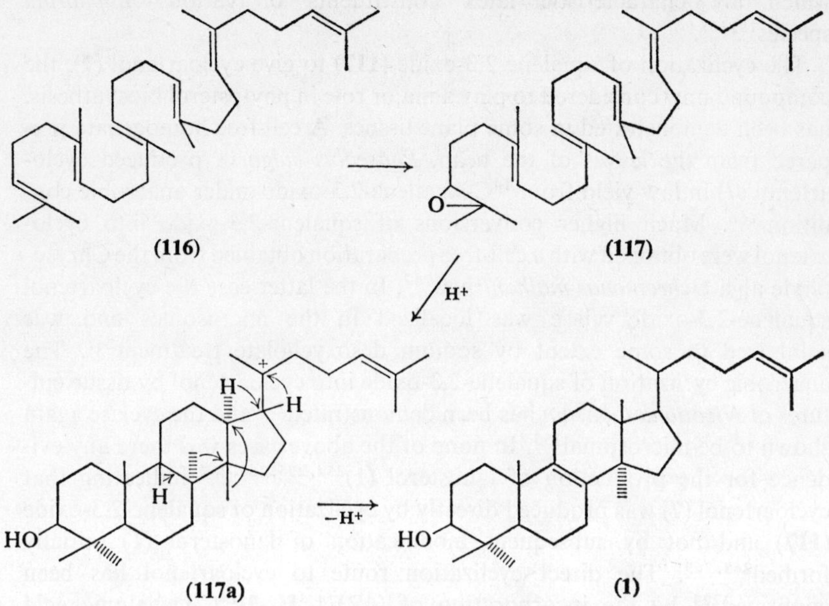

(116) → **(117)**

(117a) → **(1)**

Scheme 8.

117 → 117a

(23) → **(118)**

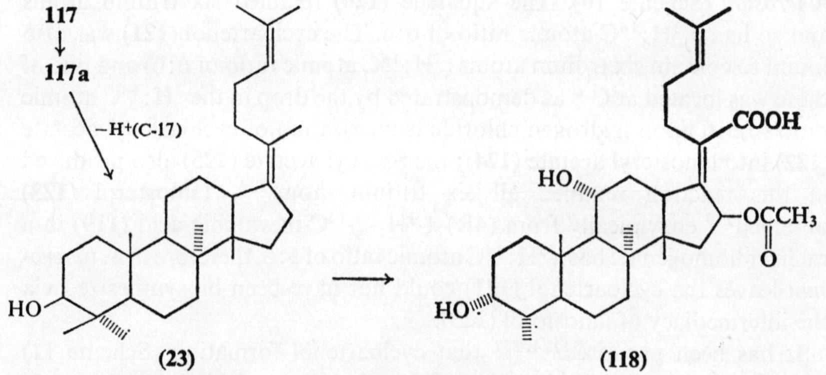

Scheme 9.

seedlings was incubated with ^{14}C-squalene-2,3-oxide[256] and the cyclase
was solubilized by treatment with sodium desoxycholate. Using the latex
of *Euphorbia cyparissias* the incorporation of 1-^{14}C-acetate into squalene-
2,3-oxide **(117)** has been demonstrated[264]. The squalene-2,3-oxide produced

6

in the latex of *Euphorbia* species is then cyclized[264] by particulate enzymes to produce triterpenes such as euphol (**15**), tirucallol (**16**) and cycloartenol (**7**) which are characteristic latex constituents of various *Euphorbia* species[18, 265].

The cyclization of squalene-2,3-oxide (**117**) to give cycloartenol (**7**), the compound now considered to play a major role in phytosterol biosynthesis, has been demonstrated in some plant tissues. A cell-free homogenate prepared from the leaves of the bean, *Phaseolus vulgaris*, produced cycloartenol (**7**) in low yield from ^{14}C-squalene-2,3-oxide under anaerobic conditions[266]. Much higher conversions of squalene-2,3-oxide into cycloartenol were obtained with a cell-free preparation obtained from the Chrysophyte alga *Ochromonas malhamensis*[267]. In the latter case the cycloartenol squalene-2,3-oxidocyclase was localized in the microsomes and was soublized to some extent by sodium desoxycholate treatment[267]. The anaerobic cyclization of squalene-2,3-oxide into cycloartenol by tissue cultures of *Nicotiana tabacum* has been demonstrated[254] and the cyclase again shown to be microsomal[255]. In none of the above cases was there any evidence for the production of lanosterol (**1**)[254, 255, 266, 267] indicating that cycloartenol (**7**) was produced directly by cyclization of squalene-2,3-oxide (**117**) and not by subsequent modification of lanosterol (**1**) initially formed[268, 269]. The direct cyclization route to cycloartenol has been confirmed[224] by the incorporation of (4R)-4-^{3}H$_1$-2-^{14}C-mevalonic acid (**119**) into squalene (**120**) and cycloartenol (**121**) by leaves of *Solanum tuberosum* (Scheme 10). The squalene (**120**) retained six tritium atoms and so had a ^{3}H:^{14}C atomic ratio of 6:6. The cycloartenol (**121**) was also found to contain six tritium atoms (^{3}H:^{14}C atomic ratio of 6:6) and one of these was located at C-8 as demonstrated by the drop in the ^{3}H:^{14}C atomic ratio to 5:6 upon hydrogen chloride isomerization of cycloartenyl acetate (**122**) into lanosteryl acetate (**124**); the parkeyl acetate (**125**) also produced in this reaction retained all six tritium atoms[224]. Lanosterol (**123**) obtained[270] enzymically from (4R)-4-^{3}H$_1$-2-^{14}C-mevalonic acid (**119**) in a rat liver homogenate has a ^{3}H:^{14}C atomic ratio of 5:6, therefore in *S. tuberosum* leaves the cycloartenol (**121**) could not have been biosynthesized via the intermediacy of lanosterol (**123**).

It has been proposed[224, 251] that cycloartenol formation (Scheme 11) involves the cyclization of chair–boat–chair–boat folded squalene-2,3-oxide (**126**) to give a cation **127**. This is then stabilized by methyl and hydrogen migrations terminated by the transfer of a hydrogen from C-9 to C-8 and loss of a proton from the C-19 methyl group with closure of the 9β,19-cyclopropane ring to give cycloartenol (**128**). Such a direct series of rearrangements would however require that the final partial migration of the C-19 methyl group be *cis* to the preceding C-9 hydrogen transfer. However this is contrary to the biogenetic isoprene rule[250] which requires that cyclization of squalene into triterpenes be accompanied by anti-planar cationic 1,2-rearrangements and 1,2-eliminations. To overcome this

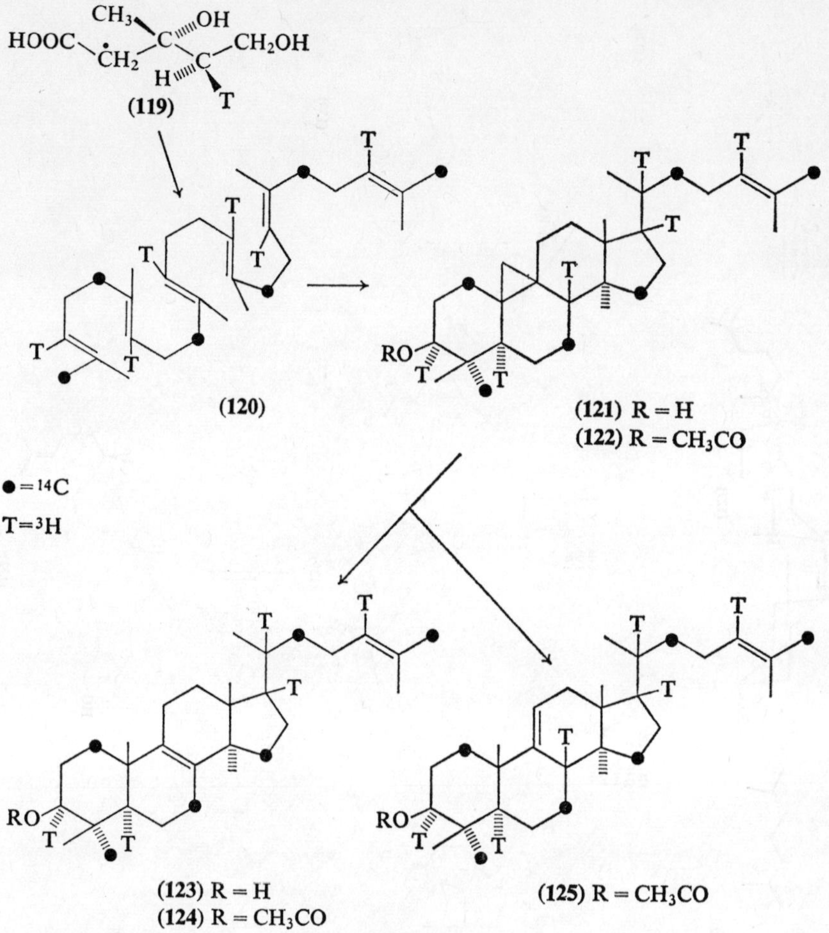

(119)

(120)

(121) R = H
(122) R = CH$_3$CO

$\bullet$ = ^{14}C

T = ^{3}H

(123) R = H
(124) R = CH$_3$CO

(125) R = CH$_3$CO

Scheme 10.

apparent anomaly it is suggested[224] that a suitable electrophilic group X$^-$, possibly on the cyclase or an associated carrier protein may transiently neutralize the charge generated at C-9 during the rearrangements of cation **127**. Withdrawal of this group would then permit the formation of the 9β,19-cyclopropane ring in a *trans* manner to leave cycloartenol as

Scheme 11.

illustrated by the sequence $126 \rightarrow 127 \rightarrow 129 \rightarrow 128$ in Scheme 11. Similarly removal of the X^- group and stabilization by loss of the C-11β proton from **129** would give parkeol (**130**) a constituent of *Butyrospermum parkii*[23] while loss of the C-8β proton from **129** would leave lanosterol (**131**), found in the latex of a few *Euphorbia* species[16-18]. Alternatively complete migration of the C-19 methyl group to C-9 after removal of the X^- group followed by hydrogen migration from C-5 to C-10 and expulsion of the C-6β proton would produce 19(10 $\rightarrow$ 9β)-*abeo*-lanosta-5,24-dien-3β-ol (**132**) which can be regarded as the parent compound of the cucurbitacin bitter principles found mainly in the Curcurbitaceae family[271,272] but also in a few species of the families Cruciferae[271,273], Scrophulariaceae[271] and Begoniaceae[274,275]. Evidence[276] for the hydrogen migration from C-5 to C-10 in the formation of **132** has been provided by the administration of $(4R)$-4-$^3H_1$-2-^{14}C-mevalonic acid to Hubbard squash seedlings to give cucurbitacin B (**133**) labelled with tritium as shown. Again the retention of a

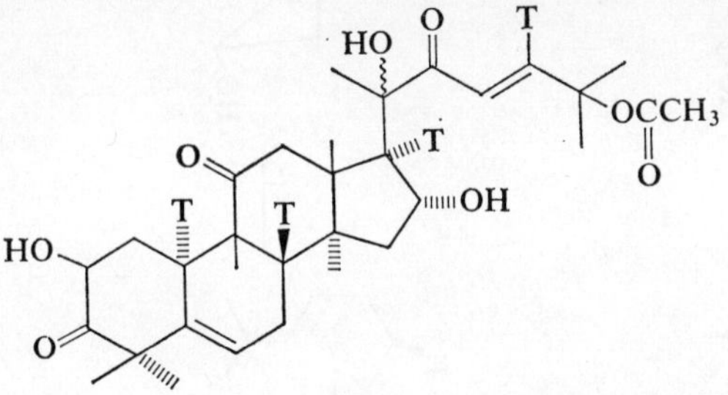

(133)

tritium at C-8 eliminates lanosterol as an intermediate but the results do not remove the possibility that either preformed parkeol or cycloartenol may be rearranged to give 19(10 $\rightarrow$ 9β)-*abeo*-lanosta-5,24-dien-3β-ol (**132**)[276].

One aspect of a cyclization mechanism such as that shown in Scheme 11 requires further comment. The transient neutralization of the positive charge at C-9 in the rearranged cation **127** or the formation of a more stable protein-bound intermediate such as **129** would mean that the cyclization process could no longer be regarded as a concerted non-stop reaction as originally postulated in the biogenetic isoprene rule[250]. Possible X^- group participation in the cyclization of squalene-2,3-oxide (**126**) to give lanosterol via the intermediate **134** (Scheme 12) has been discussed by Cornforth[277]. It is envisaged that following formation of intermediate **134** there is rotation about the C-17, C-20 bond to give **135** which can then rearrange to give lanosterol (**131**). If such a mechanism is operative for lanosterol formation it is tempting to speculate that **134** might also occur in cycloartenol formation,

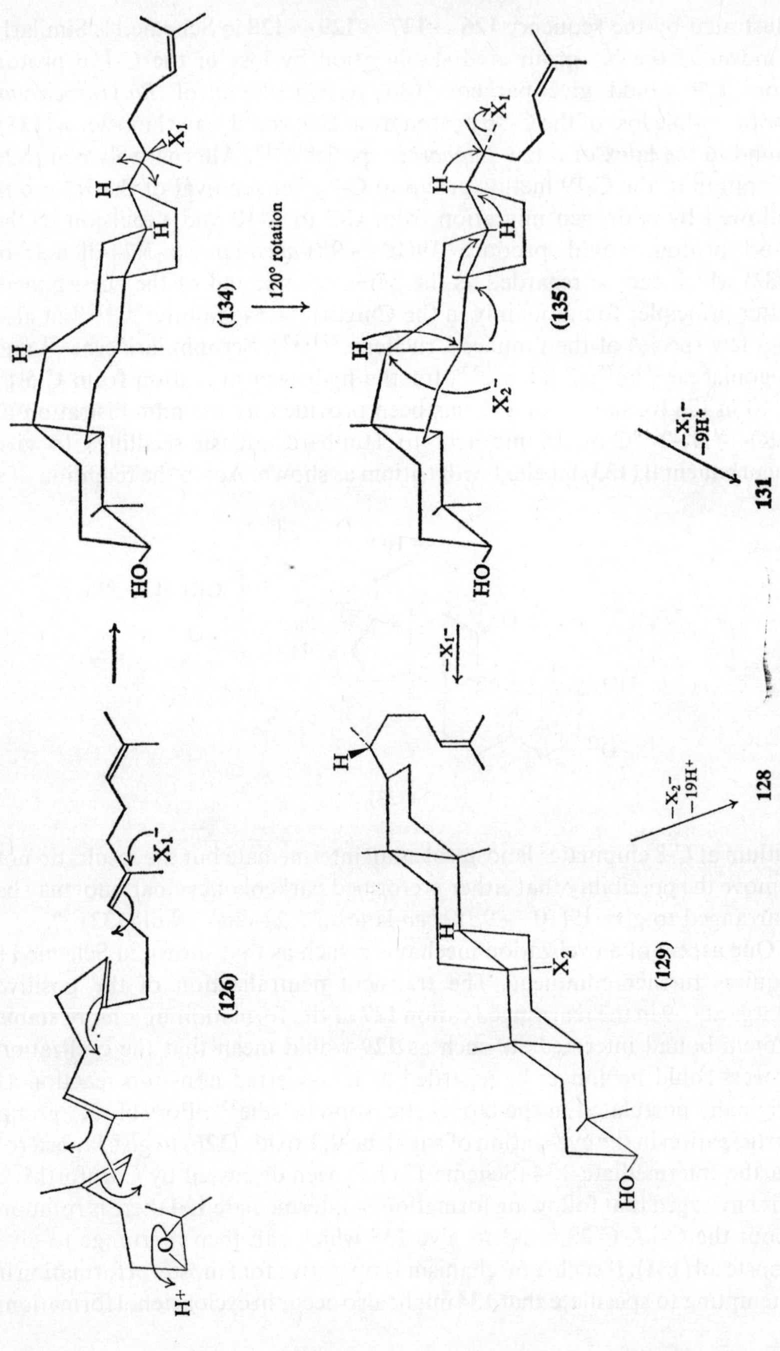

Scheme 12.

rearrangement proceeding as already discussed above, with the production of a second intermediate (**129**) as shown in Scheme 12. This mechanism is attractive since the rotation about the C-17, C-20 bond would result in the removal of the cyclized product from that portion of the enzyme which imposed the conformational requirements for cyclization upon the squalene chain and may then allow the approach of C-9 to a favourable position in relation to the second X⁻ group on either the same enzyme or an associated enzyme or carrier protein. A similar mechanism has been postulated[277] to explain β-amyrin formation.

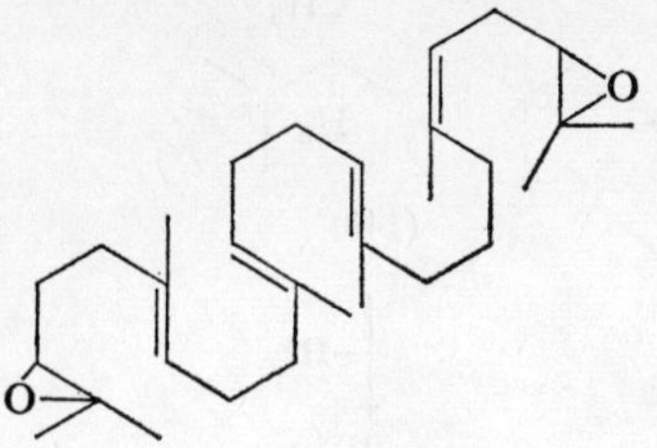

(136) 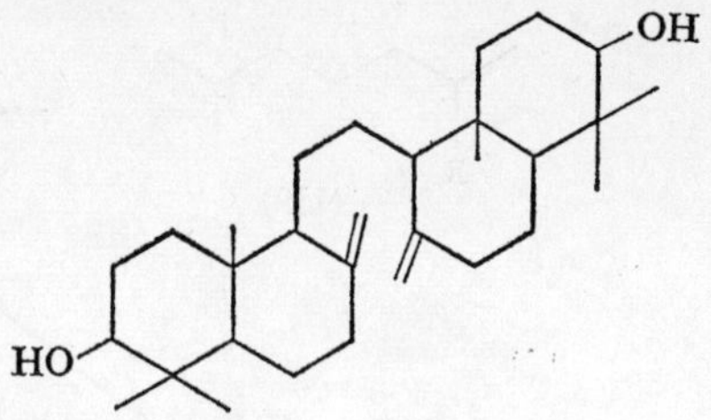   (137)

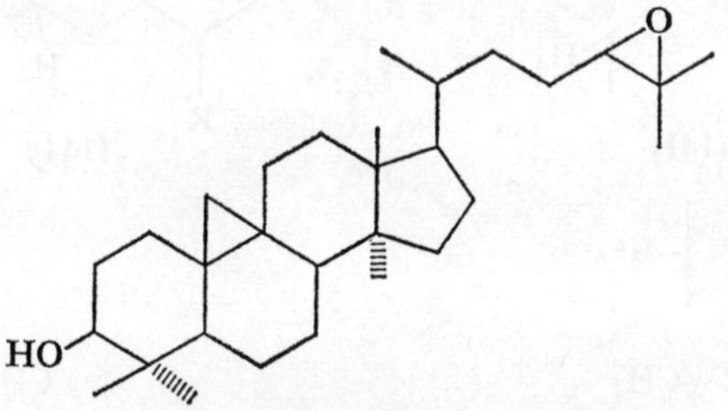

(138)

Squalene-2,3;22,23-diepoxide (**136**) has been suggested as the precursor of α-onocerin (**137**), cyclization in this case proceeding from both ends of the molecule[278]. This view has recently been substantiated by the use of a cell-free preparation obtained from *Ononis spinosa* which produced labelled α-onocerin when incubated with ¹⁴C-squalene-2,3;22,23-diepoxide[279]. However by contrast the incubation of squalene-2,3;22,23-diepoxide (**136**) with microsomes isolated from bramble (*Rubus fruticosa*) led to the formation of 24ξ,25-epoxycycloartanol (**138**)[280] which is a possible precursor of cycloart-23-en-3β,25-diol (**13**) a constituent of *Tillandsia usneoides*[101].

V. ALKYLATION OF THE PHYTOSTEROL SIDE-CHAIN

The characteristic plant sterols contain an additional group at C-24 of the side-chain which arises by a transmethylation reaction from methionine[9,281-285]. The role of methionine as the methyl donor was first demonstrated for ergosterol (**60**) biosynthesis in yeast[286-289] and subsequently for sitosterol (**80**) in *Pisum sativum*[290] and *Salvia officinalis*[291], spinasterol (**84**) in *Menyanthes trifoliata*[292] and fucosterol (**67**) in *Laminaria saccharina*[293]. The possibility that a C-24 ethyl group can arise by transethylation from ethionine has been considered and eliminated[284,294]. The actual methyl donor in the case of ergosterol (**60**) formation in yeast is *S*-adenosyl

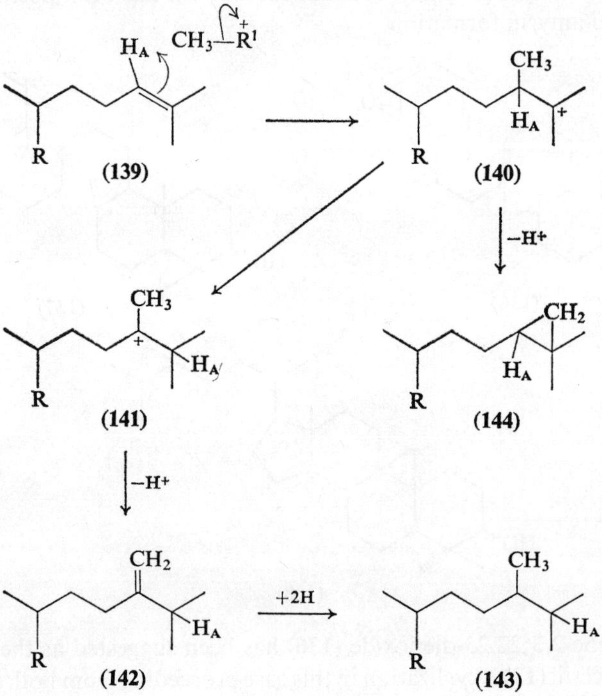

Scheme 13.

methionine[289,295] and this compound will presumably play a similar role in other organisms.

A mechanism for alkylation of the sterol side-chain (Scheme 13) was first proposed by Nes and his colleagues[290] and subsequently elaborated by other workers[6,9,281-285]. The requirement for a Δ^{24} bond in the precursor sterol (**139**) has been demonstrated[296] using a pea seedling homogenate capable of converting lanosterol (**1**), cycloartenol (**7**) and desmosterol (**44**) into the corresponding 24-methylene compounds (**3**), (**10**) and (**51**) respectively.

The conclusion that the formation of a 24-methylene compound (**142**) is the first step of the alkylation process follows from several lines of evidence. Thus a number of sterols possessing a 24-methylene group have been isolated from higher plants, algae and fungi (Table 1)[1-3,9]. The retention of only two deuteriums in the C-24 methyl group of ergosterol (**60**) biosynthesized by a methionine-less strain of *Neurospora crassa*[297,298] in the presence of trideuterated methionine ([CD$_3$-methyl]-methionine) is best explained by the formation of a 24-methylene intermediate (**142**) which is subsequently reduced to give a 24-methyl sterol (**143**) (Scheme 13). The alternative possibility involving formation of a 24,25-methylene-bridged intermediate (**144**) was eliminated by the synthesis of 24,25-methylene dihydrolanosterol (**145**)[299] and the demonstration that it is not converted into ergosterol (**60**) by yeast[300].

(145)

As shown in Scheme 13 the methylation of a Δ^{24} sterol precursor (**139**) will lead to a cation (**140**) which is stabilized by hydrogen migration from C-24 to C-25 (**141**) and loss of a proton from C-28 to give **142**. Moreover this migrated hydrogen will remain at C-25 during subsequent reduction of the 24-methylene sterol (**142**) to give the C-24 methyl sterol side-chain (**143**). The reality of this hydrogen migration during ergosterol biosynthesis has been demonstrated by the incubation of (24-^{3}H$_1$-26,27-^{14}C)-lanosterol with yeast. The tritium was retained in the isolated ergosterol and degradation showed that it was not located at C-24 but was probably at C-25[301,302].

The demonstration that labelled 24-methylene dihydrolanosterol (**3**)[50,303-305], 4α-methyl-5α-ergosta-8,24(28)-dien-3β-ol (**35**)[50], obtusifoliol (**30**)[50] and ergostatetraeneol (**55**)[306] are all incorporated into ergosterol when administered to yeast cultures further substantiates the proposed involvement of 24-methylene intermediates during ergosterol formation in yeast. Moreover a number of 24-methylene sterols have been isolated from yeast and other fungi; these include 24-methylene dihydrolanosterol (**3**) from *Fomes officinalis*[21], *Agaricus campestris*[261] and *Phycomyces blakesleeanus*[20]; 4α-methyl-5α-ergosta-8,24(28)-dien-3β-ol (**35**)[50], fecosterol

(53)[81,82], episterol (52)[78–80] and ergostatetraeneol (55)[84,85] from yeast and episterol (52) and 5-dehydroepisterol (54) from *Phycomyces blakesleeanus*[83].

The particular enzyme, *S*-adenosyl methionine-Δ^{24} sterol methyl transferase, has been obtained from yeast[307] and solubilized and purified to some extent[295]. For maximum activity this enzyme requires Mg^{2+} ions and a reducing agent such as glutathione for conversion of zymosterol (45) into fecosterol (53)[295]. The point of transmethylation in the biosynthetic pathway between lanosterol (1) and ergosterol (60) in fungi is at present uncertain and this problem has been discussed at length by Lederer[284]. The isolation from many wood-rotting fungi[1–3] of acidic triterpenes (e.g. 146

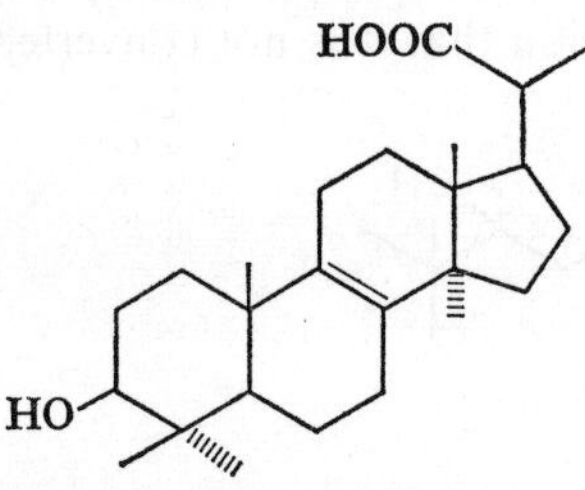

(146)

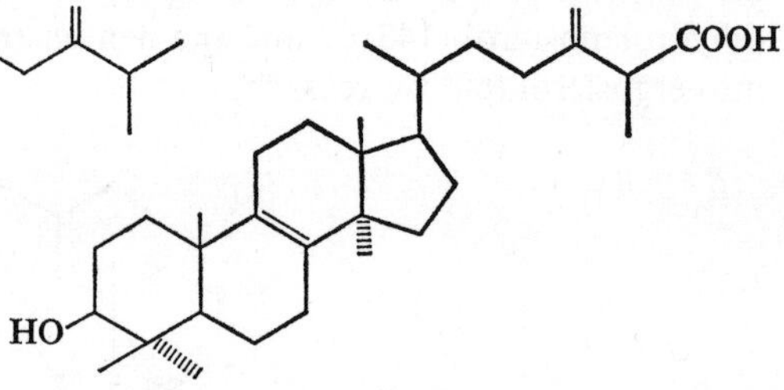

(147)

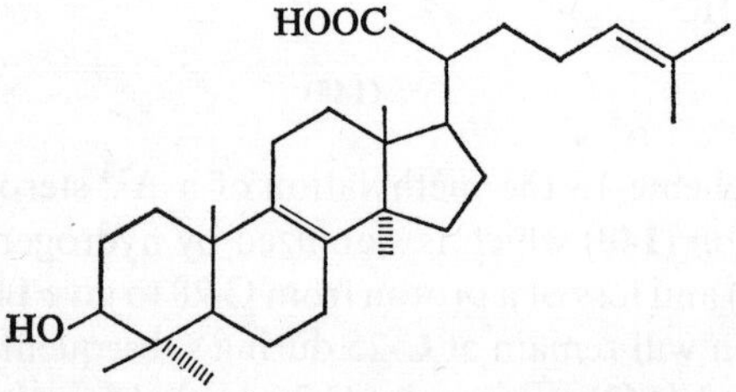

(148)

and 147) with the lanostane skeleton and a 24-methylene group shows that methylation can occur at the C_{30}-sterol level in some species. However it is not clear if C-24 alkylation occurs before or after oxidation of a side-chain methyl group in the biogenesis of these compounds. Thus it has been shown[50] that *Polyporus sulphureus* can utilize 24-methylene dihydro-lanosterol (3) added to the growth medium for the production of eburicoic acid (146) while 24-methylene dihydrolanosterol (3) has been isolated[308] from *Polyporus betulinus* which produces polyporenic acid C (147). On the other hand a C_{30} acidic triterpene has been observed in *Polyporus sulphu-reus*[309] and identified[310] as trametolenic acid (148). When compound 148 was labelled with tritium at C-3 and fed back to a culture of *P. sulphureus* it was also converted into eburicoic acid (146)[310]. With regard to the point in the biosynthetic pathway at which C-24 methylation occurs in ergosterol **(60)** production the evidence is equally ambivalent. An examination of the sterols of yeast failed to reveal any 4,4-dimethyl sterols possessing a 24-methylene group[14] although Δ^{24} sterols were identified thus indicating that methylation may perhaps be restricted to the latter stages. A similar conclusion was reached using a cell-free preparation from yeast which could incorporate ^{14}C-methyl-methionine into ergostatetraeneol **(55)** but did not apparently produce any 4,4-dimethyl or 4-monomethyl C-24 methylated derivatives[307]. Although yeast can incorporate added 24-methylene dihydrolanosterol **(3)** into ergosterol **(60)**[50, 303–305] this does not necessarily prove that **3** is an obligatory precursor, it may merely be a reflection of a lack of absolute substrate specificity by the enzymes involved in ergosterol biosynthesis. Indeed when a yeast homogenate was incubated with ^{3}H-squalene-2,3-oxide in the presence of unlabelled 24-methylene dihydrolanosterol **(3)** to act as a trap, lanosterol **(1)** was labelled but it was not possible to accumulate radioactivity in 24-methylene dihydrolanosterol **(3)**[50]. However it is not clear if *S*-adenosyl methionine-Δ^{24} sterol methyl transferase was active in the yeast homogenate used in this experiment[50]. In a recent study using the purified yeast *S*-adenosyl methionine-Δ^{24} sterol methyl transferase it was observed[311] that while zymosterol **(45)** is a good substrate, 4-methyl sterols by contrast are only very poorly utilized as substrates. This result could be interpreted as further support for the view[307] that side-chain alkylation is a terminal step in ergosterol **(60)** formation in yeast. However in an earlier study of sterol C-4 demethylase activity in yeast extracts Gaylor[312] found that the rate of demethylation of lanosterol **(1)** was enhanced by added *S*-adenosyl methionine and suggested that side-chain alkylation may precede or accompany demethylation of lanosterol **(1)**.

With a Phycomycete, *Phycomyces blakesleeanus*[19, 20], and two Basidiomycetes, *Agaricus campestris*[261] and *Polyporus betulinus*[308], which all contain ergosterol **(60)** as the predominant sterol, an examination of the 4,4-dimethyl sterol fractions has shown the presence of small amounts of lanosterol **(1)** and 24-methylene dihydrolanosterol **(3)**. Moreover the ready incorporation of ^{14}C-methyl-methionine into 24-methylene dihydrolanosterol **(3)** under conditions were ergosterol **(60)** is readily labelled has been observed with a culture of *Phycomyces blakesleeanus*[19, 20]. These results are indicative of C-24 methylation occurring at the 4,4-dimethyl sterol level in these three fungal species but clearly further information is required to determine if the 'normal' biosynthetic route proceeds predominantly via 24-methylene dihydrolanosterol **(3)**. As Lederer[284] has pointed out the alkylating enzymes of some organisms may be rather non-specific and utilize as substrate any Δ^{24} sterol available regardless of ring structure (but compare ref. 311) to give a situation similar to that previously observed with the Δ^{24} sterol reductase acting in cholesterol (**42**) biosynthesis in animals. On the experimental evidence at present available it is therefore difficult to form a definite opinion upon the major point in the biosynthetic sequence at which C-24 alkylation occurs in most fungi. However it is interesting to speculate that perhaps alkylation at an early stage will prove to be typical of some Phycomycetes and Basidiomycetes whereas with yeasts alkylation is restricted to the latter stages.

In higher plants the transmethylation mechanism outlined in Scheme 13 will lead to the conversion of cycloartenol (**7**) into 24-methylene cycloartanol (**10**) and tirucallol (**16**) into euphorbol (**18**). A cell-free system catalysing the former reaction has been obtained from pea seedlings[294] and the hydride migration from C-24 to C-25 has been demonstrated for 24-methylene cycloartanol (**10**) formation in *Nicotiana tabacum*[313] and *Fucus spiralis*[226].

The 4,4-dimethyl sterol, cyclolaudenol (**11**), found in opium[27] and *Polypodium vulgare*[28], presents an interesting alternative alkylation mechanism to that already discussed. As indicated in Scheme 14 the 24-methyl group could be introduced with concomitant formation of the 25-methylene group by three possible mechanisms. By route (*a*) elimination of a proton from C-26 of cation **140** would lead directly to the cyclolaudenol side-chain (**149**)[9,285] with retention of all three hydrogens of the introduced methyl group and H_A, the C-24 hydrogen of the precursor (**139**) (i.e. cycloartenol) remaining at C-24 in the cyclolaudenol[315]. Routes (*b*) and (*c*) both involve the intermediacy of a 24-methylene compound (**142**) (i.e. 24-methylene cycloartanol, **10**) and would therefore lead to retention of only two of the methionine methyl-hydrogens in **150** and **152**. However while route (*b*) would retain H_A at C-24 (**150**) it would be lost (**152**) by route (*c*). The use of $(4R)\text{-}4\text{-}^3H_1\text{-}2\text{-}^{14}C$-mevalonic acid has permitted certain aspects of this alkylation mechanism to be investigated[315] since H_A of **139** is derived from the 4-pro-R hydrogen of mevalonic acid (Scheme 10). Incorporation of $(4R)\text{-}4\text{-}^3H_1\text{-}2\text{-}^{14}C$-mevalonic acid into cyclolaudenol by *Polypodium vulgare* rhizomes thus demonstrated that H_A was retained at C-24 and that the 25-methylene group arose stereospecifically by loss of a proton from the terminal methyl group derived from the C-3′ position of mevalonic acid[315]. The retention of H_A therefore eliminated route (*c*) but was compatible with both routes (*a*) and (*b*). Evidence favouring route (*a*) was provided[315] by incorporating methionine labelled with radioactive carbon and tritium in the methyl group into both cyclolaudenol (**11**) and 24-methlyene cycloartanol (**10**) by *P. vulgare* rhizomes. The cyclolaudenol (**11**) had a $^3H:^{14}C$

ratio similar to the starting methionine but significantly higher than the 24-methylene cycloartanol (**10**). As pointed out by other authors[7,281] a tritium isotope effect can seriously interfere in some circumstances with the interpretation of results obtained using dual-labelled methionine. However

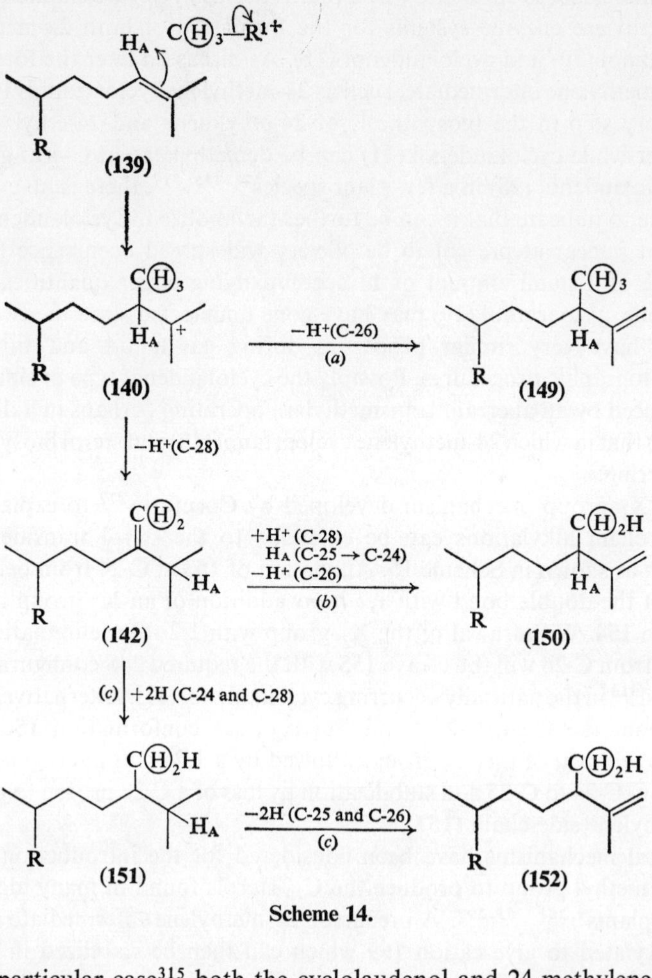

Scheme 14.

in this particular case[315] both the cyclolaudenol and 24-methylene cyclo-artanol were isolated from the same tissue during the same incubation. Thus if 24-methylene cycloartanol, which lost some tritium from C-28 during its formation, was the precursor of cyclolaudenol as required by route (*b*) then the two compounds should have $^3H:^{14}C$ ratios in close agreement irrespective of the extent of a tritium isotope effect. Since the cyclolaudenol ^{3}H: ^{14}C ratio was some 15–20 % higher than the 24-methylene cycloartanol ratio this favours the operation of route (*a*) in cyclolaudenol formation. It seems reasonable to suggest that the same mechanism will operate in the formation of carnaubadiol (**21**) from dammarenediol (**20**).

It is interesting to speculate why a plant such as *P. vulgare* should contain transmethylase enzyme systems for the formation of both 24-methylene cycloartanol (**10**) and cyclolaudenol (**11**). As discussed later the formation of a 24-methylene intermediate, such as 24-methylene cycloartanol (**10**) is an obligatory step in the biosynthesis of 24-ethylidene and 24-ethyl sterols. However while cyclolaudenol (**11**) can be demethylated at C-4 to give 31-nor-cyclolaudenol (**28**) in a few plant species[28, 315–317], there is no evidence available to indicate that it can be further metabolized. Cyclolaudenol (**11**) does not appear at present to be of very widespread occurrence but the presence of a small amount of **11** accompanying larger quantities of 24-methylene cycloartanol (**10**) may have gone undetected since the two compounds have very similar properties during gas–liquid and thin-layer chromatographic procedures. Possibly the cyclolaudenol type of side-chain is produced by an aberrant transmethylase operating perhaps in a different tissue to that in which 24-methylene cycloartanol (**10**) and sterol biosynthesis is occurring.

The 'X$^-$ group' mechanism developed by Cornforth[277] to explain biological olefin alkylations can be extended to the sterol transmethylase reaction as shown in Scheme 15. Alkylation of **153** at C-24 from below the plane of the double bond with 1,2-*trans* addition of an X$^-$ group to C-25 will give **154**. Withdrawal of the X$^-$ group with 1,2-*trans* elimination of a proton from C-26 will thus leave **155** with the required 24S configuration as reported[318] in the naturally occurring cyclolaudenol (**11**). Alternatively rotation about the C-24, C-25 bond can produce conformation **156** which permits removal of the X$^-$ group followed by a 1,2-*trans* hydrogen migration from C-24 to C-25 and stabilization by loss of a C-28 proton to give the 24-methylene side-chain (**157**).

Several mechanisms have been considered for the introduction of the second methyl group to produce the C_{29}-sterols found in many algae and higher plants[9, 281–285, 290]. A precursor 24-methylene intermediate (**158**) is first alkylated to give cation **159** which can then be stabilized in several ways as indicated in Scheme 16. Evidence for the operation of each of these routes has been obtained using various organisms.

The loss of a proton from C-28 of **159** to give a C-24 ethylidene sterol (**160** or **161**) was first proposed by Nes *et al.*[290] (Route *A*, Scheme 16). Sterols with a 24-ethylidene group have been obtained from several sources (Table 1) and indeed fucosterol (**67**) is the predominant sterol of the marine brown algae (Phaeophyceae)[60, 104, 105]. An important advance in sterol identification and the understanding of sterol biosynthesis was provided by the demonstration[108] that avenasterol (**70**) isolated from oats[107], is the isomer of fucosterol (**67**). Subsequently it was demonstrated[110, 319, 320] by nuclear magnetic resonance spectroscopy that fucosterol (**67**) and 28-isofucosterol (avenasterol, **70**) have the configurations at C-28 indicated by

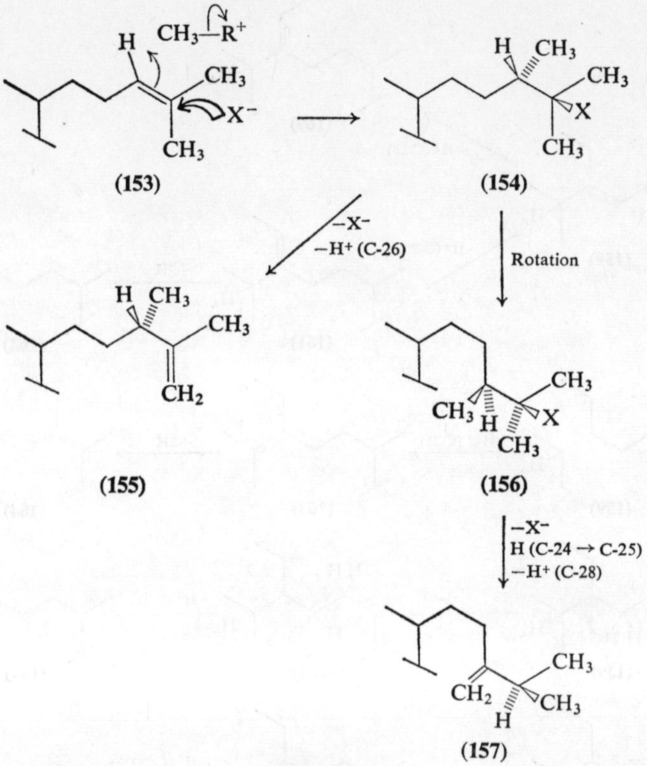

(153)   **(154)**

$-X^-$
$-H^+$ (C-26)

Rotation

(155)   **(156)**

$-X^-$
H (C-24 → C-25)
$-H^+$ (C-28)

(157)

Scheme 15.

structures **160** and **161**, respectively. 28-Isofucosterol (**70**) has been identified as the major sterol constituent in three species of marine algae[103, 109] and it occurs in small amounts in a number of higher plant tissues where it accompanies C-24 ethyl sterols such as sitosterol (**80**)[132, 231, 321–325]. Also 5α-stigmasta-7,Z-24(28)-dien-3β-ol (**71**)[77, 110, 326] and 4α-methyl-5α-stigmasta-7,Z-24(28)-dien-3β-ol (24-ethylidene lophenol, citrostadienol (**38**))[52–58, 327–329], have been identified in several plant species which contain mainly C-24 ethyl sterols.

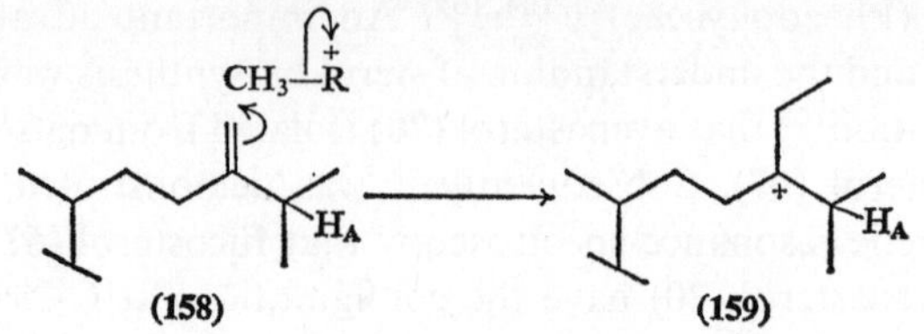

(158) → **(159)**

Route *A*

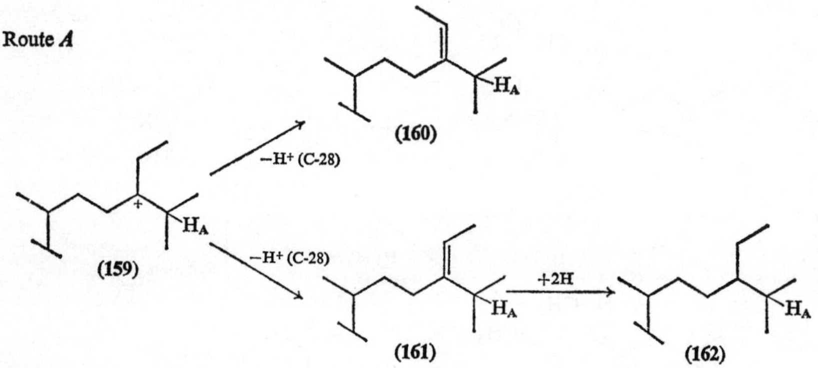

(159)

−H⁺ (C-28) → **(160)**

−H⁺ (C-28) → **(161)** +2H → **(162)**

Route *B*

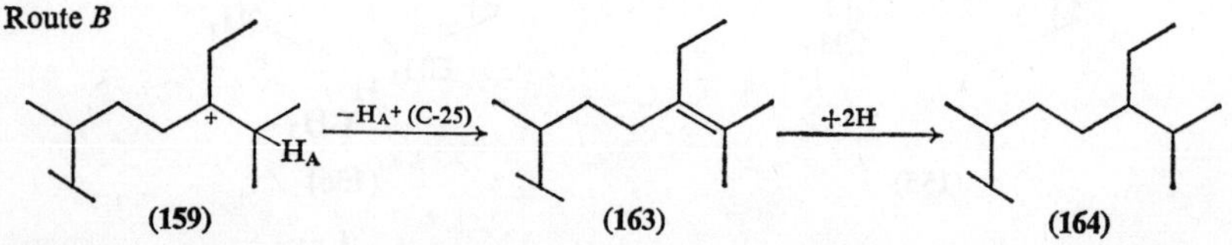

(159) −H$_A^+$ (C-25) → **(163)** +2H → **(164)**

Route *C*

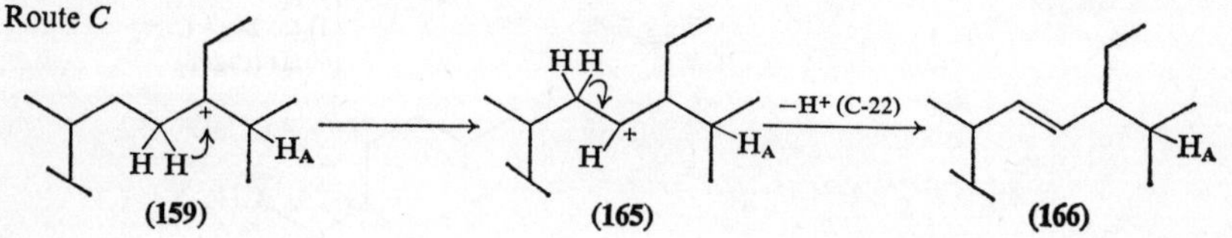

(159) → **(165)** −H⁺ (C-22) → **(166)**

Route *D*

(159) → **(167)** −H⁺ (C-26) → **(168)**

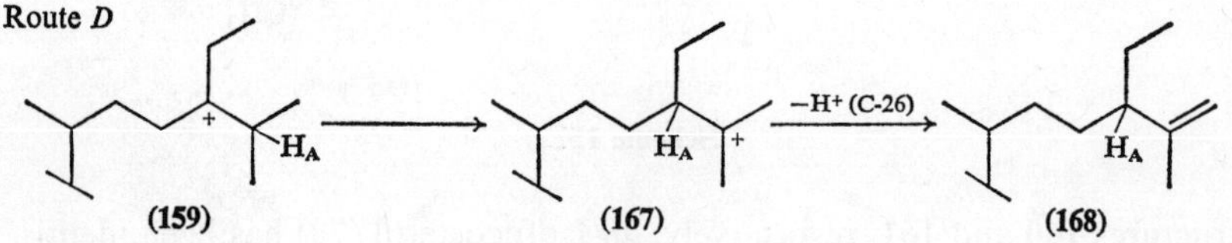

Scheme 16.

The co-occurrence of the 24-ethylidene compounds **(38)**, **(70)** and **(71)** together with 24-ethyl sterols suggests a precursor–product relationship as indicated by **161** → **162** in route *A*, Scheme 16. The incorporation of 2-¹⁴C-mevalonic acid into both 28-isofucosterol **(70)** and sitosterol **(80)** has been observed in *Pinus pinea*[323], *Larix decidua*[231], *Pisum sativum*[330] and

Euphorbia peplus[325]. A time-course study of the incorporation of 2-^{14}C-mevalonic acid into 28-isofucosterol (**70**) and sitosterol (**80**) in *E. peplus*[325] indicated the probable conversion of **70** into **80**. This conversion has been confirmed by the observation that [14]C-28-isofucosterol administered to *P. pinea* seedlings was converted into sitosterol[323].

Further evidence for the intermediacy of a 24-ethylidene sterol has been obtained by determining the number of hydrogen atoms derived from the two methionine methyl groups which are retained in the C-24 ethyl sterols of four organisms. In one study[331] [14]C[3]H$_3$-methionine was incorporated into sitosterol (**80**) by *Zea mays* and *Larix decidua* leaves. In both cases the sitosterol (**80**) had a [3]H:[14]C ratio in close agreement with the ratio of fucosterol (**67**) biosynthesized by *Fucus spiralis* from the same sample of dual-labelled methionine[331]. This comparative approach therefore indicated that only four of the hydrogen atoms of the C-24 ethyl group of sitosterol (**80**) were derived from methionine and provided some evidence for the formation of a 24-ethylidene intermediate with subsequent reduction to a 24-ethyl compound. However, as recognized at the time of this work[9, 331], results obtained using tritium-labelled methionine are open to the serious objection that the operation of a strong hydrogen isotope effect may result in the preferential retention of tritium and so seriously complicate the interpretation of results[7, 9, 281]. A preferable experimental approach to this type of problem is the use of deuterium-labelled methionine[281] followed by mass spectrometry of the biosynthesized sterol to determine the deuterium content. When C[2]H$_3$-methionine was added to a culture of the alga *Ochromonas malhamensis* the major sterol, poriferasterol (**75**), was found to contain species with one, three and four deuterium atoms located in the C-24 ethyl group[332]. Similarly the related species, *O. danica* also incorporated a maximum of four deuterium atoms into poriferasterol (**75**)[333]. Such a labelling pattern is only consistent with the intermediacy of a 24-ethylidene sterol during poriferasterol (**75**) biosynthesis in these organisms[284, 332].

No 24-ethylidene sterols have yet been isolated from *O. malhamensis* but the utilization of potential sterol precursors with a 24-ethylidene group has been examined[334, 335]. It was found that tritium-labelled 24-ethylidene lophenol (**38**)[334], 5α-stigmasta-7,Z-24(28)-dien-3β-ol (**71**)[335] and 28-isofucosterol (**70**)[335] added to the growth media were efficiently converted into poriferasterol (**75**). Labelled fucosterol (**67**) however was poorly utilized in comparison with 28-isofucosterol (**70**)[335]. Two explanations for this preferential incorporation of 28-isofucosterol (**70**) seem possible. Firstly the 24-ethylidene intermediate presumed to occur in *O. malhamensis* may have the configuration found in 28-isofucosterol (i.e. **161**) and the $\Delta^{24(28)}$ reductase might therefore be expected to show some degree of specificity towards ethylidene compounds of this configuration. Alternatively introduction of a Δ^{22} bond may precede reduction of the $\Delta^{24(28)}$ bond, the formation of a $\Delta^{22,24(28)}$ diene system would then be analogous to the formation of the $\Delta^{5,7}$ sterol in the sequence $\Delta^7 \rightarrow \Delta^{5,7} \rightarrow \Delta^5$ during ring B modification[9]. Relevant to this point are results obtained with a protozoan, *Tetrahymena pyriformis*, which has the ability to convert cholesterol (**42**) added to the growth medium into cholesta-5,7,22-trien-3β-ol[336, 337]. This protozoan can also metabolize 28-isofucosterol (**70**) to give stigmasta-5,7,22,Z-24(28)-tetraen-3β-ol (**169**) but with fucosterol (**67**) the product is stigmasta-5,7,E-24(28)-trien-3β-ol (**170**)[338, 339]. Apparently the configuration of the C-29 methyl group of fucosterol (**67**) hinders the introduction of the Δ^{22} bond. Extrapolation of these results to *O. malhamensis* could therefore also explain why fucosterol is poorly incorporated into poriferasterol if a $\Delta^{22,24(28)}$ intermediate is in fact involved.

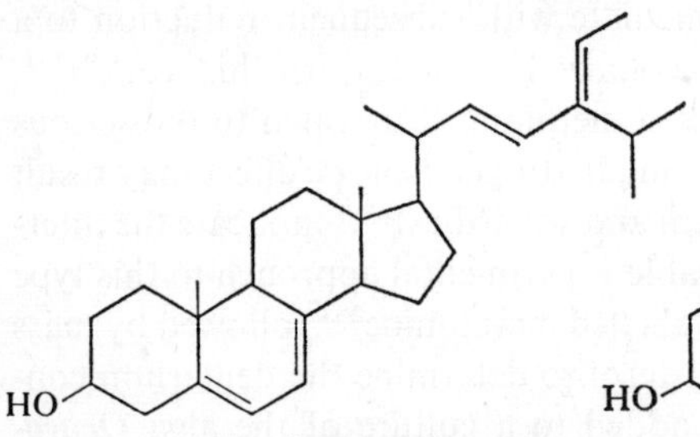

HO (169)   HO (170)

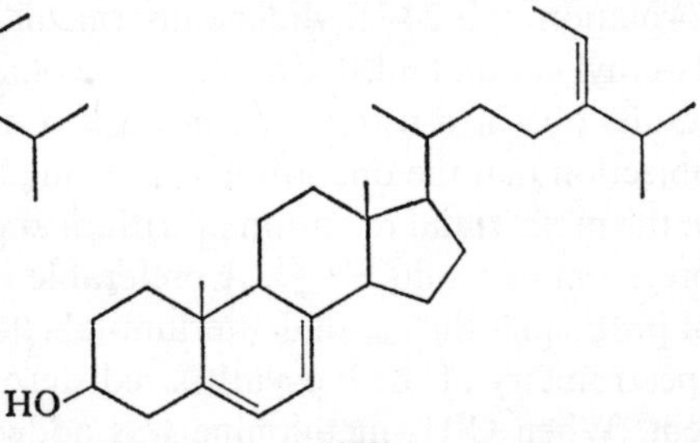

The recognition[320] that all the 24-ethylidene sterols so far isolated from higher plants have the Z-configuration (**161**) and the observation that ^{14}C-28-isofucosterol (**70**) is converted into sitosterol (**80**) by *Pinus pinea* seedlings[323] is indicative that in such plants 24-ethylidene sterols of the type (**161**) are the preferred intermediates in 24-ethyl sterol formation. The accumulation of fucosterol (**67**) as the major sterol of the Phaeophyceae could be due to its having the 'wrong' configuration for reduction or to the absence of a $\Delta^{24(28)}$ reductase. The participation of the enzyme in the alkylation reaction by the X^- group type of mechanism has been discussed in an attempt to explain the specific formation of either side-chain **160** or **161** in different classes of organism[109].

As already discussed the operation of a methylation mechanism involving a 24-methylene intermediate will result in the migration of a hydrogen from C-24 of the precursor sterol to C-25 of the 24-methylene compound (Scheme 13). At the second transmethylation step this hydrogen will remain at C-25 in the 24-ethylidene sterol (**160** or **161**) and also in any derived 24-ethyl sterol (**162**) if route *A*, Scheme 16, is operative. This hydrogen migration has been demonstrated[340] for 28-isofucosterol (**70**) biosynthesis. Incubation of 24-³H₁-lanosterol with seeds of *P. pinea* gave 28-isofucosterol (**70**) which retained the tritium. Ozonolysis to give labelled 24-ketocholesterol (**47**) followed by reduction and dehydration gave desmosterol (**44**) with loss of tritium from C-25 thereby confirming the predicted hydrogen migration[340]. Similarly the hydrogen migration has been shown to occur during fucosterol (**67**) formation by *Fucus spiralis*[225, 226]. In this case advantage was taken of the fact that (4R)-4-³H₁-mevalonic acid labels the C-24 position of cycloartenol (**7**) or lanosterol (**1**) with tritium (Scheme 10). Previous work[270] had shown that cholesterol (**42**) biosynthesized by a rat liver preparation from (4R)-4-³H₁-mevalonic acid contained three tritium atoms located at C-17, C-20 and C-24[270, 341, 342]. The fucosterol (**67**) isolated from *F. spiralis* also retained three tritium atoms[225, 226, 331], however in this case tritium could not be located at C-24. Isomerization[314] of the fucosterol (**67**) into stigmasta-5,24-dien-3β-ol resulted in the loss of one tritium atom[226] thereby confirming that the 1,2-hydrogen migration from C-24 to C-25 had again occurred during C-24 alkylation. Finally the C-24 to C-25 hydrogen migration has also been demonstrated in poriferasterol (**75**) produced by *Ochromonas malhamensis*[343]. Again three tritium atoms were incorporated into the sterol when (4R)-4-³H₁-mevalonic acid was the precursor and two were located at C-17 and C-20 respectively[343]. Ozonolysis of the poriferasterol (**75**) gave 2-ethyl-3-methylbutanol from the terminal portion of the side-chain[343]. This fragment contained one tritium atom which was not lost under basic exchange conditions[343] showing that it was not located at C-2 (corresponding to C-24 of the original poriferasterol (**75**) side-chain). By inference the tritium must therefore have been located at C-25 confirming again the 1,2-hydrogen migration, and coincidentally providing further evidence for the intermediacy of a 24-ethylidene sterol in poriferasterol formation in this organism.

The operation of alternative alkylation routes for the production of 24-ethyl sterols without the involvement of 24-ethylidene intermediates has been shown in some organisms by the use of C^2H_3-methionine. In the alga *Chlorella vulgaris* five deuterium atoms were incorporated into chondrillasterol (**77**) and 22-dihydrochondrillasterol (**78**)[344]. Similarly *C. ellipsoidea* produced poriferasterol (**75**) with retention of five deuterium atoms[345]. Route *B*, Scheme 16, has been suggested[344] to explain these results although to date there is apparently no report of a naturally occurring sterol with side-chain **163**. In support of the involvement of a $\Delta^{24(25)}$ intermediate (**163**) which is then reduced to give **164** is the observation that stigmasterol

(82) biosynthesized by *Nicotiana tabacum* and *Dioscorea tokoro* tissue cultures in the presence of $(4R)$-4-$^3H_1$-mevalonic acid retained only two tritium atoms, neither of which was located at C-24 or C-25[313]. However in this case the status of the methionine methyl hydrogens involved has not been determined but it is notable that 24-ethylidene lophenol **(38)** has been isolated from *N. tabacum* tissue culture[329] and evidence for the labelling of stigmasta-7,Z-24(28)-dien-3β-ol **(71)** from acetate by this tissue has also been presented[346].

With regard to sterol formation in *Chlorella vulgaris* in the work referred to above[345] it was found that ergost-7-en-3β-ol **(58)** isolated from the cells grown in the presence of C^2H_3-methionine contained three deuterium atoms thus eliminating a 24-methylene intermediate. However it would appear that a 24-methylene sterol must also be produced in this organism to act as substrate during the second transmethylation step to give the 24-ethyl sterols found in the alga. Thus two routes for the stabilization of cation **140** appear to operate in *C. vulgaris*.

In the slime mould, *Dictyostelium discoideum*, it has also been shown[347, 348] that five deuterium atoms from C^2H_3-methionine are incorporated into the 24-ethyl group of the principal C_{29}-sterol, stigmast-22-en-3β-ol **(81)**. This result again eliminates the involvement of a 24-ethylidene compound and route C, Scheme 16, has been proposed[348, 349]. Here the cation **159** is stabilized by a 1,2-hydrogen migration, from C-23 to C-24 to give the cation **165**, which is then followed by loss of a proton from C-22 to introduce the Δ^{22} bond **(166)** as a concerted part of the alkylation mechanism. Substantiating such a mechanism is the demonstration[349, 350] that 23-$^3H_2$-lanosterol is incorporated into stigmast-22-en-3β-ol **(81)** by *Dictyostelium discoideum* with retention of tritium at both C-23 and C-24. However the operation of an alternative pathway to stigmast-22-en-3β-ol **(81)** in this organism is indicated by the demonstration that stigmastan-3β-ol **(79)** can be converted into stigmast-22-en-3β-ol **(81)**[350, 351] thereby showing that the Δ^{22} bond can be introduced independently of C-24 alkylation[349–351].

The isolation of 24-ethyl sterols with a 25-methylene group **(89–93)** from plants of the Cucurbitaceae and Verbenaceae families suggested that a fourth alkylation mechanism, route D, **159** $\rightarrow$ **167** $\rightarrow$ **168**, might be operative in these plants. An experiment to test this mechanism has been described[352]. Leaves of *Clerodendrum campbellii*, which produces $(24S)$-24-ethylcholesta-5,22,25-trien-3β-ol **(90)** were incubated with $(4R)$-4-$^3H_1$-2-^{14}C-mevalonic acid and the isolated sterol was found to have a $^3H:^{14}C$ atomic ratio of 3:5[352]. Removal of the 25-methylene group from the side-chain **(171)** gave the 27-nor-25-one derivative **(172)** which had an unchanged $^3H:^{14}C$ atomic ratio of 3:5. Thus the methylene group of **171** was not derived from C-2 of mevalonic acid but must arise stereospecifically, probably in the alkylation reaction, from that methyl group of the precursor 24-methylene sterol which was originally derived from the C-3′ position of mevalonic acid. This is a similar situation to that already discussed above in connexion with the formation of the cyclolaudenol (**11**) side-chain. Base equilibration of **172** produced **173** with a ^{3}H:^{14}C atomic ratio of 2:5[352], one tritium was lost and it must have been located at C-24 as required if route *D* operated. Thus in plants producing sterols of the type **89–93** it appears that the first methylation reaction to give a 24-methylene sterol proceeds with a 1,2-hydrogen migration from C-24 to C-25. However during the elaboration of the 24-ethyl-25-methylene side-chain this hydrogen migrates back from C-25 to C-24. The 'X$^-$ group mechanism' can again be applied to the formation of this type of side-chain as indicated in Scheme 17. *Trans* 1,2-addition of the methyl group and the X$^-$ group to the precursor 24-methylene sterol (**174**) first gives **175**. Removal of the X$^-$ group accompanied by a 1,2-hydrogen migration and proton elimination from one terminal methyl group, which is possibly in a favourable position with regard to a proton-accepting site on the enzyme, then leads to **176**.

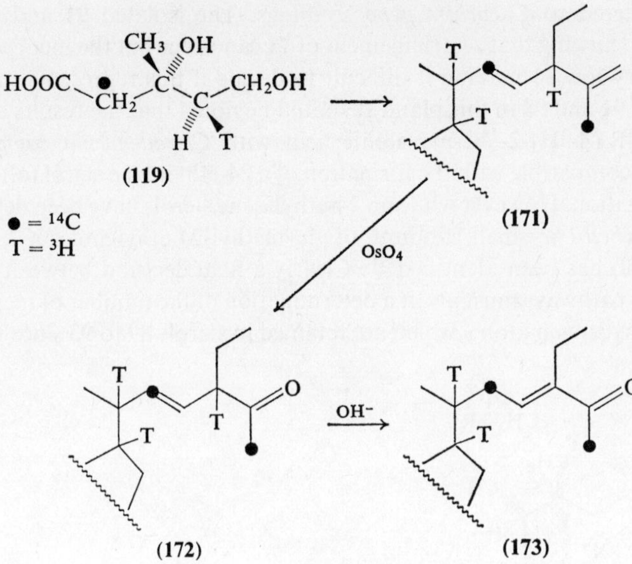

The co-occurrence of stigmasta-7,Z-24(28)-dien-3β-ol (**71**) with stigmasta-7,25-dien-3β-ol (**91**) and stigmasta-7,22,25-trien-3β-ol (**92**) in some Cucurbitaceae plants has prompted the suggestion[353] that **71** is a precursor of **91** and **92**. To test this possibility labelled **71** was synthesized[353] and administered to *Cucurbita pepo* seedlings. The isolated **91** and **92** were labelled showing that rearrangement of **71** can occur but the incorporations were very low so making it difficult to decide if this represents the major route to **91** and **92** in this plant. It should be noted that the results obtained using (4R)-4-^{3}H$_1$-2-^{14}C-mevalonic acid with *Clerodendrum campbellii*[352] are also compatible with the formation of a 24-ethylidene sterol followed by isomerization. However while no 24-ethylidene sterols have been detected in *C. campbellii* a small amount of 4α-methyl-24-ethylcholesta-7,25-dien-3β-ol (**40**) has been identified[128]. Clearly a final decision between the two possible pathways must await a determination of the number of methionine methyl hydrogen atoms which are retained in sterols **89** to **93** since route *D*,

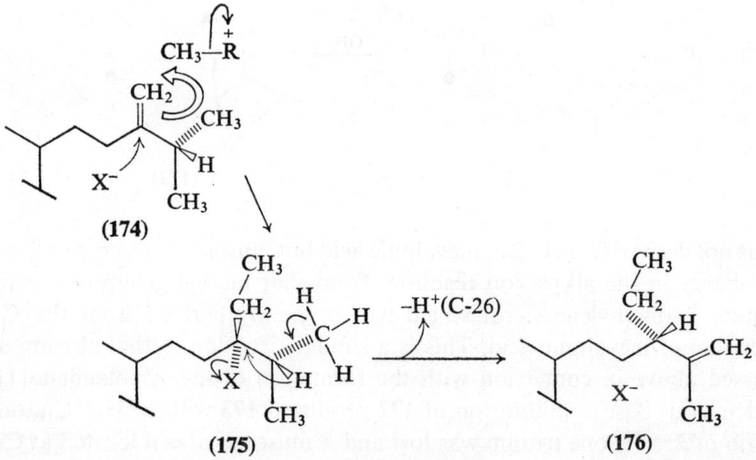

Scheme 17.

Scheme 16, would require retention of five hydrogens whereas a path involving a 24-ethylidene sterol would leave only four of the original methyl hydrogens.

There is no clear evidence to indicate at what point in the biosynthetic sequence the second alkylation step principally occurs. However at the present time no 4,4-dimethyl sterols are known which have a 24-ethyl or 24-ethylidene group indicating that the second alkylation is limited either to the 4α-methyl sterol level to give 24-ethylidene lophenol (**38**) and the related sterols **39** and **40** or to an even later stage in the biosynthetic pathway. Indeed the point of the second alkylation may be species-dependent since it has been shown that while 24-methylene cholesterol (**51**) can be alkylated to give 28-isofucosterol (**70**) in *Pinus pinea* seedlings[323, 354] this sterol (**51**) is only reduced to campesterol (**65**) by *Nicotiana tabacum* leaves[355].

The occurrence of epimeric 24-methyl and 24-ethyl sterols in nature raises the question of their biosynthetic origin[356]. In organisms in which 24-methylene and 24-ethylidene sterols occur as intermediates the presence of stereospecific $\Delta^{24(28)}$ reductases producing either 24R- or 24S-methyl and ethyl sterols can be visualized. On the present evidence of sterol identification and distribution this would seem to indicate that in most higher plants reduction of 24-methylene and 24-ethylidene compounds gives principally 24R sterols* whilst in algae and fungi 24S sterols* are the preferred products. In the formation of sterols by mechanisms not involving 24-methylene and 24-ethylidene intermediates, for example the alkyl groups of cyclolaudenol **(11)** and the sterols of *Dictyostelium discoideum* and *Chlorella vulgaris* the C-24 configuration must be imposed as a direct consequence of the actual alkylation mechanism. The occurrence in some higher plant species of brassicasterol **(57)** which has a C-24 configuration opposite to that of the other sterols in the plant presents a problem[356]. Perhaps the C-24 configuration of the C_{28}-sterol has been assigned incorrectly in this case but the possible formation of brassicasterol **(57)** from cyclolaudenol **(11)** should be considered, with the other 24R sterols arising by reduction of 24-methylene and 24-ethylidene intermediates.

VI. INTRODUCTION OF THE Δ^{22} BOND INTO PHYTOSTEROLS

There are apparently two routes in nature for the introduction of the Δ^{22}-*trans* double bond found in several phytosterols (Table 1). As already discussed, it has been shown[347–351] in the slime mould *Dictyostelium discoideum* that the Δ^{22} bond of stigmast-22-en-3β-ol **(81)** is formed as a consequence of the C-24 alkylation reactions (route C, Scheme 16). However at present this is the only case in which this mechanism is shown to operate, in all other cases so far studied it appears that the Δ^{22} bond is introduced by dehydrogenation at C-22 and C-23 and is independent of C-24 alkylation. Indeed the operation of route C, Scheme 16, has been eliminated in the case of ergosterol **(60)** formation by yeast[305, 357]. When 23-$^3H_2$-lanosterol was incorporated into ergosterol **(60)** by whole yeast cells[357] one tritium remained at C-23 but the other was eliminated, a result contrary to expectations if route C were operative. In the case of ergosterol **(60)** formation in yeast the action of a Δ^{22} dehydrogenase has been demonstrated by the conversion of 24-methyl dihydrolanosterol **(177)**[357] and ergost-7-en-3β-ol[305] **(58)** into ergosterol **(60)**. The conversion of the former compound, **(177)**, was shown to occur without prior oxidation of the 24-methyl group to a

* See footnote on page 139.

(177)

24-methylene grouping thus demonstrating that the Δ^{22} dehydrogenation reaction does not require activation of an adjacent double bond[357]. It should be noted that whilst route C, Scheme 16, is reported to occur in *D. discoideum* the presence of a Δ^{22} dehydrogenase in this mould is also clearly established because it can convert stigmastanol (**79**) efficiently into stigmast-22-enol (**81**). Since this organism contains stigmastanol (**79**) which is not produced by reduction of stigmast-22-enol (**81**)[350], it appears that two alkylation mechanisms are operating and that stigmast-22-enol (**81**) can possibly arise by two routes, either from the alkylation reaction or by desaturation of the side-chain of preformed stigmastanol (**79**)[349]. The relative importance of these alternative routes is at present obscure.

In higher plants all the evidence so far available indicates that the Δ^{22} bond is introduced by desaturation of the appropriate saturated sterol side-chain. Studies on the changes in levels of sitosterol (**80**) and stigmasterol (**82**) in various higher plants have suggested a precursor–product relationship for these two sterols[358–361]. This view is strengthened by the greater rate of incorporation of 2-^{14}C-mevalonic acid and ^{14}C-acetic acid into sitosterol (**80**) than into stigmasterol (**82**) in *Dioscorea spiculiflora*[358], *Solanum tuberosum*[359], *Nicotiana tabacum* tissue culture[362] and *Calendula officinalis*[363]. Moreover the administration of radioactive sitosterol (**80**) to leaves of *Digitalis lanata* led to the isolation of labelled stigmasterol (**82**)[364]. The observation[365] that tritiated spinasterol (**84**) is converted by *Nicotiana tabacum* leaves into both sitosterol (**80**) and stigmasterol (**82**) apparently indicates that the C-22–C-23 desaturation reaction is reversible.

The stereospecificity of hydrogen removal from C-22 and C-23 during Δ^{22} introduction is discussed in Section VIII.

VII. THE ROLES OF LANOSTEROL AND CYCLOARTENOL IN PHYTOSTEROL BIOSYNTHESIS

Cyclization of squalene-2,3-oxide can produce several tetracyclic triterpenes of the required stereochemistry for further conversion into phytosterols by demethylation at C-4 and C-14, double bond rearrangement and C-24 alkylation.

In fungi, as in animals, lanosterol (**1**) is the product of squalene-2,3-oxide cyclization (Section IV). Thus lanosterol (**1**) has been identified in a number of fungi[3, 13–15, 19–22, 50, 261–263] and its biosynthesis in yeast has been demonstrated[366, 367]. Further conversion of lanosterol (**1**) into ergosterol (**60**) is well established[302, 357, 368] but the exact sequence of intermediates is uncertain and it appears likely that several routes will be found to operate in various fungi. Scheme 18 shows possible routes based upon known fungal sterols. Thus, as already discussed (Section V), alkylation at C-24 may occur before or after C-4 and C-14 demethylation depending upon the species examined. Similarly the point of reduction of the 24-methylene group and introduction of the Δ^{22} bond may be variable. Relevant to Scheme 18 the following transformations have so far been demonstrated in fungi: $1 \rightarrow 60^{302, 357, 368}$; $3 \rightarrow 60^{50, 303-305}$; $45 \rightarrow 53^{295}$; $45 \rightarrow 55^{307}$; $58 \rightarrow 60^{305}$; $55 \rightarrow 60^{306}$; $59 \rightarrow 60^{369}$; $52 \rightarrow 54^{83}$; $52 \rightarrow 60^{83}$. The evidence for the presence of **6** in *Phycomyces blakesleeanus*[19, 20] implies that C-14 demethylation may precede that at C-4 in fungi and a $\Delta^{8,14}$ diene sterol is implicated in this demethylation reaction in yeast[370] as found previously in cholesterol biosynthesis in rat liver[8, 371, 372]. A microsomal preparation has been obtained from yeast which catalyses C-4 and C-14 demethylation and requires glutathione, NAD^+ and Mg^{2+} for full activity[312]. The sequence of removal of the 4α- and 4β-methyl groups from lanosterol (**1**) in its conversion into ergosterol (**60**) remains to be determined but it is interesting to note that the demethylation of 3β-hydroxyprotosta-17(20),24-diene (**23**) in its conversion into fusidic acid (**118**) by *Fusidium coccineum* apparently involves the loss of the 4β-methyl group[373, 374]. This is opposite to the sequence in animals[375, 376] and higher plants (see later) where the 4α-methyl group is the first to be removed.

Sterol formation in algae and higher plants was originally assumed to proceed via lanosterol (**1**) as in animals and fungi. However there are very few reports of lanosterol (**1**) in higher plants and at present the latex of some species of the Euphorbiaceae seems to be the only authentic higher plant source of lanosterol (**1**)[16–18, 377]. A report of lanosterol (**1**) and 24,25-dihydrolanosterol in coffee beans[378] is now known to be incorrect[379], a reinvestigation[52, 379] showed that the 4,4-dimethyl sterols of coffee oil are cycloartenol (**7**) and 24-methylene cycloartanol (**10**). Cycloartenol (**7**) is of very wide distribution in higher plants[3, 8, 9, 94], and this led to the suggestion[380–384] that cycloartenol (**7**) rather than lanosterol (**1**) is the first cyclic precursor of the phytosterols. The rapid incorporation of acetate and mevalonate into cycloartenol (**7**) has been demonstrated in a number of

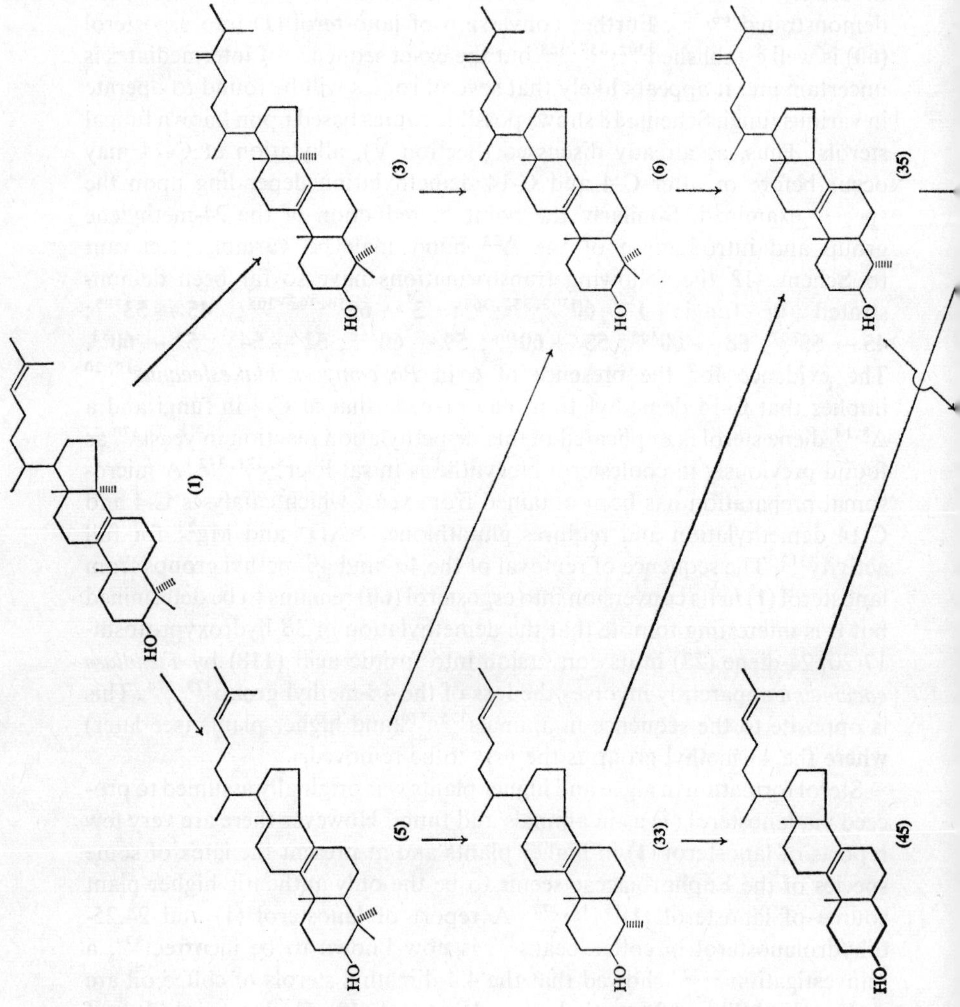

(1)

(3)

(5)

(6)

(33)

(35)

(45)

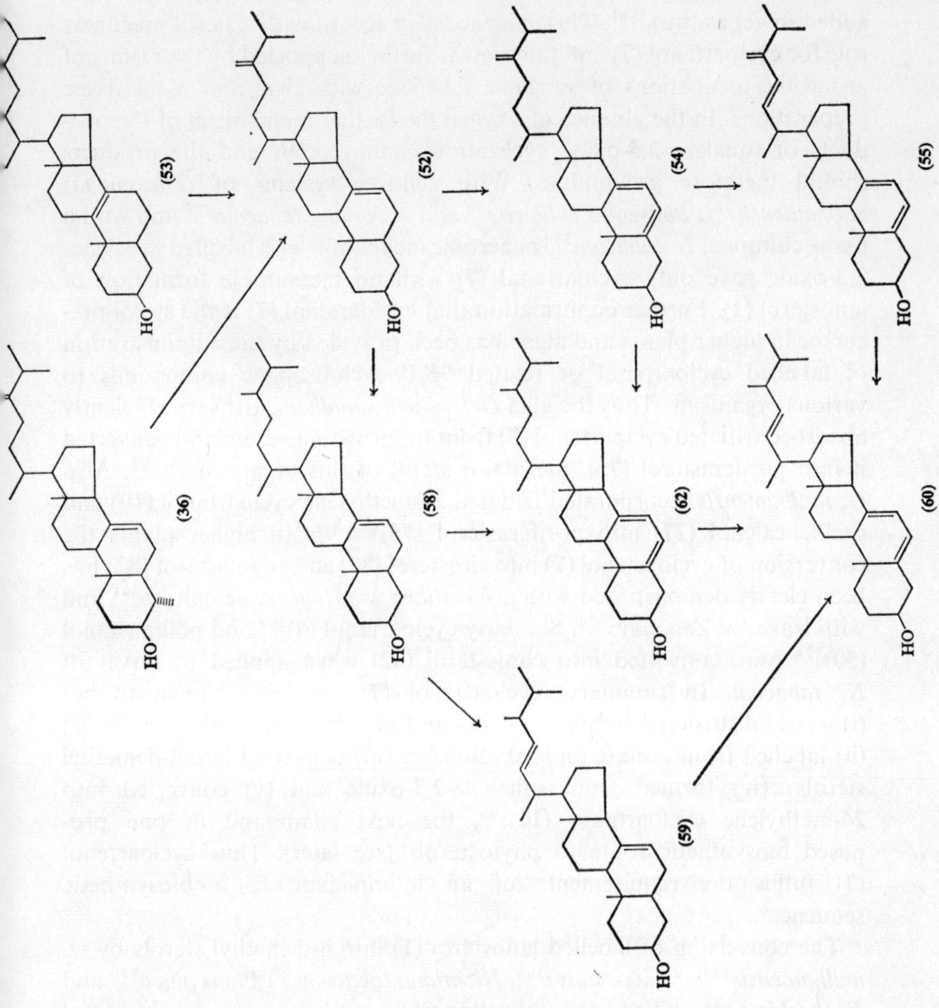

(53)

(52)

(54)

(55)

(36)

(58)

(62)

(60)

(59)

Scheme 18.

plants including *Nicotiana tabacum*[382], *Agave toumeyana*[385], *Dioscorea composita*[385] bramble[385], endive[385] carrot[385], *Pisum sativum*[383], *Larix decidua*[162,231], *Zea mays*[386–388], *Solanum tuberosum*[224], *Fucus spiralis*[226], and *Musa sapientum*[389]. However in no case has lanosterol (**1**) been found labelled[162,224,226,382,385,387–390] even where unlabelled lanosterol has been added to act as 'trap'[391]. This evidence is in accord with a sterol precursor role for cycloartenol (**7**) and this view is further supported by the results of anaerobic incubations of squalene-2,3-oxide with algal and plant tissue preparations. In the absence of oxygen the further metabolism of the products of squalene-2,3-oxide cyclization cannot occur and the products should therefore accumulate. With cell-free systems of *Ochromonas malhamensis*[267], *Phaseolus vulgaris*[266] and *Nicotiana tabacum*[255] and with a tissue culture of *N. tabacum*[254] anaerobic incubation with labelled squalene-2,3-oxide gave only cycloartenol (**7**) with no measurable formation of lanosterol (**1**). Further confirmation that cycloartenol (**7**) is the sterol precursor in higher plants and algae has been provided by the administration of labelled cycloartenol or related $9\beta,19$-cyclopropane compounds to various organisms. Thus the alga *Ochromonas malhamensis* very efficiently absorbed tritiated cycloartenol (**7**) from the growth medium and converted it into poriferasterol (**75**), the major sterol of this organism[334,392]. Also *O. malhamensis* incorporated tritiated 24-methylene cycloartanol (**10**) and cycloeucalenol (**27**) into poriferasterol (**75**)[334,392]. In higher plants the conversion of cycloartenol (**7**) into sitosterol (**80**) and stigmasterol (**82**) has been clearly demonstrated with a *Nicotiana tabacum* tissue culture[391] and with leaves of *Zea mays*[388]. Similarly cycloartanol (**8**)[393] and pollinastanol (**50**)[394] were converted into cholesterol (**42**) when applied to leaves of *N. tabacum*. In summary, cycloartenol (**7**) has been shown to be: (i) widely distributed in higher plants and in some algae[87,109,226,395,396]; (ii) labelled from acetate and mevalonate; (iii) converted into 4-demethyl sterols; (iv) formed from squalene-2,3-oxide and (v) converted into 24-methylene cycloartanol (**10**)[296], the next compound in one proposed biosynthetic route to phytosterols (see later). Thus cycloartenol (**7**) fulfils the requirements of an intermediate in a biosynthetic sequence[6].

The conversion of labelled lanosterol (**1**) into 4-demethyl sterols by *O. malhamensis*[334,392], *Zea mays*[388], *Nicotiana tabacum*[391], *Pinus pinea*[340] and *Euphorbia peplus*[397] and the utilization of 24-methylene dihydrolanosterol (**2**) by *N. tabacum*[355] and *Spinacea oleracea*[398] can be readily explained by a lack of substrate specificity shown by the various enzyme systems involved in sterol formation. The possibility that parkeol (**4**) can also act as a phytosterol precursor has been explored[399] but it was found that a *N. tabacum* tissue culture did not convert this compound into sterols, the only metabolite identified was the 24,25-epoxide of parkeol.

The lanosterol (**1**) found in the latex of some *Euphorbia* species[16-18], can arise either directly from the cyclization of squalene-2,3-oxide or by opening the 9β,19-cyclopropane ring of preformed cycloartenol (**7**). The latter route is perhaps favoured by the observations that latex from *E. helioscopia* incorporated 1-^{14}C-acetate into both lanosterol (**1**) and cycloartenol (**7**)[400] while latex of *E. lathyris* isomerized 25-^{14}C-cycloartenol (**7**) into lanosterol (**1**)[401].

Any discussion of the route from cycloartenol (**7**) to 4-demethyl sterols must take into account which of the several possible C-24 alkylation mechanisms is operating in the organism under consideration. To date the only route considered in any detail is one involving 24-methylene and 24-ethylidene intermediates and it was based upon the sterols which are known to co-occur in various higher plant species[8, 9, 45, 162, 381-383]. The first step (Scheme 19) is envisaged to be alkylation at C-24 to give 24-methylene cycloartanol (**10**) which is widely distributed in higher plants[3, 9, 94]. Demethylation at C-4 will then lead to cycloeucalenol (**27**) another 9β,19-cyclopropane compound which is now being recognized as widely distributed in plants[3, 9, 402, 403]. At this point it seems probable that the 9β,19-cyclopropane ring is opened[9, 383] and this view gained support when the structure of obtusifoliol (**30**), isolated originally from *Euphorbia obtusifolia*, was elucidated[48] and it was subsequently shown to occur in several plants together with cycloeucalenol (**27**)[52, 386, 402, 404]. Demethylation of obtusifoliol (**30**) at C-14 and rearrangement of the nuclear double bond to Δ^7 yields 24-methylene lophenol (**36**) which can follow two possible routes, either C-4 demethylation to give the various C_{28}-sterols of plants, or alternatively a second alkylation at C-28 and thence to the 4-demethyl C_{29}-sterols. It is at this point, when the second transmethylation reaction probably occurs, that divergences in the biosynthetic pathway occur depending upon which alkylation mechanism (see Section V) is operating in the plant species under study. For example, in *Clerodendron* species and members of the Cucurbitaceae family the operation of route *D*, Scheme 16, will lead to 4α-methyl-(24S)-24-ethyl-5α-cholesta-7,25-dien-3β-ol (**40**), isolated from leaves of *Clerodendrum campbellii*[128], which can then undergo C-4 demethylation to lead to the 25-methylene sterols characteristic of these plants. Alternatively by route *A*, Scheme 16, 24-ethylidene lophenol (**38**) will be obtained and this sterol seems to be of fairly wide distribution amongst higher plants[8, 9, 94]. Removal of the 4α-methyl group from **38** gives stigmasta-7,Z-24(28)-dien-3β-ol (**71**)[77, 110, 325] which can then progress via a $\Delta^{5,7}$ intermediate to 28-isofucosterol (**70**) another compound found in a number of higher plants[321–325]. Finally reduction of the 24-ethylidene group of **70** will produce sitosterol (**80**). However, reduction of the 24-ethylidene group at earlier stages in the sequence is probable and would give for example stigmast-7-en-3β-ol (**83**) from which spinasterol (**84**) or sitosterol (**80**) can be obtained[405]. In the alga *Ochromonas malhamensis* the principal sterol is poriferasterol (**75**), the C-24 epimer of stigmasterol (**82**),

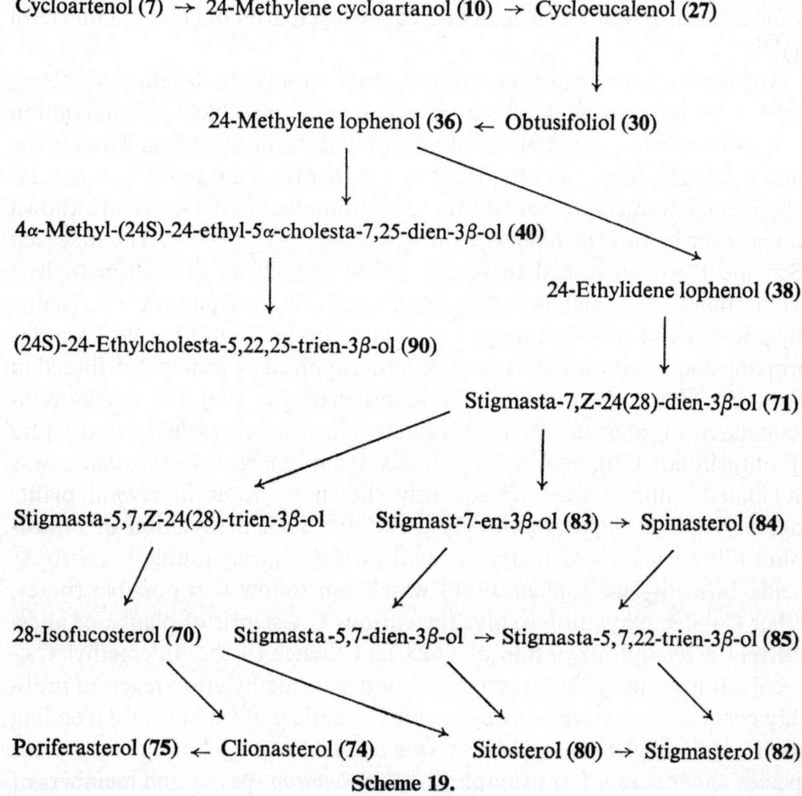

Scheme 19.

and a 24-ethylidene intermediate has been implicated in its biosynthesis[332]. This suggests that a pathway such as that shown in Scheme 19 is operative and this is supported by the demonstration[334, 335, 392] that *O. malhamensis* can efficiently convert tritium-labelled cycloartenol (**7**), 24-methylene cycloartanol (**10**), cycloeucalenol (**27**), 24-methylene lophenol (**36**), 24-ethylidene lophenol (**38**), stigmasta-7,Z-24(28)-dien-3β-ol (**71**) and 28-isofucosterol (**70**) into poriferasterol (**75**).

Regarding the sequence of removal of the three methyl groups from C-4 and C-14 it appears from the structures of known phytosterols that one of the C-4 groups is usually lost prior to C-14 demethylation. This compares with cholesterol (**42**) biosynthesis in animals where the C-14 methyl group is usually the first to be lost by oxidative decarboxylation[4-8]. Delayed C-14 demethylation is probably a consequence of the $9\beta,19$-cyclopropane ring present in the early plant sterol intermediates. In such compounds the 14α-methyl group is rather hindered (see **128**) and this may prevent attack by the demethylase enzyme system. Moreover C-14 decarboxylation is facilitated by a Δ^8 (or Δ^7) bond[4-6,251,375,406] and this only appears in the proposed biosynthetic route as a result of the opening of the cyclopropane ring. The order of removal of the C-4 methyl groups has been investigated[316] in the fern *Polypodium vulgaris* during the conversion of cycloartanol (**8**) into 31-nor-cyloartanol (**26**). It was shown[316] using $(4R)$-4-$^3H_1$-2-^{14}C-mevalonic acid that the 4α-methyl group of cycloartenol (**7**) is derived from C-2 of mevalonic acid and that the 4α-methyl group is lost from cycloartanol (**8**) while the 4β-methyl group is epimerized to the 4α-position in the production of 31-nor-cycloartanol (**26**). Similarly the loss of the 4α-methyl group of 24-methylene cycloartanol (**10**) has been shown in the formation of cyclo-leucalenol (**27**) and in the conversion of cyclolaudenol (**11**) into 31-nor-cyclolaudenone (**178**)[317] by the banana, *Musa sapientum*[407,408]. The same

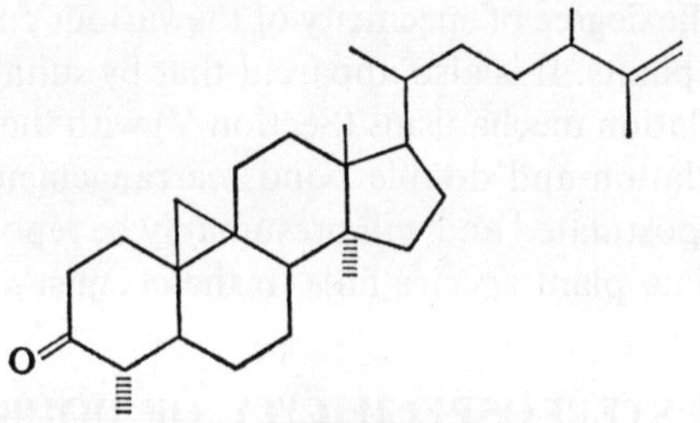

(178)

order of C-4 removal occurs in the demethylation of cholesterol precursors in rat liver[375,376]. A further similarity between the C-4 demethylation reactions in rat liver[409] and several plants[226,316,407,408,410] is shown by the exchange of the 3α-hydrogen atom which has been explained by the involvement of a 3-ketone in the demethylation reaction[5-8,408,409]. This latter finding renders obscure the significance of the rapid labelling of esterified 4,4-dimethyl and 4α-methyl sterols when 2-^{14}C-mevalonic acid is administered to plant tissues[386,389].

Evidence has been obtained for the involvement of the supernatant and microsomal fractions of the plant cell in sterol biosynthesis[411] but the preparation of active cell-free preparations for the conversion of cyclo-artenol (**7**) into sterols is proving elusive. The recognition of water-soluble forms of sterol in *Euglena gracilis*[412] and in yeast[413, 414] and of squalene in *Kalanchoë blossfeldiana*[415] suggests that such 'soluble' sterol conjugates should perhaps be considered as playing an important role in phytosterol production[415a].

It must be stressed that the biosynthetic routes outlined in Scheme 19 are only intended as a guide to the manner in which the plant sterols can be elaborated from cycloartenol (**7**). Consideration of the large array of plant sterols now characterized (Table 1) reveals that many divergent pathways are possible. Thus while Scheme 19 shows a linear pathway from cyclo-artenol (**7**) to 24-ethylidene lophenol (**38**) lack of specificity by the C-24 alkylation and C-4 and C-14 demethylation enzymes and by the enzyme responsible for the opening of the 9β,19-cyclopropane ring could give diverg-ing and converging pathways. Indeed when tobacco tissue cultures were incubated with ^{14}C-acetic acid for increasing periods of time the specific radioactivities of compounds **7**, **10**, **27**, **30** and **36** were not apparently in accord with the simple linear route[416]. Sterols **10** and **27** had specific radio-activities lower than **7** and **30** indicating that they lie on one of the several possible parallel routes from cycloartenol (**7**) to obtusifoliol (**30**)[416]. Possibly the relative importance of the various routes will vary from species to species and depend upon the degree of specificity of the various enzyme systems in-volved in different plants. It is also apparent that by suitable permutations of the various alkylation mechanisms (Section V) with the processes of C-4 and C-14 demethylation and double bond rearrangement that numerous new sterols can be postulated and will presumably be reported in the future when the appropriate plant species falls to the chemist's soxhlet.

VIII. THE STEREOSPECIFICITY OF DOUBLE BOND INTRODUCTION

In the course of sterol formation there is a movement of the double bond[5-8] through the sequence $\Delta^8 \rightarrow \Delta^7 \rightarrow \Delta^{5,7} \rightarrow \Delta^5$. The stereospecificity of hydrogen eliminations at C-6 and C-7 during these steps has been examined in cholesterol (**42**) biosynthesis by rat liver[8]. The isomerization of the Δ^8 bond to the Δ^7 position occurs with loss of the 7β-hydrogen atom[417-421] while introduction of the Δ^5 bond proceeds with elimination of the 5α- and 6α-hydrogen atoms[231, 421-424].

The stereochemistry of hydrogen atom elimination from C-7 in plant sterol elaboration has been examined by using mevalonic acid labelled stereospecifically at C-2 with tritium. As shown in Scheme 20 (2R)-2-^{3}H$_1$-

2-^{14}C-mevalonic acid (179) will lead via squalene (181) (see Section III) to cycloartenol (183) which has a 7α-tritium atom; similarly from (2S)-2-^{3}H$_1$-2-^{14}C-mevalonic acid (180) and the derived squalene (182) there will be a 7β-tritium atom in the labelled cycloartenol (184). When the 2R and 2S tritiated mevalonates were fed separately to the alga *Ochromonas malhamensis* and the biosynthesized poriferasterols (185 and 186 respectively)

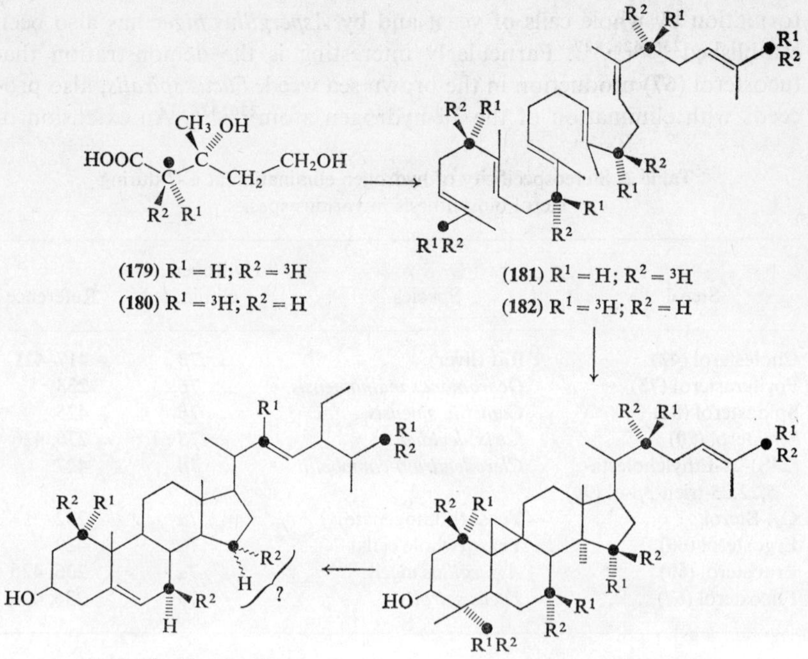

(179) R^1 = H; R^2 = ^{3}H

(180) R^1 = ^{3}H; R^2 = H

(181) R^1 = H; R^2 = ^{3}H

(182) R^1 = ^{3}H; R^2 = H

(185) R^1 = H; R^2 = ^{3}H

(186) R^1 = ^{3}H; R^2 = H

(183) R^1 = H; R^2 = ^{3}H

(184) R^1 = ^{3}H; R^2 = H

Scheme 20.

were isolated and subjected to suitable degradation it was found that the 7β-hydrogen of cycloartenol had been lost[233]. Using the same technique it has been demonstrated that the 7β-hydrogen atom of cycloartenol is eliminated during the formation of the 4-demethyl sterols in the leaves of three species of higher plants (Table 2). This shows a mechanistic similarity between the $\Delta^8 \rightarrow \Delta^7$ isomerase reaction in these plant species and the analogous reaction which occurs in mammalian sterol synthesis. However by contrast in the production of fungal sterols it was found that the 7α-hydrogen atom is eliminated. This was first demonstrated[232] with a yeast

7

homogenate which although it failed to produce ergosterol (**60**) gave a C_{27}-sterol labelled from 2R- or 2S- tritiated mevalonic acid. It was found that this sterol retained the hydrogen atom at C-7 derived from the 2-pro-S position of mevalonic acid but lost the C-7 hydrogen atom arising from the 2-pro-R hydrogen atom of mevalonic acid[232]. This corresponds to the loss of the 7α-hydrogen atom of lanosterol (**1**), the precursor of fungal sterols (see Section VI). The elimination of the 7α-hydrogen during ergosterol (**60**) formation by whole cells of yeast and by *Aspergillus niger* has also been established[236, 426, 429]. Particularly interesting is the demonstration that fucosterol (**67**) production in the brown sea weed, *Fucus spiralis*, also proceeds with elimination of the 7α-hydrogen atom[236, 426]. An extension of

Table 2. Stereospecificity of hydrogen elimination at C-7 during sterol biosynthesis in various species

Sterol	Species	Hydrogen atom eliminated	Reference
Cholesterol (**42**)	Rat (liver)	7β	417–421
Poriferasterol (**75**)	*Ochromonas malhamensis*	7β	233
Spinasterol (**84**)	*Camellia sinensis*	7β	425
Sitosterol (**80**)	*Larix decidua*	7β	236, 426
(24S)-24-Ethylcholesta-5,22,25-trien-3β-ol (**90**)	*Clerodendrum campbellii*	7β	427
C_{27}-Sterol	Yeast (homogenate)	7α	232
Ergosterol (**60**)	Yeast (whole cells)	7α	429
Ergosterol (**60**)	*Aspergillus niger*	7α	236, 426
Fucosterol (**67**)	*Fucus spiralis*	7α	236, 426

these studies on biosynthetic stereospecificity to other classes of alga, fungi and higher plants would perhaps yield interesting phylogenetic data.

The introduction of the Δ^5 bond into fungal, algal and higher plant sterols proceeds in all cases examined (Table 3) by the elimination of the 5α- and 6α-hydrogen atoms revealing that the same mechanism of dehydrogenation is probably common to all these organisms. Some of these studies were facilitated by the use of (5R)-5-$^3H_1$-2-^{14}C-mevalonic acid (**187**) which, via squalene (**188**) introduces a 6β-tritium into cycloartenol (**189**) (Scheme 21). Thus when (5R)-5-$^3H_1$-2-^{14}C-mevalonic acid (**187**) was incorporated into the culture medium of *Ochromonas malhamensis*, the isolated poriferasterol (**190**) retained a tritium atom at C-6 thereby demonstrating abstraction of the 6α-hydrogen in Δ^5 bond introduction[430]. In the case of ergosterol (**60**) formation in yeast cells the loss of the 5α- and 6α-hydrogen atoms was

Table 3. Stereochemistry of hydrogen elimination from C-6 during
sterol biosynthesis in various species

Sterol	Species	Hydrogen atom eliminated	Reference
Cholesterol (42)	Rat (liver)	6α	231, 421–424
Poriferasterol (75)	*Ochromonas malhamensis*	6α	430
Sitosterol (80)	*Larix decidua*	6α	231
Ergosterol (60)	Yeast	6α	369
Ergosterol (60)	*Aspergillus fumigatus*	6α	234, 236
	Blakeslea trispora	6α	236
(24S)-24-Ethylcholesta-5,22,25-trien-3β-ol (90)	*Clerodendrum campbelli*	6α	427

demonstrated using $[5\alpha,6\alpha\text{-}^3H_2]$-ergosta-7,22-dien-3β-ol[369]. The details
of the introduction of the Δ^5 bond are not clear at present. The isolation and
purification of a 5α-hydroxy sterol dehydrase from yeast has been
reported[431, 432] which will promote the anaerobic dehydration of ergosta-
7,22-dien-3β,5α-diol to give ergosterol (60). However involvement of a
5α-hydroxylation–dehydration reaction in ergosterol biosynthesis in yeast

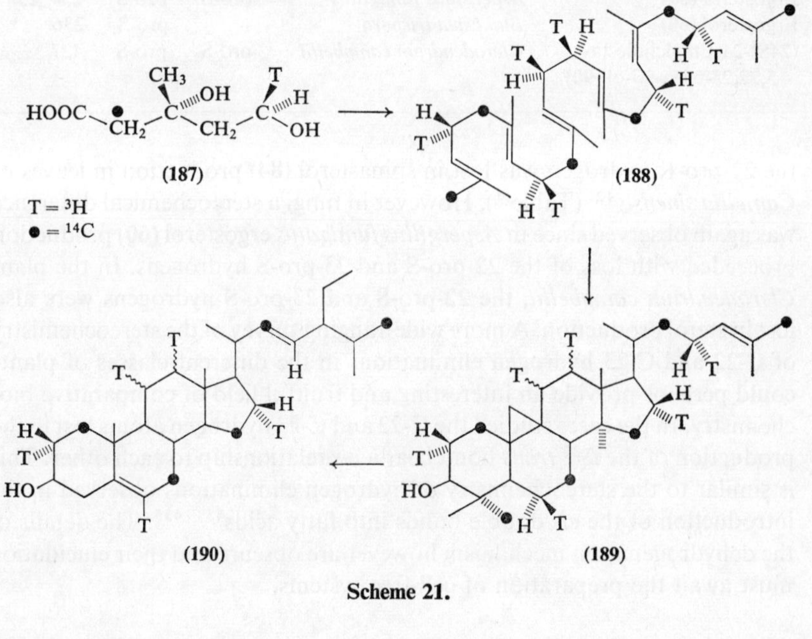

$T = {}^3H$

● $= {}^{14}C$

(187)   (188)

(190)   (189)

Scheme 21.

has been contested[369] since under anaerobic conditions whole yeast cells were unable to convert ergosta-7,22-dien-3β,5α-diol into ergosterol (**60**) although this transformation did occur under aerobic conditions.

The use of 2R-, 2S- and 5R-tritiated mevalonic acid species has permitted the determination of the stereochemistry of hydrogen atom eliminations at C-22 and C-23 in the formation of Δ^{22} sterols (Schemes 20 and 21). The overall result of the C-22–C-23 dehydrogenation mechanism in poriferasterol (**75**) formation by an alga, *Ochromonas malhamensis*, showed the removal of the 22-pro-R and 23-pro-R hydrogens[233, 430]. Similarly the 23-pro-R hydrogen atom was eliminated in the formation of ergosterol (**60**) and 7-dehydroporiferasterol (**76**) by the related organism *O. danica*[343] and

Table 4. Stereospecificity of hydrogen elimination from C-22 and C-23
during sterol biosynthesis in various species

| | | Hydrogen atoms eliminated | | |
Sterol	Species	C-22	C-23	Reference
Poriferasterol (**75**)	*Ochromonas malhamensis*	pro-R	pro-R	233, 430
7-Dehydroporiferasterol (**76**)	*O. danica*		pro-R	343
Ergosterol (**60**)	*O. danica*		pro-R	343
Spinasterol (**84**)	*Camellia sinensis*	pro-R		425
Ergosterol (**60**)	*Aspergillus fumigatus*	pro-S	pro-S	234, 236
Ergosterol (**60**)	*Blakeslea trispora*		pro-S	236
(24S)-24-Ethylcholesta-5,22,25-trien-3β-ol (**90**)	*Clerodendrum campbellii*	pro-S	pro-S	427

the 22-pro-R hydrogen was lost in spinasterol (**84**) production in leaves of *Camellia sinensis*[425] (Table 4). However in fungi a stereochemical difference was again observed since in *Aspergillus fumigatus* ergosterol (**60**) production proceeded with loss of the 22-pro-S and 23-pro-S hydrogens. In the plant *Clerodendrum campbellii*, the 22-pro-S and 23-pro-S hydrogens were also lost in sterol production. A more wide-ranging survey of the stereochemistry of C-22 and C-23 hydrogen eliminations in the different classes of plants could perhaps provide an interesting and fruitful field of comparative biochemistry. In the cases studied the C-22 and C-23 hydrogen atoms lost in the production of the Δ^{22} *trans* bond bear a *cis* relationship to each other. This is similar to the stereochemistry of hydrogen eliminations observed in the introduction of the *cis* double bonds into fatty acids[433–435]. The details of the dehydrogenation mechanism however are obscure and their elucidation must await the preparation of cell-free systems.

IX. APPENDIX: THE ASSIGNMENT OF CONFIGURATION AT C-24 OF THE STEROL SIDE-CHAIN

The introduction of an alkyl group at C-24 of the sterol side-chain produces chirality at this carbon. An extension of the Plattner convention[438-440] by Fieser and Fieser[1] enabled C-24 alkyl groups to be assigned as 24α or 24β as indicated by **191** and **192** respectively. This convention has been used

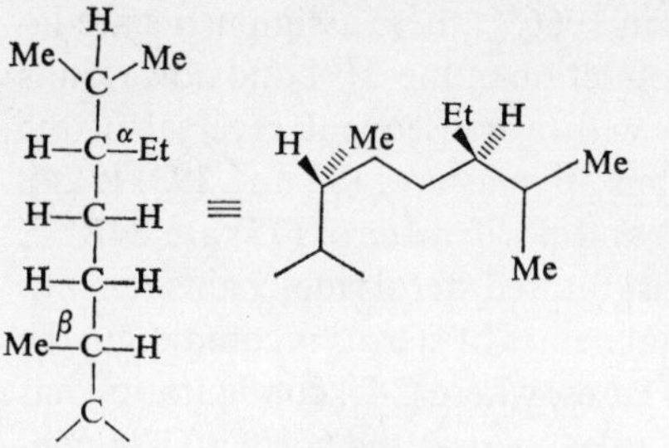

(191) 24α, 24R

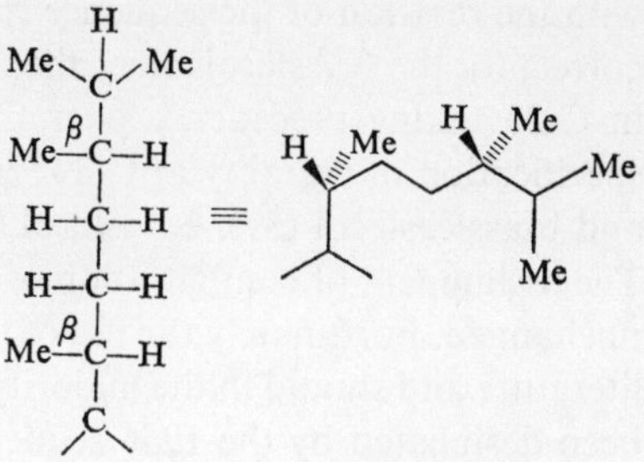

(192) 24β, 24S

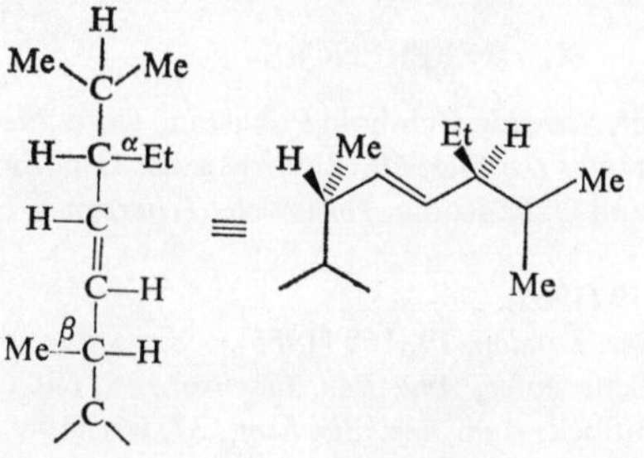

(193) 24α, 24S

extensively in the literature and the evidence for the assignments of C-24 configuration of naturally occurring C_{28}- and C_{29}-sterols has been reviewed[1,2,441]. Recently the IUPAC/IUB Revised Tentative Rules for Steroid Nomenclature[442] recommended that chirality at C-24 should be designated as either R or S by the sequence rule procedure proposed by Cahn, Ingold and Prelog[443-445]. Thus **191** is 24R and (**192**) is 24S and it therefore follows that 5α-stigmastanol (**79**), sitosterol (**80**) and campesterol (**65**) which are 24α-sterols have the 24R configuration while clionasterol (**74**) and ergost-5-enol (**56**) which are 24β-sterols have the 24S configuration. However the application of the sequence rule to the assignment of C-24 configuration in sterols with a Δ^{22} bond has produced an ambiguous situation in the literature owing to a revision of the sequence rule in relation to the priority of double-bonded carbons[444]. In stigmasterol (**82**, side-chain **193**) the absolute configuration at C-24 is the same as sitosterol (**80** and **191**) and by the original 1956 rules of Cahn *et al.*[443] stigmasterol was designated the 24R configuration since the C-25 secondary group was regarded to take precedence over C-23 of the Δ^{22} double bond[87,356]. Similarly ergost-5-enol (**56**), brassicasterol (**57**), ergosterol (**60**), clionasterol (**74**) and poriferasterol (**75**) all have the same configuration at C-24 (24β) and according to the 1956 sequence rules[443] all were assigned the 24S configuration[87,149,356]. However with the revision of the sequence rules in 1966[444] these assignments are incorrect for the Δ^{22} sterols since the introduction of the Δ^{22} bond now results in C-23 taking precedence over C-25 with a consequent reversal in the specification of chirality at C-24[441]. Thus stigmasterol (**82** and **193**) is 24S and brassicasterol (**57**), ergosterol (**60**) and poriferasterol (**75**) are 24R[441]. The assignment of configuration of the saturated sterol side-chains remains unchanged. Fortunately the use of trivial names of sterols is common in the literature and should in the majority of cases, where C-24 configuration has been designated by the 1956 sequence rules, permit the correct C-24 configuration to be deduced.

X. REFERENCES

1. L. F. Fieser and M. Fieser, *Steroids*, Reinhold Publishing Corp. New York, 1959.
2. C. W. Shoppee, *Chemistry of the Steroids*, Butterworths, London, 1964.
3. G. Ourisson, P. Crabbé and O. R. Rodig, *Tetracyclic Triterpenes*, Hermann, Leeds, 1964.
4. K. Bloch, *Science*, **150**, 19 (1965).
5. R. B. Clayton, *Quart. Rev. London*, **19**, 168 (1965).
6. I. D. Frantz and G. J. Schroepfer, *Ann. Rev. Biochem.*, **36**, 691 (1967).
7. C. J. Sih and H. W. Whitlock, *Ann. Rev. Biochem.*, **37**, 661 (1968).
8. L. J. Goad, in *Naturally Occurring Compounds formed Biologically from Mevalonic Acid* (Ed. T. W. Goodwin), Academic Press, London, 1970, p. 45.
9. L. J. Goad, in *Terpenoids in Plants* (Ed. J. B. Pridham), Academic Press, London, 1967, p. 159.
10. E. Heftmann, *Lloydia*, **30**, 209 (1967).
11. E. Heftmann, *Lloydia*, **31**, 293 (1968).
12. T. W. Goodwin, *Biochem. J.*, **123**, 293 (1971).
13. L. Ruzicka, R. Denss and O. Jeger, *Helv. Chim. Acta*, **28**, 759 (1945).
14. G. Ponsinet and G. Ourisson, *Bull. Soc. Chim. France*, 3682 (1965).
15. R. S. Ludwiczak and U. Wrzeciono, *Rozn. Chem.*, **34**, 77 (1960).
16. A. G. González and M. C. G. Mora, *Ann. Roy. Soc. Esp. Fis. Quim.*, **48B**, 475 (1952).
17. A. G. González and A. H. Toste, *Ann. Roy. Soc. Esp. Fis. Quim.*, **48B**, 487 (1952).
18. G. Ponsinet and G. Ourisson, *Phytochemistry*, **7**, 89 (1968).
19. G. Goulston, *Ph.D. Thesis*, University College of Wales, Aberystwyth, 1969.
20. G. Goulston, L. J. Goad and T. W. Goodwin, *Biochem. J.*, **102**, 15C (1967).
21. W. W. Epstein and G. Van Lear, *J. Org. Chem.*, **31**, 3434 (1966).
22. M. Lenfant, M. F. Lecompte and G. Farugia, *Phytochemistry*, **9**, 2529 (1970).

23. M. C. Dawson, T. G. Halsall, E. R. H. Jones and P. A. Robbins, *J. Chem. Soc.*, 586 (1953).

24. H. R. Bentley, J. A. Henry, D. S. Irvine and F. S. Spring, *J. Chem. Soc.*, 3673 (1953).

25. M. Shimizu, F. Uchimaru and G. Ohta, *Chem. Pharm. Bull. (Tokyo)*, **12**, 74 (1964).

26. G. Ohta and M. Shimizu, *Chem. Pharm. Bull. (Tokyo)*, **6**, 325 (1958).

27. H. R. Bentley, J. A. Henry, D. S. Irvine, D. Mukerji and F. S. Spring, *J. Chem. Soc.*, 596 (1955).

28. G. Berti, F. Bottari, B. Macchia, A. Marsili, G. Ourisson and H. Piotrowska, *Bull. Soc. Chim. France*, 2359 (1964).

29. H. Pinhas, *Bull. Soc. Chim. France*, 2037 (1969).

30. G. T. Newbold and F. S. Spring, *J. Chem. Soc.*, 249 (1944).

31. D. W. Haines and F. C. Warren, *J. Chem. Soc.*, 2554 (1949).

32. J. B. Barbour, F. C. Warren and D. A. Wood, *J. Chem. Soc.*, 2537 (1951).

33. I. M. Heilbron, G. L. Moffet and F. S. Spring, *J. Chem. Soc.*, 1583 (1934).

34. D. S. Irvine, W. Lawrie, A. S. McNab and F. S. Spring, *J. Chem. Soc.*, 2029 (1956).

35. K. G. Lewis, *J. Chem. Soc.*, 73 (1959).

36. G. Dupont, M. Julia and W. R. Wragg, *Bull. Soc. Chim. France*, 643 (1951).

37. J. S. Mills and A. E. A. Werner, *J. Chem. Soc.*, 3132 (1955).

37a. J. S. Mills, *J. Chem. Soc.*, 2196 (1956).

38. D. Shiengthong, A. Verasarn, P. Nanonggai-Suwanrath and E. W. Warnhoff, *Tetrahedron*, **21**, 917 (1965).

39. E. Ritchie, R. G. Senior and W. C. Taylor, *Australian J. Chem.*, **22**, 2371 (1969).

40. M. Devys, A. Alcaide, F. Pinte and M. Barbier, *Tetrahedron Letters* No. 53, 4621 (1970).

41. G. Berti, F. Bottari, A. Marsili, I. Morelli, M. Polvari and A. Mandelbaum, *Tetrahedron Letters* No. 2, 125 (1967).

42. K. N. N. Ayengar and S. Rangaswami, *Tetrahedron Letters* No. 37, 3567 (1967).

43. J. S. G. Cox, F. E. King and T. J. King, *J. Chem. Soc.*, 1384 (1956).

44. L. Amores-Marin, W. I. Torres and C. F. Asenjo, *J. Org. Chem.*, **24**, 411 (1959).

45. B. L. Williams, L. J. Goad and T. W. Goodwin, *Phytochemistry*, **6**, 1137 (1967).

46. G. Charles, T. Njimi, G. Ourisson, J. D. Ehrhardt, C. Conreur, A. Cavé and R. Goutarel, *C.R. Acad. Sci.*, **268C**, 2105 (1969).

47. H. H. Rees, L. J. Goad and T. W. Goodwin, *Phytochemistry*, **7**, 1875 (1968).

48. J. B. Barrera, J. L. Breton, J. D. Martin and A. G. Gonzalez, *Ann. Roy. Soc. Esp. Fis. Quim.*, **63B**, 191 (1967).

49. C. Djerassi, G. W. Krakower, A. J. Lemin, L. H. Lin and J. S. Mills, *J. Am. Chem. Soc.*, **80**, 6284 (1958).

50. D. H. R. Barton, D. M. Harrison, G. P. Moss and D. A. Widdowson, *J. Chem. Soc. C*, 775 (1970).

51. G. Osske and K. Schreiber, *Tetrahedron*, **21**, 1559 (1965).

52. B. A. Nagasampagi, J. W. Rowe, R. Simpson and L. J. Goad, *Phytochemistry* **10**, 1101 (1971).

53. Y. Mazur, A. Weizmann and F. Sondheimer, *J. Am. Chem. Soc.*, **80**, 1007 (1958).

54. Y. Mazur, A. Weizmann and F. Sondheimer, *J. Am. Chem. Soc.*, **80**, 6293 (1958).

55. Y. Mazur, A. Weizmann and F. Sondheimer, *J. Am. Chem. Soc.*, **80**, 6296 (1958).

56. A. Weizmann and Y. Mazur, *J. Org. Chem.*, **23**, 832 (1958).

57. K. Schreiber and G. Osske, *Tetrahedron*, **20**, 2575 (1964).

58. J. Bergman, B. O. Lindgren and C. M. Svahn, *Acta. Chem. Scand.*, **19**, 1661 (1965).

59. K. Tsuda, S. Akagi and Y. Kishida, *Chem. Pharm. Bull. (Tokyo)*, **6**, 101 (1958).

60. K. Tsuda, S. Akagi, Y. Kishida, R. Hayatsu and K. Sakai, *Chem. Pharm. Bull. (Tokyo)*, **6**, 724 (1958).
61. G. F. Gibbons, L. J. Goad and T. W. Goodwin, *Phytochemistry*, **6**, 677 (1967).
62. K. Tsuda, K. Sakai, K. Tanabe and Y. Kishida, *J. Am. Chem. Soc.*, **82**, 1442 (1960).
63. M. F. Hügel, W. Vetter, H. Audier, M. C. Barbier and E. Lederer, *Phytochemistry*, **3**, 7 (1964).
64. H. Wieland and Y. Kanaoka, *Ann. Chem.*, **530**, 146 (1937).
65. A. Stabursvik, *Acta Chem. Scand.*, **7**, 1220 (1953).
66. H. Mori, K. Shibata, K. Tsuneda, M. Sawai and K. Tsuda, *Chem. Pharm. Bull. (Tokyo)*, **16**, 1407 (1968).
67. A. M. Motzfeldt, *Acta Chem. Scand.*, **24**, 1846 (1970).
68. B. A. Knights, *Phytochemistry*, **9**, 903 (1970).
69. C. Djerassi, D. H. Murray and R. Villotti, *Proc. Chem. Soc.*, 450 (1961).
70. C. Djerassi, J. C. Knight and H. Brockmann, *Chem. Ber.*, **97**, 3118 (1964).
71. C. Djerassi, J. C. Knight and D. I. Wilkinson, *J. Am. Chem. Soc.*, **85**, 835 (1963).
72. M. F. Hügel, M. Barbier and E. Lederer, *Bull. Soc. Chim. France*, 2012 (1964).
73. M. Devys and M. Barbier, *Bull. Soc. Chim. Biol.*, **49**, 865 (1967).
73a. M. Barbier, *Progr. Phytochem.*, **2**, 1 (1970).
74. M. Devys, A. Alcaide, F. Pinte and M. Barbier, *C.R. Acad. Sci.*, **269**, 2033 (1969).
75. M. Barbier, M. F. Hügel and E. Lederer, *Bull. Soc. Chim. Biol.*, **42**, 91 (1960).
76. L. N. Standifer M. Devys and M. Barbier, *Phytochemistry*, **7**, 1361 (1968).
77. B. A. Knights and W. Laurie, *Phytochemistry*, **6**, 407 (1967).
78. H. Wieland and G. A. Gough, *Ann. Chem.*, **482**, 36 (1930).
79. D. H. R. Barton, *J. Chem. Soc.*, 813 (1945).
80. D. H. R. Barton, *J. Chem. Soc.*, 512 (1946).
81. H. Wieland, F. Rath and H. Hesse, *Ann. Chem.*, **548**, 54 (1941).
82. H. Morimoto, T. Imada, T. Murata and N. Matsumoto, *Ann. Chem.*, **708**, 230 (1967).
83. G. Goulston and E. I. Mercer, *Phytochemistry*, **8**, 1945 (1969).
84. O. N. Breivik, S. L. Owades and R. F. Light, *J. Org. Chem.*, **19**, 1734 (1954).
85. K. Petzoldt, M. Kuhne, E. Blanke, K. Kieslich and E. Kasper, *Ann. Chem.*, **709**, 203 (1967).
86. G. W. Patterson and R. W. Kraus, *Plant Cell Physiol.*, **6**, 211 (1965).
87. M. C. Gershengorn, A. R. H. Smith, G. Goulston, L. J. Goad, T. W. Goodwin and T. H. Haines, *Biochemistry*, **7**, 1698 (1968).
88. E. Fernholz and H. E. Stavely, *J. Am. Chem. Soc.*, **62**, 428, 1875 (1940).
89. H. Wieland and G. Coutelle, *Ann. Chem.*, **548**, 270 (1941).
90. R. K. Callow, *Biochem. J.*, **25**, 87 (1931).
91. C. Tanret, *Ann. Chim.*, **15**, 313 (1908).
92. F. H. Milazzo, *Can. J. Botany*, **43**, 1347 (1965).
93. A. Stoll and E. Jucker, in *Modern Methods of Plant Analysis*, Vol. 3 (Eds. K. Paech and M. V. Tracey), Springer-Verlag, Berlin, 1955, p. 141.
94. Reference to *Chemical Abstracts* will give many reports of the natural occurrence of these compounds.
95. D. H. R. Barton and T. Brunn, *J. Chem. Soc.*, 2728 (1951).
96. P. Singh and S. Rangaswami, *Current Sci.*, **35**, 515 (1966).
97. W. Fuerst, *Ann. Chem.*, **699**, 206 (1966).
98. N. Entwistle and A. D. Pratt, *Tetrahedron*, **24**, 3949 (1968).
99. N. Entwistle and A. D. Pratt, *Tetrahedron*, **25**, 1449 (1969).
100. A. N. Starratt, *Phytochemistry*, **5**, 1341 (1966).

101. C. Djerassi and R. McCrindle, *J. Chem. Soc.*, 4034 (1962).
102. R. J. Kemp and E. I. Mercer, *Biochem. J.*, **110**, 111 (1968).
103. K. Tsuda and K. Sakai, *Chem. Pharm. Bull. (Tokyo)*, **8**, 554 (1960).
104. P. W. Carter, I. M. Heilbron and B. Lythgoe, *Proc. Roy. Soc.*, **128B**, 82 (1939).
105. S. Ito, T. Tamura and T. Matsumoto, *J. Res. Inst. Technol. Nihon University*, No. 13, 99 (1956).
106. K. Tsuda, R. Hayatsu, Y. Kishida and S. Akagi, *J. Am. Chem. Soc.*, **80**, 921 (1958).
107. D. R. Idler, S. W. Nicksic, D. R. Johnson, V. W. Meloche, H. A. Scheutte and C. A. Baumann, *J. Am. Chem. Soc.*, **75**, 1712 (1953).
108. B. A. Knights, *Phytochemistry*, **4**, 857 (1965).
109. G. F. Gibbons, L. J. Goad and T. W. Goodwin, *Phytochemistry*, **7**, 983 (1968).
110. D. J. Frost and J. P. Ward, *Tetrahedron Letters*, 3779 (1968).
111. D. J. Frost and J. P. Ward, *Rec. Trav. Chim.*, **89**, 1054 (1970).
112. J. A. Fioretti, M. G. Kolor and R. P. McNaught, *Tetrahedron Letters*, No. 34, 2971 (1970).
113. R. Hanna, *Lloydia*, **27**, 40 (1964).
114. W. Bergmann and R. J. Feeney, *J. Org. Chem.*, **15**, 812 (1950).
115. E. M. Low, *J. Mar. Res.*, **14**, 199 (1955).
116. G. W. Patterson, *Plant Physiol.*, **42**, 1457 (1967).
117. J. W. Cook and M. F. C. Paige, *J. Chem. Soc.*, 336 (1944).
118. J. W. Clark-Lewis and I. Dainis, *Australian J. Chem.*, **20**, 1961 (1967).
119. E. Heftmann, B. E. Wright and G. U. Liddel, *Arch. Biochem. Biophys.*, **91**, 266 (1960).
120. K. Takeda, T. Kubota and Y. Matsui, *Chem. Pharm. Bull. (Tokyo)*, **6**, 437 (1958).
121. M. C. Hart and F. W. Heye, *J. Biol. Chem.*, **95**, 311 (1932).
122. E. Fernholz and M. L. Moore, *J. Am. Chem. Soc.*, **61**, 2467 (1939).
123. I. Nishioka, *J. Pharm. Soc. Japan*, **78**, 1432 (1958).
124. S. Rakhit, M. M. Dhar, H. Anand and M. L. Dhar, *J. Sci. Ind. Res. (India)*, **18B**, 422 (1959).
125. L. H. Zalkow, G. A. Cabat, G. L. Chetty, M. Ghosal and G. Keen, *Tetrahedron Letters*, No. 55, 5727 (1968).
126. M. Manzoor-I-Khuda, *Tetrahedron*, **22**, 2377 (1966).
127. W. Sucrow, *Chem. Ber.*, **99**, 2765 (1966).
128. L. M. Bolger, H. H. Rees, E. L. Ghisalberti, L. J. Goad and T. W. Goodwin, *Tetrahedron Letters*, No. 35, 3043 (1970).
129. W. Sucrow, *Chem. Ber.*, **99**, 3559 (1966).
130. B. R. Gonzalez and F. M. Panizo, *Anal. Fis. Quim*, **63B**, 1123 (1967).
131. N. Ikekawa, K. Tsuda and N. Norisaki, *Chem. Ind. (London)* 1179 (1966).
132. J. St. Pyrek, *Chem. Commun.*, 107 (1969).
133. E. Y. Levin and K. Bloch, *Nature*, **202**, 4927 (1964).
134. R. C. Reitz and J. G. Hamilton, *Comp. Biochem. Physiol.*, **25**, 401 (1968).
135. N. J. Souza and W. R. Nes, *Science*, **162**, 363 (1968).
136. N. Ikekawa, N. Morisaki, K. Tsuda and T. Yoshida, *Steroids*, **12**, 41 (1968).
137. G. W. Patterson, *Comp. Biochem. Physiol.*, **24**, 501 (1968).
138. D. R. Idler, A. Saito and P. Wiseman, *Steroids*, **11**, 465 (1968).
139. D. R. Idler and P. Wiseman, *Comp. Biochem. Physiol.*, **35**, 679 (1970).
140. A. Alcaide, M. Devys and M. Barbier, *Phytochemistry*, **7**, 329 (1968).
141. A. Alcaide, M. Barbier, P. Potier, A. M. Magueur and J. Teste, *Phytochemistry*, **8**, 2301 (1969).
142. R. P. Collins and K. Kalnins, *Comp. Biochem. Physiol.*, **30**, 779 (1969).

143. I. M. Heilbron, *J. Chem. Soc.*, 79 (1942).
144. L. Avivi, O. Iaron and S. Halevy, *Comp. Biochem. Physiol.*, **21**, 321 (1967).
145. S. Halevy, L. Avivi, and H. Katan, *J. Protozool.*, **12**, 293 (1966).
146. S. Huneck, *Phytochemistry*, **8**, 1313 (1969).
147. D. M. Orcutt and B. Richardson, *Steroids*, **16**, 429 (1970).
148. G. W. Patterson, *Comp. Biochem. Physiol.*, **31**, 391 (1969).
149. G. W. Patterson, *Lipids*, **6**, 120 (1971).
150. T. W. Goodwin, in *Lipids and Biomembranes of Eucaryotic Microorganisms* (Ed. J. A. Erwin), Academic Press, London, 1972, in press.
151. N. J. McCorkindale, S. A. Hutchinson, B. A. Pursey, W. T. Scott and R. Wheeler, *Phytochemistry*, **8**, 861 (1969).
152. C. G. Elliot, M. R. Hendrie, B. A. Knights and W. Parker, *Nature*, **203**, 427 (1964).
153. J. W. Hendrix, *Ann. Rev. Phytopathol.*, **8**, 111 (1970).
154. R. Ellouz and M. Lenfant, *Tetrahedron Letters*, 609 (1969).
155. S. Yoshida, R. Takasaki and H. Sueyoshi, *J. Pharm. Soc. Japan*, **76**, 1335 (1956).
156. C. S. Barnes, M. N. Galbraith, E. Ritchie and W. C. Taylor, *Australian J. Chem.*, **18**, 1411 (1965).
157. A. Marsili and I. Morelli, *Phytochemistry*, **7**, 1705 (1968).
158. A. Marsili and I. Morelli, *Phytochemistry*, **9**, 651 (1970).
159. G. Berti and F. Bottari, *Progr. Phytochem.*, **1**, 589 (1968).
160. R. S. Ludwiczak and K. Stackowiak, *Rozn. Chem.*, **37**, 575 (1963).
161. J. W. Rowe, *Phytochemistry*, **4**, 1 (1965).
162. L. J. Goad and T. W. Goodwin, *Eur. J. Biochem.*, **1**, 357 (1967).
163. A. K. Barua, P. K. Sanyal and P. Chakrabarti, *J. Ind. Chem. Soc.*, **44**, 549 (1967).
164. N. Kawano, H. Miura and Y. Kamo, *Yakugaku Zasshi*, **87**, 1146 (1967).
165. W. Sucrow and A. Reimerdes, *Z. Naturforsch.*, **23b**, 42 (1968).
166. W. Sucrow and B. Girgensohn, *Chem. Ber.*, **103**, 750 (1970).
167. W. Sucrow and P. Polyzou, *Tetrahedron Letters*, (21), 1883 (1971).
168. J. W. Cornforth, *J. Lipid Res.*, **1**, 3 (1959).
169. T. W. Goodwin, *The Biosynthesis of Vitamins and Related Compounds*, Academic Press, London, 1963.
170. T. T. Tchen, *J. Biol. Chem.*, **233**, 1100 (1958).
171. K. Folkers, C. H. Shunk, B. O. Linn, F. M. Robinson, P. E. Wittreich, J. W. Huff, J. L. Gilfillan and H. R. Skeggs, in *Biosynthesis of Terpenes and Sterols* (Eds. G. E. W. Wolstenholme and M. O'Connor), Churchill, London, 1959, p. 20.
172. F. Lynen, in *Biosynthesis of Terpenes and Sterols* (Eds. G. E. W. Wolstenholme and M. O'Connor), Churchill, London, 1959, p. 95.
173. G. Popják and J. W. Cornforth, *Advan. Enzymol.*, **22**, 281 (1960).
174. K. Bloch, S. Chaykin, A. H. Phillips and A. DeWaard, *J. Biol. Chem.*, **234**, 2595 (1959).
175. U. Henning, E. M. Moslein and F. Lynen, *Arch. Biochem. Biophys.*, **83**, 259 (1959).
176. H. R. Levy and G. Popják, *Biochem. J.*, **75**, 417 (1960).
177. F. Lynen, H. Eggerer, U. Henning and I. Kessel, *Angew. Chem.*, **70**, 783 (1958).
178. A. DeWaard, A. H. Phillips and K. Bloch, *J. Am. Chem. Soc.*, **81**, 2913 (1959).
179. S. Chaykin, J. Law, A. H. Phillips, T. T. Tchen and K. Bloch, *Proc. U.S. Natl. Acad. Sci.*, **44**, 998 (1958).
180. B. W. Agranoff, H. Eggerer, U. Henning and F. Lynen, *J. Biol. Chem.*, **235**, 326 (1960).

181. F. Lynen, J. Knappe, H. Eggerer, U. Henning and B. W. Agranoff, *Fed. Proc.*, **18**, 278 (1959).
182. D. H. Shah, W. W. Cleland and J. W. Porter, *J. Biol. Chem.*, **240**, 1946 (1965).
183. B. W. Agranoff, H. Eggerer, U. Henning and F. Lynen, *J. Am. Chem. Soc.*, **81**, 1254 (1959).
184. F. Lynen, B. W. Agranoff, H. Eggerer, U. Henning and E. M. Moslein, *Angew. Chem.*, **71**, 657 (1959).
185. D. S. Goodman and G. Popják, *J. Lip. Res.*, **1**, 286 (1960).
186. G. Popják, in *Biosynthesis of Terpenes and Sterols* (Eds. G. E. W. Wolstenholme and M. O'Connor), Churchill, London, 1959, p. 93.
187. G. Popják, J. W. Cornforth, R. H. Cornforth, R. Ryhage and D. S. Goodman, *J. Biol. Chem.*, **237**, 56 (1962).
188. G. Popják, G. Schroepfer and J. W. Cornforth, *Biochem. Biophys. Res. Commun.*, **6**, 438 (1962).
189. J. W. Cornforth and R. H. Cornforth, Biochemical Soc. Symposia No. 29, *Natural Substances formed biologically from Mevalonic acid* (Ed. T. W. Goodwin), Academic Press, London, 1970, p. 5.
190. G. Ryback, in *Terpenoids in Plants* (Ed. J. B. Pridham), Academic Press, London, 1967, p. 48.
191. J. W. Cornforth, *Quart. Rev., London*, **23**, 125 (1969).
192. G. Popják and J. W. Cornforth, *Biochem. J.*, **101**, 553 (1968).
193. J. W. Cornforth, R. H. Cornforth, C. Donninger and G. Popják, *Proc. Roy. Soc.*, **163B**, 492 (1966).
194. J. W. Cornforth, R. H. Cornforth, G. Popják and L. Yengoyan, *J. Biol. Chem.*, **241**, 3970 (1966).
195. C. Donninger and G. Popják, *Proc. Roy. Soc.*, **163B**, 465 (1966).
196. E. Capstack, D. J. Baisted, W. W. Newschwander, G. A. Blondin, N. L. Rosin and W. R. Nes, *Biochemistry*, **1**, 1178 (1962).
197. E. Capstack, N. L. Rosin, G. A. Blondin and W. R. Nes, *J. Biol. Chem.*, **240**, 3258 (1965).
198. W. D. Loomis and J. Battaile, *Biochim. Biophys. Acta*, **67**, 54 (1963).
199. K. Oshima-Oba and I. Uritani, *Plant Cell Physiol.*, **10**, 827 (1969).
200. C. J. Pollard, J. Bonner, A. J. Haagen-Smit and C. C. Nimmo, *Plant Physiol.*, **41**, 66 (1966).
201. V. H. Potty and J. H. Bruemmer, *Phytochemistry*, **9**, 99 (1970).
202. D. R. Thomas and A. K. Stobart, *Phytochemistry*, **9**, 1443 (1970).
203. L. J. Rogers, S. P. J. Shah and T. W. Goodwin, *Biochem. J.*, **99**, 381 (1966).
204. I. P. Williamson and R. G. O. Keckwick, *Biochem. J.*, **96**, 862 (1965).
205. D. N. Skilleter, I. P. Williamson and R. G. O. Keckwick, *Biochem. J.*, **98**, 27P (1966).
206. E. Beytia, P. Valenzuela and O. Cori, *Arch. Biochem. Biophys.*, **129**, 346 (1969).
207. P. Valenzuela, E. Beytia, O. Cori and A. Yudelevich, *Arch. Biochem. Biophys.*, **113**, 536 (1966).
208. P. Valenzuela, O. Cori and A. Yudelevich, *Phytochemistry*, **5**, 1005 (1966).
209. L. J. Rogers, S. P. J. Shah and T. W. Goodwin, *Biochem. J.*, **100**, 14C (1966).
210. C. J. Chesterton and R. G. O. Keckwick, *Arch. Biochem. Biophys.*, **125**, 76 (1968).
210a. K. Ogura, T. Nishano, T. Koyama and S. Seto, *Phytochemistry*, **10**, 779 (1971).
211. J. E. Graebe, *Science*, **157**, 73 (1967).
212. J. E. Graebe, *Phytochemistry*, **7**, 2003 (1968).
213. P. Benveniste, G. Ourisson and L. Hirth, *Phytochemistry*, **9**, 1073 (1970).

214. C. George-Nascimento, E. Beytia, A. R. Aedo and O. Cori, *Arch. Biochem. Biophys.*, **132**, 470 (1969).
215. D. J. Baisted, *Phytochemistry*, **6**, 93 (1967).
216. R. T. van Aller and W. R. Nes, *Phytochemistry*, **7**, 85 (1968).
217. J. Battaile and W. D. Loomis, *Phytochemistry*, **5**, 423 (1966).
218. D. A. Beeler, D. G. Anderson and J. W. Porter, *Arch. Biochem. Biophys.*, **102**, 26 (1963).
219. B. Holland, *M.Sc. Thesis*, University College of Wales, Aberystwyth, 1966.
220. J. E. Graebe, D. T. Dennis, C. D. Upper and C. A. West, *J. Biol. Chem.*, **240**, 1847 (1965).
221. T. W. Goodwin and R. J. H. Williams, *Proc. Roy. Soc.*, **163B**, 515 (1966).
222. R. J. H. Williams, G. Britton and T. W. Goodwin, *Biochem. J.*, **105**, 99 (1967).
223. H. H. Rees, E. I. Mercer and T. W. Goodwin, *Biochem. J.*, **99**, 726 (1966).
224. H. H. Rees, L. J. Goad and T. W. Goodwin, *Biochem. J.*, **107**, 417 (1968).
225. L. J. Goad and T. W. Goodwin, *Biochem. J.*, **96**, 79P (1965).
226. L. J. Goad and T. W. Goodwin, *Eur. J. Biochem.*, **7**, 502 (1969).
227. T. W. Goodwin and R. J. H. Williams, *Biochem. J.*, **94**, 5C (1965).
228. K. J. Stone and F. W. Hemming, *Biochem. J.*, **96**, 14C (1965).
229. K. J. Stone and F. W. Hemming, *Biochem. J.*, **109**, 877 (1968).
230. R. J. H. Williams, G. Britton, J. M. Charlton and T. W. Goodwin, *Biochem. J.*, **104**, 767 (1967).
231. L. J. Goad, G. F. Gibbons, L. M. Bolger, H. H. Rees and T. W. Goodwin, *Biochem. J.*, **114**, 885 (1969).
232. E. Caspi and P. J. Ramm, *Tetrahedron Letters*, No. 3, 181 (1969).
233. A. R. H. Smith, L. J. Goad and T. W. Goodwin, *Chem. Commun.*, 926 (1968).
234. T. Bimpson, L. J. Goad and T. W. Goodwin, *Chem. Commun.*, 297 (1969).
235. G. Britton, personal communication.
236. T. Bimpson, *Ph.D. Thesis*, University of Liverpool, 1970.
237. P. W. Holloway and G. Popják, *Biochem. J.*, **106**, 835 (1968).
238. L. Canonica, A. Fiecchi, M. G. Kienle, A. Scala, G. Galli, E. G. Paoletti and R. Paoletti, *J. Am. Chem. Soc.*, **90**, 3597 (1968).
239. E. Caspi, J. B. Greig, P. J. Ramm and K. R. Varma, *Tetrahedron Letters*, 3829 (1968).
240. G. F. Gibbons, L. J. Goad and T. W. Goodwin, *Chem. Commun.*, 1212 (1968).
241. E. J. Corey, W. E. Russey and P. R. Ortiz de Montellano, *J. Am. Chem. Soc.*, **88**, 4750 (1966).
242. E. E. van Tamelen, J. D. Willet, R. B. Clayton and K. E. Lord, *J. Am. Chem. Soc.*, **88**, 4752 (1966).
243. E. E. van Tamelen, J. D. Willet and R. B. Clayton, *J. Am. Chem. Soc.*, **89**, 3371 (1967).
244. E. J. Corey, P. R. Ortiz de Montellano, K. Lin and P. D. G. Dean, *J. Am. Chem. Soc.*, **89**, 2797 (1967).
245. E. J. Corey and W. E. Russey, *J. Am. Chem. Soc.*, **88**, 4751 (1966).
246. P. D. G. Dean, P. R. Ortiz de Montellano, K. Bloch and E. J. Corey, *J. Biol. Chem.*, **242**, 3014 (1967).
247. S. Yamamoto, K. Lin and K. Bloch, *Proc. U.S. Nat. Acad. Sci.*, **63**, 110 (1969).
248. E. J. Corey, P. R. Ortiz de Montellano and H. Yamamoto, *J. Am. Chem. Soc.*, **90**, 6254 (1968).
249. S. Yamamoto and K. Bloch, in *Natural Substances formed Biologically from Mevalonic Acid* (Ed. T. W. Goodwin), Academic Press, London, 1970, p. 35.
250. L. Ruzicka, *Proc. Chem. Soc.* (*London*), 341 (1959).

251. L. Richards and J. B. Hendrickson, *Biosynthesis of Sterols, Terpenes and Aceto-genins*, W. A. Benjamin Inc., New York, 1964.
252. P. Benveniste and R. A. Massy-Westrop, *Tetrahedron Letters*, No. 37, 3553 (1967).
253. W. W. Reid, *Phytochemistry*, **7**, 451 (1968).
254. U. Eppenberger, L. Hirth and G. Ourisson, *Eur. J. Biochem.*, **8**, 180 (1969).
255. R. Heintz and P. Benveniste, *Phytochemistry*, **9**, 1499 (1970).
256. E. J. Corey and P. R. Ortiz de Montellano, *J. Am. Chem. Soc.*, **89**, 3362 (1967).
257. W. O. Godtfredson, H. Loreh, E. E. van Tamelen, J. D. Willet and R. B. Clayton, *J. Am. Chem. Soc.*, **90**, 208 (1968).
258. T. Hattori, H. Igarashi, S. Iwasaki and S. Okuda, *Tetrahedron Letters*, No. 13, 1023 (1969).
259. S. Okuda, Y. Sato, T. Hattori, H. Igarashi, T. Tsuchiya and N. Wasad, *Tetrahedron Letters*, No. 46, 4769 (1968).
260. A. Kawaguchi and S. Okuda, *Chem. Commun.*, 1012 (1970).
261. W. Sach and L. J. Goad, unpublished results.
262. D. H. R. Barton, K. F. Gosden, G. Mellows and D. A. Widdowson, *Chem. Commun.*, 1067 (1968).
263. E. I. Mercer and M. W. Johnson, *Phytochemistry*, **8**, 2329 (1969).
264. G. Ponsinet and G. Ourisson, *Phytochemistry*, **7**, 757 (1968).
265. G. Ponsinet and G. Ourisson, *Phytochemistry*, **6**, 1235 (1967).
266. H. H. Rees, L. J. Goad and T. W. Goodwin, *Tetrahedron Letters*, 723 (1968).
267. H. H. Rees, L. J. Goad and T. W. Goodwin, *Biochim. Biophys. Acta*, **176**, 894 (1969).
268. D. Lavie, Y. Shoo, O. R. Gottlieb and E. Glotter, *J. Org. Chem.*, **28**, 1790 (1963).
269. F. Khuong-Huu-Laine, A. Milliet, N. G. Bisset and R. Goutarel, *Bull. Soc. Chim. France*, 1216 (1966).
270. J. W. Cornforth, R. H. Cornforth, C. Donninger, G. Popják, Y. Schimizu, S. Ichii, E. Forchielli and E. Caspi, *J. Am. Chem. Soc.*, **87**, 3224 (1965).
271. G. P. Moss, *Planta Med.*, **14** (Suppl.), 86 (1966).
272. T. W. Goodwin and L. J. Goad, in *The Biochemistry of Fruits and their Products*, Vol. 1 (Ed. A. C. Hulme), Academic Press, London, 1970, p. 305.
273. R. Gmelin, *Planta Med.*, **14** (Suppl.), 119 (1966).
274. R. W. Doskotch, M. Y. Malik and J. L. Beal, *Lloydia*, **32**, 115 (1969).
275. R. W. Doskotch and C. D. Hufford, *Can. J. Chem.*, **48**, 1787 (1970).
276. J. M. Zander and D. C. Wigfield, *Chem. Commun.*, 1599 (1970).
277. J. W. Cornforth, *Angew Chem. Intern. Ed. Engl.*, **7**, 903 (1968).
278. Y. Kitahara, T. Kata and M. Kishi, *Chem. Pharm. Bull.* (*Tokyo*), **16**, 2216 (1968).
279. M. Rowan, P. D. G. Dean and T. W. Goodwin, *FEBS Letters*, **12**, 229 (1971).
280. R. Heintz, P. C. Schaefer and P. Benveniste, *Chem. Commun.*, 946 (1970).
281. E. Lederer, *Biochem. J.*, **93**, 449 (1964).
282. E. Lederer, *Experientia*, **20**, 473 (1964).
283. E. Lederer, *Israel J. Med. Sci.*, **1**, 1129 (1965).
284. E. Lederer, *Quart Rev. London*, **23**, 453 (1969).
285. R. B. Clayton, *Quart. Rev. London*, **19**, 201 (1965).
286. G. J. Alexander and E. Schwenk, *J. Biol. Chem.*, **232**, 611 (1958).
287. G. J. Alexander, A. M. Gold and E. Schwenk, *J. Am. Chem. Soc.*, **79**, 2967 (1957).
288. G. J. Alexander, A. M. Gold and E. Schwenk, *J. Biol. Chem.*, **232**, 599 (1958).
289. L. W. Parks, *J. Am. Chem. Soc.*, **80**, 2023 (1958).
290. M. Castle, G. A. Blondin and W. R. Nes, *J. Am. Chem. Soc.*, **85**, 3306 (1963).
291. H. J. Nicholas and S. Moriarty, *Fed. Proc.*, **22**, 529 (1963).

292. S. Bader, L. Guglielmetti and D. Arigoni, *Proc. Chem. Soc.*, 16 (1964).
293. V. Villanueva, M. Barbier and E. Lederer, *Bull. Soc. Chim. France*, 1423 (1964).
294. M. Castle, G. A. Blondin and W. R. Nes, *J. Biol. Chem.*, **242**, 5796 (1967).
295. J. T. Moore and J. L. Gaylor, *J. Biol. Chem.*, **244**, 6334 (1969).
296. P. T. Russell, R. T. van Aller and W. R. Nes, *J. Biol. Chem.*, **242**, 5802 (1967).
297. G. Jauréguiberry, J. H. Law, J. A. McCloskey and E. Lederer, *C.R. Acad. Sci.*, **258**, 3587 (1964).
298. G. Jauréguiberry, J. H. Law, J. A. McCloskey and E. Lederer, *Biochemistry*, **4**, 347 (1965).
299. R. Toubiana and E. Lederer, *Bull. Soc. Chim. France*, 2563 (1965).
300. R. Toubiana, unpublished data quoted in ref. 284.
301. M. Akhtar, P. F. Hunt and M. A. Parvez, *Chem. Commun.*, 565 (1966).
302. M. Akhtar, P. F. Hunt and M. A. Parvez, *Biochem. J.*, **103**, 616 (1967).
303. M. Akhtar, M. A. Parvez and P. F. Hunt, *Biochem. J.*, **100**, 38C (1966).
304. D. H. R. Barton, D. M. Harrison and G. P. Moss, *Chem. Commun.*, 595 (1966).
305. M. Akhtar, M. A. Parvez and P. F. Hunt, *Biochem. J.*, **113**, 727 (1969).
306. D. H. R. Barton, T. Shioiri and D. A. Widdowson, *Chem. Commun.*, 939 (1970).
307. H. Katsuki and K. Bloch, *J. Biol. Chem.*, **242**, 222 (1967).
308. E. L. Ghisalberti and H. H. Rees, unpublished results.
309. V. R. Villanueve, M. Barbier and E. Lederer, *Bull. Soc. Chim. Biol.*, **49**, 389 (1967).
310. M. Devys and M. Barbier, *Bull. Soc. Chim. Biol.*, **51**, 925 (1969).
311. J. T. Moore and J. L. Gaylor, *J. Biol. Chem.*, **245**, 4684 (1970).
312. J. T. Moore and J. L. Gaylor, *Arch. Biochem. Biophys.*, **124**, 167 (1968).
313. Y. Tomita and A. Uomori, *Chem. Commun.*, 1416 (1970).
314. K. Sakai, *Chem. Pharm. Bull.* (*Tokyo*), **11**, 752 (1963).
315. E. L. Ghisalberti, N. J. de Souza, H. H. Rees, L. J. Goad and T. W. Goodwin, *Chem. Commun.*, 1401 (1969).
316. E. L. Ghisalberti, N. J. de Souza, H. H. Rees, L. J. Goad and T. W. Goodwin, *Chem. Commun.*, 1403 (1969).
317. F. F. Knapp and H. J. Nicholas, *Steroids*, **16**, 329 (1970).
318. J. A. Henry, D. S. Irvine and F. S. Spring, *J. Chem. Soc.*, 1607 (1955).
319. D. J. Frost and J. P. Ward, *Rec. Trav. Chim.*, **89**, 186 (1970).
320. R. B. Bates, A. D. Brewer, B. A. Knights and J. W. Rowe, *Tetrahedron Letters*, No. 59, 6163 (1968).
321. D. S. Ingram, B. A. Knights, I. J. McEvoy and P. McKay, *Phytochemistry*, **7**, 1241 (1968).
322. B. A. Knights, *Phytochemistry*, **7**, 1707 (1968).
323. R. T. van Aller, H. Chikamatsu, N. J. de Souza, J. P. John and W. R. Nes, *Biochem. Biophys. Res. Commun.*, **31**, 842 (1968).
324. B. N. Bowden, *Phytochemistry*, **10**, 3135 (1971).
325. D. J. Baisted, *Phytochemistry*, **8**, 1697 (1969).
326. W. Sucrow, *Tetrahedron Letters*, No. 20, 2443 (1968).
327. M. Fracheboud, J. W. Rowe, R. W. Scott, S. M. Fanega, A. J. Buhl and J. K. Toda, *Forest Products J.*, **18**, 37 (1968).
328. B. O. Lindgren and C. M. Svahn, *Acta Chem. Scand.*, **20**, 1763 (1966).
329. P. Benveniste, L. Hirth and G. Ourisson, *Phytochemistry*, **5**, 31 (1966).
330. L. M. Bolger, H. H. Rees, L. J. Goad and T. W. Goodwin, *Biochem. J.*, **114**, 892 (1969).

331. L. J. Goad, A. S. A. Hamman, A. Dennis and T. W. Goodwin, *Nature*, **210**, 1322 (1966).
332. A. R. H. Smith, L. J. Goad, T. W. Goodwin and E. Lederer, *Biochem. J.*, **104**, 56C (1967).
333. M. Lenfant, unpublished work quoted in ref. 284.
334. J. L. Lenton, J. Hall, A. R. H. Smith, E. L. Ghisalberti, H. H. Rees, L. J. Goad and T. W. Goodwin, *Arch. Biochem. Biophys.*, **143**, 664 (1971).
335. F. F. Knapp, J. B. Greig, L. J. Goad and T. W. Goodwin, *Chem. Commun.*, **707** (1971).
336. F. B. Mallory, R. L. Conner, J. R. Landrey and C. W. L. Iyengar, *Tetrahedron Letters*, 6103 (1968).
337. R. L. Conner, F. B. Mallory, J. R. Landrey and C. W. L. Iyengar, *J. Biol. Chem.*, **244**, 2325 (1969).
338. W. R. Nes, *J. Am. Oil Chem. Soc.* **47**, 85A (1970).
339. W. R. Nes, P. A. G. Malya, F. B. Mallory, K. A. Ferguson, J. R. Landrey and R. L. Conner, *J. Biol. Chem.*, **246**, 561 (1971).
340. K. H. Raab, N. J. de Souza and W. R. Nes, *Biochim. Biophys. Acta*, **152**, 742 (1968).
341. E. Caspi, K. R. Varma and J. B. Greig, *Chem. Commun.*, 45 (1969).
342. E. Caspi and L. J. Mulheirn, *Chem. Commun.*, 1423 (1969).
343. A. R. H. Smith, *Ph.D. Thesis*, University of Liverpool, 1969.
344. Y. Tomita, A. Uomori and H. Minato, *Phytochemistry*, **9**, 555 (1970).
345. Y. Tomita, A. Uomori and E. Sakurai, *Phytochemistry*, **10**, 573 (1971).
346. P. Benveniste, H. J. E. Hewlins and B. Fritig, *Eur. J. Biochem.*, **9**, 526 (1969).
347. M. Lenfant, E. Zissman and E. Lederer, *Tetrahedron Letters*, 1049 (1967).
348. M. Lenfant, R. Ellouz, B. C. Das, E. Zissman and E. Lederer, *Eur. J. Biochem.*, **7**, 159 (1969).
349. R. Ellouz and M. Lenfant, *Tetrahedron Letters*, 3967 (1970).
350. R. Ellouz and M. Lenfant, *Tetrahedron Letters*, 609 (1969).
351. R. Ellouz and M. Lenfant, *Tetrahedron Letters*, 2655 (1969).
352. L. M. Bolger, H. H. Rees, E. L. Ghisalberti, L. J. Goad and T. W. Goodwin, *Biochem. J.*, **118**, 197 (1970).
353. W. Sucrow and B. Radüchel, *Phytochemistry*, **9**, 2003 (1970).
354. R. T. van Aller, R. T. H. Chikamatsu, N. J. de Souza, J. P. John and W. R. Nes, *J. Biol. Chem.*, **244**, 6645 (1969).
355. A. Alcaide, M. Devys, J. Bottin, M. Fetizon, M. Barbier and E. Lederer, *Phytochemistry*, **7**, 1773 (1968).
356. B. A. Knights and A. M. M. Berrie, *Phytochemistry*, **10**, 131 (1971).
357. M. Akhtar, M. A. Parvez and P. F. Hunt, *Biochem. J.*, **106**, 623 (1968).
358. R. D. Bennett, E. Heftmann, W. H. Preston and J. R. Haun, *Arch. Biochem. Biophys.*, **103**, 74 (1963).
359. D. F. Johnson, E. Heftmann and G. V. C. Houghland, *Arch. Biochem. Biophys.*, **104**, 102 (1964).
360. R. J. Kemp, L. J. Goad and E. I. Mercer, *Phytochemistry*, **6**, 1609 (1967).
361. G. M. Jacobsohn and M. J. Frey, *Arch. Biochem. Biophys.*, **127**, 655 (1968).
362. P. Benveniste, M. J. E. Hewlins and B. Fritig, *Eur. J. Biochem.*, **9**, 526 (1969).
363. Z. Kasprzyk and Z. Wojciechowski, *Phytochemistry*, **8**, 1921 (1969).
364. R. D. Bennett and E. Heftmann, *Steroids*, **14**, 403 (1969).
365. A. Alcaide, M. Devys and M. Barbier, *Phytochemistry*, **9**, 1553 (1970).
366. E. Schwenk, G. J. Alexander, C. A. Fish and T. H. Stodt, *Fed. Proc.*, **14**, 752 (1955).

367. E. Kodicek, in *Ciba Foundation Symposium on the Biosynthesis of Terpenes and Sterols* (Eds. G. E. W. Wolstenholme and N. O'Connor), Churchill, London, 1959, p. 173.

368. E. Schwenk and G. J. Alexander, *Arch. Biochem. Biophys.*, **76**, 65 (1958).

369. M. Akhtar and M. A. Parvez, *Biochem. J.*, **108**, 527 (1968).

370. M. Akhtar, W. A. Brooks and I. A. Watkinson, *Biochem. J.*, **115**, 135 (1969).

371. A. Fiecchi, N. Galli Kienle, A. Scala, G. Galli, E. Grossi Paoletti, F. Cattobeni and R. Paoletti, *Proc. Roy. Soc.*, **180B**, 147 (1972).

372. G. J. Schroepfer, B. N. Lutsky, J. A. Martin, S. Huntoon, B. Fourcans, W. H. Lee and J. Vermilion, *Proc. Roy. Soc.*, **180B**, 125 (1972).

373. D. Arigoni, quoted in ref. 374.

374. E. Caspi and L. J. Mulheirn, *J. Am. Chem. Soc.*, **92**, 404 (1970).

375. K. B. Sharpless, T. E. Snyder, T. A. Spencer, K. K. Maheshwari, J. A. Nelson and R. B. Clayton, *J. Am. Chem. Soc.*, **91**, 3394 (1969).

376. R. Rahman, K. B. Sharpless, T. A. Spencer and R. B. Clayton, *J. Biol. Chem.*, **245**, 2667 (1970).

377. A. G. Gonzalez, *Chem. Abstr.*, **60**, 5585e (1964).

378. H. P. Kaufmann and A. K. Sen Gupta, *Fette Seifen*, **66**, 461 (1964).

379. A. Alcaide, M. Devys, M. Barbier, H. P. Kaufmann and A. K. Sen Gupta, *Phytochemistry*, **10**, 209 (1971).

380. K. Schreiber and G. Osske, *Kulturpflänze*, **10**, 372 (1962).

381. N. von Ardenne, G. Osske, K. Schreiber, K. Steinfelder and R. Tümmler, *Kulturpflänze*, **13**, 101 (1965).

382. P. Benveniste, L. Hirth and G. Ourisson, *Phytochemistry*, **5**, 45 (1966).

383. L. J. Goad and T. W. Goodwin, *Biochem. J.*, **99**, 735 (1966).

384. M. Barbier, *Rev. Franc. Corps. Gras.*, **13**, 331 (1966).

385. J. D. Ehrhardt, L. Hirth and G. Ourisson, *Phytochemistry*, **6**, 815 (1967).

386. R. J. Kemp, A. S. A. Hamman, L. J. Goad and T. W. Goodwin, *Phytochemistry*, **7**, 447 (1968).

387. G. F. Gibbons, *Ph.D. Thesis*, University of Liverpool (1968).

388. G. F. Gibbons, L. J. Goad, T. W. Goodwin and W. R. Nes, *J. Biol. Chem.*, **246**, 3967 (1971).

389. F. F. Knapp and H. J. Nicholas, *Phytochemistry*, **10**, 85 (1971).

390. R. T. Aexel, S. Evans, M. Kelly and H. J. Nicholas, *Phytochemistry*, **6**, 511 (1967).

391. M. J. E. Hewlins, J. D. Ehrhardt, L. Hirth and G. Ourisson, *Eur. J. Biochem.*, **8**, 184 (1969).

392. J. Hall, A. R. H. Smith, L. J. Goad and T. W. Goodwin, *Biochem. J.*, **112**, 129 (1969).

393. M. Devys, A. Alcaide and M. Barbier, *Bull. Soc. Chim. Biol,*, **51**, 133 (1969).

394. M. Devys, A. Alcaide and M. Barbier, *Phytochemistry*, **8**, 1441 (1969).

395. G. H. Beastall, H. H. Rees and T. W. Goodwin, *Tetrahedron Letters*, (52), 4935 (1971).

396. A. Alcaide, M. Devys and M. Barbier, *Phytochemistry*, **7**, 329 (1968).

397. D. J. Baisted, R. L. Gardner and L. A. McReynolds, *Phytochemistry*, **7**, 945 (1968).

398. M. Devys, A. Alcaide, M. Barbier and E. Lederer, *Phytochemistry*, **7**, 613 (1968).

399. P. C. Schaefer, F. de Reinach and G. Ourisson, *Eur. J. Biochem.*, **14**, 284 (1970).

400. G. Ponsinet and G. Ourisson, *Phytochemistry*, **6**, 1235 (1967).

401. G. Ponsinet and G. Ourisson, *Phytochemistry*, **7**, 757 (1968).

402. L. J. Goad, B. L. Williams and T. W. Goodwin, *Eur. J. Biochem.*, **3**, 232 (1967).

403. F. F. Knapp and H. J. Nicholas, *Phytochemistry*, **8**, 207 (1969).

404. P. Benveniste, *Phytochemistry*, **7**, 951 (1968).
405. M. Devys, A. Alcaide and M. Barbier, *Bull. Soc. Chim. Biol.*, **50**, 1751 (1968).
406. K. B. Sharpless, T. E. Snyder, T. A. Spencer, K. K. Maheshwari, G. Guhn and R. B. Clayton, *J. Am. Chem. Soc.*, **90**, 6874 (1968).
407. F. F. Knapp and H. J. Nicholas, *Chem. Commun.*, 399 (1970).
408. F. F. Knapp and H. J. Nicholas, *Phytochemistry*, **10**, 97 (1971).
409. M. Lindberg, F. Gautchi and K. Bloch, *J. Biol. Chem.*, **238**, 1661 (1963).
410. H. H. Rees, E. I. Mercer and T. W. Goodwin, *Biochem, J.*, **99**, 726 (1966).
411. F. F. Knapp, R. T. Aexel and H. J. Nicholas, *Plant Physiol.*, **44**, 442 (1969).
412. R. D. Brandt, R. J. Pryce, C. Andig and G. Ourisson, *Eur. J. Biochem.*, **17**, 344 (1970).
413. B. G. Adams and L. W. Parks, *J. Cell. Physiol.*, **70**, 161 (1967).
414. B. G. Adams and L. W. Parks, *J. Lipid Res.*, **9**, 8 (1968).
415. R. J. Pryce, *Phytochemistry*, **10**, 1303 (1971).
415a. C. Andig, R. D. Brandt, G. Ourisson, R. J. Pryce and M. Rohmer, *Proc. Roy. Soc.*, **180B**, 115 (1972).
416. P. Benveniste, M. J. E. Hewlins and B. Fritig, *Eur. J. Biochem.*, **9**, 526 (1969).
417. E. Caspi, J. B. Greig, P. J. Ramm and K. R. Varma, *Tetrahedron Letters*, 3829 (1968).
418. L. Canonica, A. Fiecchi, M. G. Kienle, A. Scala, G. Galli, E. G. Paoletti and R. Paoletti, *Steroids*, **11**, 749 (1968).
419. L. Canonica, A. Fiecchi, M. G. Kienle, A. Scala, G. Galli, E. G. Paoletti and R. Paoletti, *Steroids*, **12**, 445 (1968).
420. G. F. Gibbons, L. J. Goad and T. W. Goodwin, *Chem. Commun.*, 1212 (1968).
421. M. Akhtar, A. D. Rahimtula, I. A. Watkinson, D. C. Wilton and K. A. Munday, *Eur. J. Biochem.*, **9**, 107 (1969).
422. A. M. Paliokas and G. J. Schroepfer, *Biochem. Biophys. Res. Commun.*, **26**, 736 (1967).
423. A. M. Paliokas and G. J. Schroepfer, *J. Biol. Chem.*, **243**, 453 (1968).
424. M. Akhtar and S. Marsh, *Biochem. J.*, **102**, 462 (1967).
425. R. K. Sharma, *Chem. Commun.*, 543 (1970).
426. T. Bimpson, L. J. Goad and T. W. Goodwin, unpublished observations, (1971).
427. L. M. Bolger, *Ph.D. Thesis*, University of Liverpool (1970).
428. W. Sucrow, P. Polyzou, M. Slopinuka and G. Snatzke, *Tetrahedron Letters*, No. 35, 3237 (1971).
429. M. Akhtar, A. D. Rahimtula and D. C. Wilton, *Biochem. J.*, **117**, 539 (1970).
430. A. R. H. Smith, L. J. Goad and T. W. Goodwin, *Chem. Commun.*, 1259 (1968).
431. R. W. Topham and J. L. Gaylor, *Biochem. Biophys. Res. Commun.*, **27**, 644 (1967).
432. R. W. Topham and J. L. Gaylor, *J. Biol. Chem.*, **245**, 2319 (1970).
433. G. J. Schroepfer and K. Bloch, *J. Biol. Chem.*, **240**, 54 (1965).
434. L. J. Morris, R. V. Harris, W. Kelly and A. T. James, *Biochem. J.*, **109**, 673 (1968).
435. L. J. Morris, *Biochem. J.*, **118**, 681 (1970).
436. N. Sugiyama, M. Yamamoto and K. Yamada, *Nippon Kagaku Zasshi*, **89**, 710 (1968).
437. A. B. Sen and A. R. Chowdhury, *J. Indian Chem. Soc.*, **47**, 1063 (1970).
438. P. A. Plattner, *Chem. Ind. London*, SNI June 23 (1951).
439. P. A. Plattner, *Helv. Chim. Acta*, **34**, 1680 (1951).
440. IUPAC Nomenclature of Organic Chemistry (1957), Butterworths, London, 1958.
441. C. J. W. Brooks, *Rodd's Chemistry of Carbon Compounds*, Vol. IID (Ed. S. Coffey), Elsevier Publ. Co., 1970, p. 1.

442. IUPAC/IUB Revised Tentative Rules for Steroid Nomenclature in: *Rodd's Chemistry of Carbon Compounds*, Vol. IID (Ed. S. Coffey), Elsevier Publ. Co., 1970, p. 422.

443. R. S. Cahn, C. Ingold and V. Prelog, *Experientia*, **12**, 81 (1956).

444. R. S. Cahn, C. Ingold and V. Prelog, *Angew. Chem. Intern. Ed. Engl.*, **5**, 385 (1966).

445. IUPAC Tentative Rules for the Nomenclature of Organic Chemistry. Section E, Fundamental Stereochemistry, in *Eur. J. Biochem.*, **18**, 151 (1971).

446. W. W. Epstein and H. C. Rilling, *J. Biol. Chem.*, **245**, 459 (1970).

447. E. E. van Tamelen and M. A. Swartz, *J. Am. Chem. Soc.*, **93**, 1780 (1971).

448. J. Edmond, G. Popjak, S. M. Wong and V. P. Williams, *J. Biol. Chem.*, **246**, 6254 (1971).

The Use of Comparative Amino Acid Sequence Data in Evolutionary Studies of Higher Plants

D. BOULTER   DEPARTMENT OF BOTANY, UNIVERSITY OF DURHAM, ENGLAND

I. INTRODUCTION

Phylogenetic relationships between organisms are usually based on a comparison of their anatomical and morphological features with special emphasis on fossil evidence. However, from the biochemical standpoint, organisms differ from one another in that each has its own complement of proteins, whose activities are the basis of function and give rise to its morphological characteristics. The primary structure of a protein is defined as the sequence of amino acids of which it is composed, and this is specified for each protein by the sequence of nucleotides in a part of the nucleic acid which constitutes the genetic material of the organism. Crick[27] and Zuckerkandl and Pauling[137] have pointed out therefore, that the primary structure of nucleic acids and proteins contains much information about their own origins and about the evolution of the organisms in which they occur. In this connexion, a distinction should be made between these macromolecules, called semantides by Zuckerkandl and Pauling[137], and the micromolecules of intermediary metabolism, which result from enzyme action and whose use in systematics of higher plants has been discussed by various authors[2, 7, 52, 123]; micromolecules resemble morphological characters in that both are the result of the action of many genes.

Whilst comparisons of nucleic acid structure are potentially useful in evolutionary studies, very little higher plant data have been accumulated[71]. The use of protein data however, does have the disadvantage that owing to the degeneracy of the genetic code, some differences in alleleic DNA may not be translated into differences in the resultant polypeptide chain (isosemantic heterozygosity). There are also significant stretches of informational DNA which are not transcribed into polypeptide chains[130]; both these phenomena lead to some loss of phylogenetic information when considering differences in protein sequences.

It is now clear that organisms as diverse as human beings and fungi have some proteins in common. These proteins, although they occur in many different organisms, are similar in structure and function; for example, cyto-chromes c from a wide range of eukaryotic organisms have been shown to have the same function[82] and identical amino acids in about a third of their sequences[30]. One explanation for these findings is that part of the genetic material, that coding for the cytochrome c molecule in these organisms, has a common ancestry. An alternative explanation is that the similarity of these sequences is a consequence of convergent evolution, and that the constant features of the sequence have been selected since they are necessary for the correct functioning of the cytochrome c molecule. Whilst either or both explanations are possible, the results presented in this review make it un-likely that convergent evolution can explain all the similarities which have been shown to exist between cytochromes c and between similar sets of some other proteins.

By providing a complicated phylogenetically homologous structure which can be compared from organism to organism, the amino acid sequence of homologous proteins should throw light on the relationships and order of origin of the major groups of organisms. Since there are several thousand different proteins to choose from, there is the potentiality for a large in-crease in the material available for evolutionary studies; this data can in theory be objectively quantified.

Several other aspects of the structure of proteins have been used in attempts to establish the phylogenetic relationships between major groups, for example, amino acid composition, the pattern of peptides from enzymic digestion, serological, catalytic, chromatographic and electro-phoretic properties. In the past these methods have not proved very success-ful for this purpose, since the structural feature compared did not reflect with sufficient accuracy or resolution the genetical differences between the species investigated. As methods improve, and in the area of serology this is already the position[3, 5, 21, 41, 88, 104], this objection may be overcome. At the present time, however, although there are difficulties in acquiring data and interpreting it, the sequence method is the most reliable available.

The use of sequence data in animal phylogeny has been reviewed several times[11, 15, 30, 34, 35, 79, 83, 89, 93, 107, 137], and results from animals will be used only where plant data are not available to illustrate a principle. The purpose of this review is, firstly, to give background information necessary for an understanding of the problems involved in the use of sequence data in evolutionary studies, and secondly, to indicate the current state of our knowledge with reference to higher plant evolution. Thus, amino acid sequence data of plant proteins not relevant to this purpose will not be in-cluded.

II. METHODOLOGY

A. Sequence Determinations

Methods of protein purification have been the subject of many reviews[1,23,25] and sequence methods have also been well documented[10,24,26,55,75,90]. All that will be mentioned here are the main considerations of strategy.

1. *Protein purification*

One of the main disadvantages of the sequence determination method, compared with some serological and gel electrophoretic studies, is that the

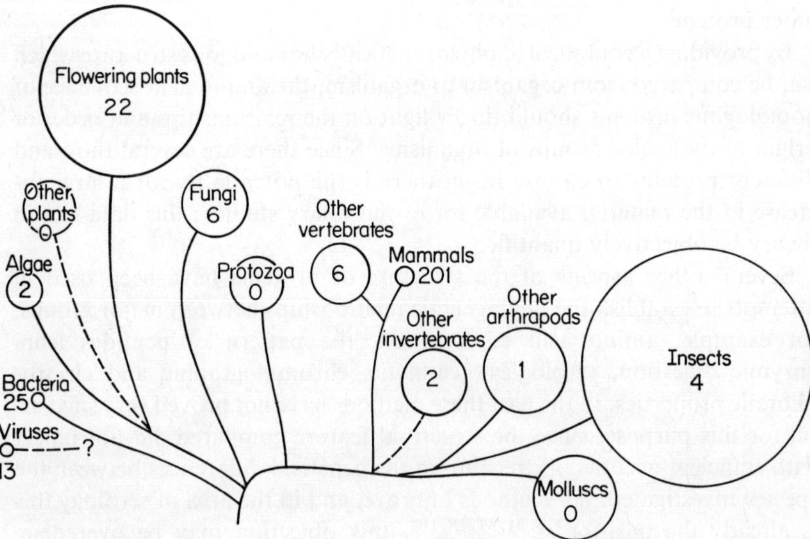

Figure 1. The number of sequences determined as modified from Dayhoff[30]. The size of the circles gives an indication of the number of species within a group and the numbers inside the circles refer to the number of determined sequences.

former requires a pure protein. In phylogenetic studies using sequence data, protein purification is, and is liable to remain, the major limiting factor. Thus, one suspects that it is the ease of obtaining sufficient amounts of protein which often dictates the choice of organism investigated (Fig. 1) rather than phylogenetic considerations. With the cytochrome *c* studies on animals, for example[30] by choosing vertebrates and pteryogotous insects, the most important problems of animal phylogeny have been by-passed[61,62]. Often the most important species phylogenetically are those from which it is difficult to obtain sufficient pure protein, and sequence methods which require small amounts of protein for the complete determination of a sequence are, therefore, to be preferred.

2. Digestion with enzymes

Once purified, the protein is cleaved into specific peptides with proteolytic enzymes, and by far the most commonly used, due to their specificity and yield of peptides, are trypsin and chymotrypsin. The identification of suitable proteolytic enzymes in addition to those presently used, would be an important advance in this field. Chemical methods[133] which hydrolyse the protein at specific residues, in particular the cyanogen bromide method, are also used.

3. Separation of peptides

Peptides in mixtures may be separated one from another by ion-exchange chromatography on cation- or anion-exchange resins and celluloses, by gel filtration, or by electrophoresis in various media[75]. Less generally used are partition chromatography, counter current distribution and immuno-chemical methods. The complete separation of the peptides of an enzymic digest is often a major limiting factor in determining a sequence. Generally speaking, present methods are normally feasible with proteins with molecular weights of up to 40,000 daltons. With small amounts of material, peptide separation by high-voltage paper electrophoresis is often preferable since it is very sensitive. Another important advantage of this method is that confirmatory evidence of the amino acid composition of a peptide can be obtained from its mobility[94].

4. Sequential chemical degradation

Sequential analysis of proteins and peptides is normally carried out by the procedure devised by Edman in 1950. This method involves the step-wise coupling of the N-terminal amino acid with phenyl isothiocyanate, cleavage and separation of the thiazolinone derivative and its conversion to the more stable isomeric phenylthiohydantoin (PTH amino acid) which can then be identified; chemical details of these processes are discussed by Ilse and Edman[58]. Gas–liquid chromatography is now more or less satisfactory for the identification of PTH amino acids[53, 126], but the majority of workers use two indirect systems of identification, i.e. the subtractive method[56], or the 'dansyl' method of Gray and Hartley[51]. Both methods have the disadvantage that side-chain amide groups cannot be identified. The dansyl-Edman method uses very small amounts of material, and for this reason is preferable to the subtractive-Edman method when material has to be conserved; although it is only semi-quantitative, in the hands of an experienced worker the method is very satisfactory. Having established the amino acid sequence of the peptides separated from two different enzymic digests, it is normally possible from a consideration of overlaps to deduce the complete amino acid sequence of a protein[103].

Since purification of proteins and the separation of peptide mixtures resulting from their digestion are the two major bottlenecks in sequencing, 'short-cuts', such as comparisons of peptide maps, are of limited use because of the attendant possibility of missing sequence differences. Nolan and Margoliash[93] quote, for example, the case of the homologous C-terminal peptide of cytochrome *c*, which is: thr,ala,ala,lys (pigeon); ala,thr,ala,lys (duck); and, lys,thr,ala,ala (rattlesnake).

With small proteins such as cytochrome *c*, ferredoxin, plastocyanin etc., $1\ \mu\text{mol} \equiv c\ 12$ mg or even less is sufficient protein to determine a complete sequence using dansyl-Edman analysis of peptides separated by high-voltage paper electrophoresis. A protein chemist with the help of an experienced technician could be expected to determine a sequence in two months.

5. *Future developments*

a. Automatic sequencers. Automatic sequencers are now available commercially, and their operation is satisfactory; they are however, expensive to buy and to operate. They are capable, at least in some instances, of determining thirty to sixty residues from the N-terminal end of a protein[19, 37, 92, 127], but they will not deal with proteins which have a blocked N-terminus. For the near future, such machines will almost certainly be used in conjunction with manual methods, and it should be borne in mind that the main bottleneck in a phylogenetic programme is usually the purification of the protein in the first place. With time we may expect perhaps a change in the strategy of sequencing, the aim then being to generate large peptides, for instance by using cyanogen bromide cleavage at methionine, a relatively rare amino acid in proteins, or by other limited digestion procedures. The large peptides would be separated by gel filtration and the sequence of the major part of each determined by automated methods; this approach will become especially valuable now automated methods deal with shorter peptides as well.

b. Mass spectrometry. Recent developments in the determination of the amino acid sequence of peptides by mass spectrometry have been impressive, although it is unlikely that peptides very much greater than ten amino acids in length will be amenable to this method. By exploiting the differential volatility of peptides using different temperature regimes during mass spectrometry, it may be possible to eliminate the peptide separation step and determine simultaneously the sequences of small peptides directly in mixtures. In the foreseeable future mass spectrometer methods are likely to be used in conjunction with manual methods.

B. Phylogenetic Trees

1. Comparison of sequences

If two amino acid sequences are compared pair-wise over their entire length, one arrives at the total number of amino acid differences between them. However, the substitution of one amino acid for another at a particular position may have required more than one mutational event in the nucleotides coding for these particular amino acids. Thus, the mutation distance between two proteins is defined as the minimum number of nucleotides that would need to be changed, in order for the gene for one protein to code for the other[45]. This is arrived at by making a pair-wise comparison of the two sequences over their entire length, and summing the mutation value (the minimum number of nucleotide changes required to convert the coding triplet for one amino acid to that of the other) for each changed position. Tables of the mutation values for all possible amino acid pairs are given in Fitch and Margoliash[45]. However, the genetic code is degenerate and there are often several codons for a particular amino acid. Thus, substitution of one amino acid for another in two proteins may require different numbers of nucleotide changes (one to three) depending on which codons of a degenerate set coded for the amino acids in question, and in computing minimum mutation distances, it is always the minimum number of nucleotides which is used. In order to compare sequences, it is necessary to align them and computer methods are now available for this purpose[43, 102]. However, it is often possible to align two sequences by some common feature, for example with cytochrome *c*, by the two cysteinyl residues 4 positions apart which bind the heme prosthetic group in all mitochondrial-type cytochromes *c*. Sometimes aligning methods clearly indicate that additions or deletions have occurred in the sequences of homologous proteins during the course of evolution, and these must be taken into account when comparisons are made. Also, if the sequences differ in their overall lengths, suitable corrections must be made for this difference, in addition to any adjustments made for gaps etc.[45]

Similarities between sequences could have arisen, (*a*) purely by chance (parallel or coincident change); (*b*) by convergence of unrelated proteins during evolution, certain amino acid sequences being more efficient for a common biological function (analogy); or (*c*) by descent from a common ancestor (true homology). There are several approaches to the problem of distinguishing whether similarities between sequences are greater than can be accounted for by random processes, i.e. of distinguishing between the first and the other two possibilities[18, 93]. Fitch[42] has used a statistical approach, which he modified subsequently[44], in which the frequencies of occurrence of individual mutation distances are plotted as a function of the mutation distances, using peptide segments of fixed length from two sequences and computing the minimum mutation distances for all such possible pairs. These procedures, however, do not distinguish whether the occurrence of similarities unaccountable by chance, have arisen by homology or convergence. Even when departure from randomness has been demonstrated therefore, it is still necessary to establish how this has arisen, and to estimate the number of residues which are the same in different sequences because of their biological function. Fitch and Margoliash[46] have presented a semi-rigorous statistical approach to this problem when there is a set of amino acid sequences available with many invariant residues. This method determines the number of invariants, irrespective of how many more sequences are examined, and if the invariant positions in two sequences exceed this number this indicates that the proteins are ancestrally related.

Dickerson[32] has pointed out that the use of minimum mutational distances in these methods has, due to the degeneracy of the genetic code, an inbuilt error and that, in any case, the nature of the code is to minimize the effect of random mutations on the chemical properties of the polypeptide chain subsequently produced[32, 124]. Particularly difficult cases are those where additions or deletions may have occurred, and when it is necessary to 'optimize' the similarity between the sequences by making allowances for these 'gaps'[93]. Several authors[18, 36, 91] have proposed homologies between *Pseudomonas* c_{551}, *Rhodospirillum* c_2, and eukaryotic cytochrome *c*. However, their proposals for alignment of the sequences disagree with one another. Dickerson[32] has been able to identify constraints in the 3-dimensional structure of cytochrome *c* from the results of X-ray analysis of horse and bonito cytochromes *c*, and by using this information he has arrived at much more meaningful suggestions for the matched regions in the bacterial and eukaryotic sequences. He concludes that any homology demonstrated solely on the basis of minimum mutation distance considerations, must be accepted with reservations until checked by X-ray data (see also Gibbs and McIntyre[50]).

Thus, the present position is that, whilst most workers accept the explanation of a common ancestry for the similarities in sets of sequences such as those of the forty or so cytochromes examined to date, a convergent evolutionary explanation for their similarity cannot be rigorously ruled out.

2. Construction of trees

Replacement distances between proteins either using amino acid differences or nucleotide mutation values, are the basic data which are used to construct phylogenetic trees. Given n species, there are $n(n-1)/2$ comparisons at each stage, and the number of possible trees that can be constructed from these, which would subsequently have to be evaluated, is so large that the task is not feasible, even with a computer. Several methods have been devised, therefore, to reduce effectively the number of trees to be evaluated but the choice of method is a matter of judgement since statistical procedures require the examination of too many trees to be feasible.

However, before considering the methods used to reduce the number of trees to be evaluated, it must be appreciated that there are several ways, (*a*) of comparing different sequences (see previous section); (*b*) of estimating similarities or dissimilarities (doublet matrixes, diagonal match, etc.); and (*c*) of generating classifications and trees from these estimates (sorting strategies). For the purpose of this review the limitations of, and the assumptions involved in the use of two methods will be described. One of these, the Fitch and Margoliash method, is an example of a matrix method, the other, the Dayhoff and Eck method, is an example of an ancestral sequence method. For a full discussion and comparison of the various methods which have been used, see refs. 17, 38, 49, 50, 63, 72, 73, 110, 111.

a. Fitch and Margoliash[45, 46]. Trees are constructed by assigning single amino acid sequences to sub-sets which are then combined by selecting from among all the possible pairings that assignment of protein sub-sets which provides the lowest average mutation distance between them. A phylogenetic tree is a graphical representation of the order in which sub-sets are joined. The assignments at every level of sub-set combination are assumed correct, although the way in which sequences are assigned to sub-sets may give rise to errors which are not corrected by the method. For this reason several trees are constructed in turn by initially combining several alternative pairs of sequence sub-sets. Fitch and Margoliash[45] for example, constructed forty phylogenetic trees in their examination of cytochromes *c*. The best tree was selected from these on the following basis: estimates were made of the replacement distances along segments of the tree between each pair of species, and these distances were summed. The tree selected was that in which these reconstructed replacement distances most closely agreed with the original minimum replacement distances for the species pairs.

This method therefore, apart from not examining all possible trees, also assumes that the differences observed between sequences of homologous proteins of present-day organisms, have been established by the minimum number of mutations during evolution. This hypothesis has been questioned by Inger[59] and Rogers, Fleming and Estabrook[101].

Pair-wise dissimilarities of sequences of present-day species (populations) will only be ultrametric, i.e. correspond to an hierarchy (evolutionary tree), if the evolutionary rate has been constant with respect to the measure of dissimilarities[60]. Thus, a constant rate of evolution has been assumed in constructing trees from sequence data by the Fitch and Margoliash method. This assumption appears reasonable for some proteins, since it has been shown to be the case for cytochrome *c*, haemoglobin and fibrinopeptides provided sufficiently long periods are considered[30, 33, 68, 93]. (Also see later in this review.)

   b. Dayhoff and Eck[31]. This method constructs trees using inferred ancestral sequences[31, 137]. Trees consist of a closed topology of nodal points from each of which three branches lead off either to sequences or to another node. The ancestral sequences at the nodes are inferred from the amino acid sequences by using a computer programme. Sequences are added to the growing tree in any order and the 'best' tree, selected from those generated, is that which involves the smallest number of amino acid replacements over the whole tree[31]. Thus, again, a minimum evolution hypothesis is assumed. As in the other method, it is not possible to generate and evaluate all possible trees, and an attempt is made to compensate for the assumption that intermediate trees are correct in spite of the arbitrary way sequences are added in any order to the growing tree. Thus, a computer programme is used which cuts off in turn each side-branch of the 'completed' tree and grafts it into all other possible positions on the tree, evaluating each of these trees and again selecting that requiring the least number of amino acid replacements as the final tree. This method gives a closed topology, and biological considerations are used to establish the point of earliest time.

Whilst there are limitations in both methods, in the author's opinion the latter method is to be preferred, since in the Fitch and Margoliash method it is gross overall sequence changes which are considered, whereas in the latter method actual sequences at the nodal branch points are inferred from a consideration of all the sequences which make up the tree.

In spite of the theoretical objections outlined above, the most striking aspect of the results obtained so far using sequence data is the fact that these trees agree in considerable detail with those derived from fossil, morphological and cytological evidence[30]. This not only lends considerable confidence to the sequence method, but also illustrates the fact that changes in a single gene are in these instances (cytochromes *c*, globins, fibrinopeptides and immunoglobulins[30]), a statistically valid expression of the evolutionary history of the genome of a species as a whole.

3. *Species and gene phylogenetic trees*

It is now apparent that gene duplications involving part of a linkage group, followed by the separate evolution of the two or more genes, has occurred frequently during both plant and animal evolution[6, 9, 76, 95, 128, 129]. Possible mechanisms such as unequal exchange between chromatids, unequal homologous crossover, or regional replication, all of which could lead to an increase in the number of copies of a gene, have been discussed[95, 128, 129]. Genes originally identical may, after duplication, diverge subsequently to the extent that they code for proteins with different functions. Such genes are homologous, but are called 'parologous' by Fitch and Margoliash[47] to distinguish them from the other class of homologous genes which they call 'orthologous' genes. The products of parologous genes cannot be used to establish a species phylogeny, since the point of divergence of two parologous genes is not the same as the point of divergence of the species in which they occur, and species phylogenies are established using the products of orthologous genes. Examples of gene phylogenies using the products of parologous genes are given in Fitch and Margoliash[47].

One of the main difficulties in establishing gene phylogenies is the identification of parologous proteins, since these may have diverged functionally and structurally so as to obscure their homologous relationship. Even proteins with a similar function may not necessarily be homologous. Thus, the class of protease enzymes has examples of parologous proteins as well as non-homologous (analogous) proteins, the latter having a different structure but a common function. Thus, one group of proteases with an active serine residue at the catalytic site, consists of animal trypsin, chymotrypsin and elastase; these molecules are probably ancestrally related, whereas the bacterial subtilisins, which also contain active serine in the catalytic site, are probably unrelated to the animal proteins. Other types of proteases have an active cysteine residue in the catalytic site, e.g. papain, bromelain, animal carboxypeptidases, pepsin and renin, and their relationship to one another and to the other proteases is at present uncertain[93]. A possible partial solution to the problem of identifying parologous proteins may lie in the reconstruction of the ancestral sequence of each of a set of different proteins using the results from species phylogenies[47]. A comparison of these ancestral sequences may reveal homology between different protein types which an examination of their present-day representatives might fail to reveal. Originally, it is possible that there was only one gene coding for one protein, and that the array of different proteins found in present-day organisms has arisen by divergence of duplicated genes during the course of evolution.

4. *Time-scales*

A phylogenetic tree as such need not have an attached time-scale, but when it does the relationships between groups can be related to geological time. Comparisons of protein sequences obtained from a wide range of organisms using several proteins including haemoglobin, cytochrome *c* and fibrinopeptides, have led to the suggestion that elapsed time is a parameter related to the number of different residues accumulated during the course of evolution since the divergence of homologous proteins from a common ancestor (see Table III in ref. 93). Thus, providing sufficiently long periods of divergence are considered, i.e. >100 million years, homologous genes in different lines of descent will show an approximately constant number of mutations when compared with each other relative to a common ancestor[81, 136]. Accepting this generalization it is possible to calculate an average period required for a single substitution to occur between two diverging lines of descent (a unit evolutionary period), and hence to establish a time-scale.

Fossil estimates of evolutionary distance place the time of the divergence of mammals and birds at 280 million years ago. Using this point as a reference, the average minimum mutation distance of any bird to any mammal is approximately 13.5 million years[93], so that the unit evolutionary period calculated on this data will be approximately 21×10^6 years. Using this value and knowing the minimum mutational distance between various groups calculated from the observed sequence data, it is possible to make a plot of minimum mutation distance against elapsed time, from which can be read off the time of divergence of the various groups, e.g. mammals and birds to reptiles, vertebrates to invertebrates, etc. Statistical methods can be used to correct for the probability of multiple effective mutations within the same codon, and using this method calculations of the times of divergence of the major groups agree with those established by classical methods[93].

III. THE RELATIONSHIPS OF MOLECULAR AND CLASSICAL BIOLOGICAL INFORMATION

A. The Integration of Molecular and Classical Phylogenetic Data

Having constructed a phylogenetic tree without using biological information other than the identification of the original species from which proteins were extracted, the implication is not that the depicted relationships are correct, but that this is the best estimate of them at present possible with the existing sequence data. However, it is a basic principle of taxonomy that all available information should be used in attempting to establish phylogenetic relationships, and there are many examples in the history of phylogenetic studies where reliance on only part of the evidence has led to an incorrect interpretation. Because morphological characteristics are the result of the action of many genes, they are often not homologous in different taxa and, in any case, do not occur ubiquitously. Thus, many morphological characters have to be considered in constructing phylogenetic schemes, whereas a single protein can be used; this is a reflection of the direct relationship of proteins and genes.

However, information from sequence trees must eventually be integrated with other biological information. In the case of the animal cytochromes there are some small differences in the details of the protein data tree, according to which method is used to estimate similarities between sequences[50]. Furthermore, the findings obtained with the protein data do not agree in some details with those arrived at by classical methods[45, 50], underlining the need to consider conclusions derived from sequence data alongside all the other biological data available. It is of interest to note that the large amount of work carried out by Margoliash, Smith and others (see ref. 30) in determining the sequences of cytochrome *c* of 27 animals, did not lead to any new phylogenetic insights; ironically this is a strength rather than a weakness, as it means that the sequence method can be checked against an established phylogeny, and thus, its overall usefulness established. This situation can be compared with that of the evolution of plants, where adequate fossil evidence is not available and where the objective sequence method becomes extremely important[12]. It is also a reasonable hope that as more sequence data accumulate and as different proteins are sequenced from the same organism, reliable phylogenetic trees will be generated by this method.

B.   The Relationship of Phylogenetic and Phenetic Classifications

The statement, 'that the green algae gave rise to the ferns', means that the ancestors of the ferns were a group of organisms which had the major characteristics of the present-day green algae, and which, if alive today, would be placed in that group; in the absence of fossilized ancestors however, it is extremely difficult to know how closely they would resemble the present-day green algae.

Natural classifications of present-day organisms are based on the sum total of resemblances or differences in an indefinite number of characters, i.e. are phenetic[16]. The naturalness of such a classification stems from the fact that evolution has taken place; but since the classification depends upon the maximum correlation of characters of present-day organisms, how far it will correlate with a phylogenetic classification, such as one established from sequence data, remains to be seen[2, 109, 122, 125].

C. Molecular Versus Morphological Evolution

1. *Neutral mutations*

The concept of the overall biochemical unity of organisms established by biochemical research during the last 60 years, means that the evolution of the basic biochemical processes seen in present-day organisms, was completed before the evolutionary changes which have led to the morphological diversity of the multi-cellular organisms of today had got underway. In considering the evolution of these, it is selection and other evolutionary processes acting within a population on variation at the whole organism level, which have determined the course of evolution.

If we consider the main changes which have taken place during the 'morphological' period of evolution, i.e. the acquisition of new structures, new organs, changes in size, behaviour patterns and so forth, it is not easy to cite examples where a change in metabolism is likely to have initiated a new evolutionary trend[87]. A possible exception is the diversification in the types of lactic dehydrogenase found in birds, since evolution of the muscle-type could extend their range and therefore, their evolutionary possibilities[131]. Isoenzymes, which are of widespread occurrence in plants[105], may in some instances, have similar effects, although there is little evidence for this at the present time.

Morphological characters evolve at variable rates[87, 106], and this is in accord with the major role of natural selection in evolutionary change, since selection pressures have not been constant. In contrast, as mentioned earlier, the average rate of evolution of several proteins, although different, would appear to have been constant over long periods[93]. Various tests of this linearity hypothesis (called regularity tests by Wilson and Sarich[132]) have been carried out[4, 33, 66, 68]. This constant rate of evolution applies to proteins with the same function in similar environments, i.e. they show linearity because they are so well adapted to their roles in an environment which is shielded from the direct effects of selection. For example, when the evolutionary trend leading to the seed habit was initiated in a fern-like ancestor of modern seed plants, what direct selective effect was this likely to have had on the structure of cytochrome *c*, a well-established mitochondrial protein involved in the respiratory process? Again, how far can one expect to relate the few differences in dog and human cytochromes c to their being specially adjusted to the different internal environments of these organisms? On the other hand, one would not expect linearity to apply to enzymes undergoing rapid change, such as might occur during a period in which a protein acquired a new function. Dickerson[33] has suggested that in following amino acid substitutions in proteins, we are seeing a more or less direct manifestation of mutations in DNA, modified by the overall requirements for a functional molecule. The more exactly specified a molecule must be for function, the less likely will random change in its amino acids be acceptable, and the longer the 'unit evolutionary period'. Thus, the rate of evolution of different proteins differs considerably[30]; perhaps the most striking example of a conservative protein being histone IV, which has changed little since the plant and animal kingdoms diverged[108].

The enzymes involved in the metabolism of most secondary plant constituents, e.g. pigments, phenolics, alkaloids, terpenoids, uncommon amino acids etc., are largely unknown. Since the products of these enzymes are responsible for flower pigmentation, scent, etc., they may be of considerable importance in the evolution of plant groups, particularly the angiosperms, whose evolution has involved interrelationships with animals especially insects; natural selection at the whole organism level might well play an important role in the evolution of these proteins. This may also be true of the key proteins which determine characteristics at the whole organism level, i.e. control differentiation, cell division, etc., and about which little is known also. It is important therefore, not to generalize the present findings with a few proteins to the whole of protein evolution which involves molecules with very different levels of function.

An experimental test of the non-linear rate of morphological evolution in contrast to the linear rate of protein evolution, has been suggested[66]; namely to use one of those living species which have ancient fossil replicas. He suggests comparing the amino acid sequences of homologous proteins from a so-called living fossil and a closely related species, with these same proteins in another major group. If the distinction is correct, one would expect the same level of amino acid difference in the homologous proteins of the two former species relative to those of the other group. Thus, primitiveness in morphological characters would not be reflected in the primary structure of the proteins.

One explanation of the constant rate of protein evolution is that most substitutions found in proteins with a clearly established function (i.e. surviving mutational events) are selectively neutral. If selectively neutral substitutions are commonplace then it is inevitable that some of these

8

would spread at random and become fixed in natural populations during evolution; mathematical models for this fixation process have been described[65, 67, 68]. Kimura[65] and Kimura and Ohta[67] have argued also, that the general occurrence of polymorphism in proteins[54, 77] is further evidence that most mutations at the relevant loci are neutral, although this conclusion has not been accepted by Maynard Smith[86]. At present, observations are restricted to a few organisms in which the protein evidence fits the mathematical model. What is needed to establish the theory however, is direct evidence that the different forms of a protein in natural populations are not subject to strong selection pressures. In animals this has been shown not to be the case by Yarbrough and Kojima[135], who have demonstrated a frequency-dependent selective effect on two esterases in *Drosophila melanogaster*, and by Koehn, Perez and Merritt[70], who have good evidence for selection affecting esterases in natural populations of *Notropus straminens*.

Ohno[95] suggests that natural selection acts primarily on larger cistrons whose products, because they are larger, have more sites at which an amino acid substitution is 'tolerable'. These amino acid substitutions at functionally less critical sites, change the kinetic properties of an enzyme, thus allowing natural selection to favour a 'tolerable' mutation. Natural selection acts on these larger loci, which having accumulated tolerable mutations give rise to a larger variety of alleles by recombination. Thus, it is quite likely that a living fossil and its more advanced relatives may have accumulated the same number of neutral mutations if one considers only the products of short cistrons, e.g. amino acid sequences of cytochrome *c*.

In Ohno's view, natural selection is too conservative a process to allow the basic character of genes to change; this requires the accumulation of 'forbidden' mutations. Evolutionary leaps are the result of redundant copies of genes escaping the pressure of natural selection, accumulating mutations which would not be allowable in a single gene, and so giving rise to functionally new genes. Thus, whilst not denying the importance of natural selection, he suggests that relief from selection pressure by gene duplication is equally important.

D. Animal Versus Plant Evolution

Plants differ from animals in that in the former polyploidy is commonplace, having become well established during evolution. Vegetative reproduction, which is also of widespread occurrence in plants, often stabilizes most forms of polyploidy, and hermaphroditism can simplify problems associated with sex determination[112].

A further difference between plants and animals is that in vertebrates there is a good correlation between major taxonomic groups and ecological niches; evolutionary pressure has been directed towards adaptation to a niche[28]. In contrast, in angiosperms, a single family may occupy many different ecological niches, and evolutionary barriers between them are minimized. Thus, characters which are not closely related to survival value and ecological niches may prove to be a better guide to relationships in angiosperms than characters directly related to selection. Woolhouse[134] on the other hand, considers it possible that the major constraints determining the amino acid sequence of proteins of plant families which have diverse habitats and/or life forms, may be demands inherent in these life forms and environments, rather than a common ancestry. This assumes that selection for these diverse habitats has played a major role in the direction of the evolution of proteins, a viewpoint for which there is no evidence, although this may reflect the choice of proteins so far investigated. The fact that plants from diverse habitats can be clearly assigned to their families, means that the characters related to the habitat have not been generally used in devising classifications[28]. A possible exception is the use of 'woodiness' by Hutchinson[57]; his classification however, has not been accepted by other taxonomists on the grounds that it conflicts with other biological information[29]. The sequence method has already been successfully applied to animal data, and, whilst it is clear that differences exist in the modes of nutrition and breeding mechanisms of plants and animals, there is no reason to suppose that these differences should invalidate the application of the sequence method to plants.

IV.   SEQUENCE DATA

A.   Ferredoxin

Ferredoxins are non-heme iron-containing, low molecular weight proteins, which transfer electrons at low redox potentials in photosynthesis. They occur in primitive anaerobic bacteria, and in all photosynthetic organisms including bacteria, algae and higher plants, providing therefore, a possible way of establishing pathways of evolution of photosynthetic organisms from nonphotosynthetic bacteria[85]. The fact that ferredoxins are of low molecular weight, can be purified in milligram quantities relatively easily, and are of widespread occurrence, makes them particularly suitable proteins for evolutionary studies. Table 1 gives the sequences of the four higher plant ferredoxins so far established. These are all type 4 ferredoxins, have molecular weights of about 11,500, and contain two atoms of iron and sulphur[40]. Too few sequences are known to construct a meaningful

Table 1. Comparison of higher plant ferredoxins

(10)   (20)   (30)

Taro[98]	A	T	Y	K	V	K	L	V	T	P	S	G	Q	Q	E	F	Q	C	P	D	D	V	Y	I	L	D	Q	A	E	E	V	G	I
Spinach[84]	A	A	Y	K	V	T	L	V	T	P	T	G	N	V	E	F	Q	C	P	D	D	V	Y	I	L	D	A	A	E	E	E	G	I
Alfalfa[64]	A	S	Y	K	V	K	L	V	T	P	E	G	T	Q	E	F	E	C	P	D	D	V	Y	I	L	D	H	A	E	E	E	G	I
Leucaena glauca[8]	-	A	F	K	V	K	L	L	T	P	D	G	P	K	E	F	E	C	P	D	D	V	Y	I	L	D	Q	A	E	E	L	G	I
Scenedesmus[115]	A	T	Y	K	V	T	L	K	T	P	S	G	D	Q	T	I	E	C	P	D	D	T	Y	I	L	D	A	A	E	E	A	G	L

(40)   (50)   (60)

Taro[98]	D	L	P	Y	S	C	R	A	G	S	C	S	S	C	A	G	K	V	K	V	G	D	V	D	Q	S	D	G	S	F	L	D	D
Spinach[84]	D	L	P	Y	S	C	R	A	G	S	C	S	S	C	A	G	K	L	K	T	G	S	L	N	Q	D	D	Q	S	F	L	D	D
Alfalfa[64]	V	L	P	Y	S	C	R	A	G	S	C	S	S	C	A	G	K	V	A	A	G	E	V	N	Q	S	D	G	S	F	L	D	D
Leucaena glauca[8]	D	L	P	Y	S	C	R	A	G	S	C	S	S	C	A	G	K	L	V	E	G	D	L	D	Q	S	D	Q	S	F	L	D	D
Scenedesmus[115]	D	L	P	Y	S	C	R	A	G	A	C	S	S	C	A	G	K	V	E	A	G	T	V	D	Q	S	D	Q	S	F	L	D	D

(70)   (80)   (90)

Taro[98]	E	Q	I	G	E	G	W	V	L	T	C	V	A	Y	P	V	S	D	G	T	I	E	T	H	K	E	E	E	L	T	A
Spinach[84]	D	Q	I	D	E	G	W	V	L	T	C	A	A	Y	P	V	S	D	V	T	I	E	T	H	K	E	E	E	L	T	A
Alfalfa[64]	D	Q	I	E	E	G	W	V	L	T	C	V	A	Y	A	K	S	D	V	T	I	E	T	H	K	E	E	E	L	T	A
Leucaena glauca[8]	E	Q	I	E	E	G	W	V	L	T	C	A	A	Y	P	R	S	D	V	V	I	E	T	H	K	E	E	E	L	T	A
Scenedesmus[115]	S	Q	M	D	G	G	F	V	L	T	C	V	A	Y	P	T	S	D	C	T	I	A	T	H	K	E	E	D	L	F	-

The single-letter code used is that given in the tentative rules of the IUPAC—IUB Commission on Biochemical Nomenclature—see Appendix.

phylogenetic tree, but the higher plant sequences and that of *Scenedesmus* (green alga) are compared in Table 2.

Table 2. Comparison of complete sequences of 5 green plant ferredoxins (in minimum base differences per codon). (H. Matsubara and T. H. Jukes— personal communication, 1970)

	Taro[b]	Spinach	Alfalfa	*Leucaena glauca*[a]
Spinach	0.26			
Alfalfa	0.21	0.27		
Leucaena glauca[a]	0.23	0.29	0.26	
Scenedesmus	0.37	0.41	0.43	0.47

[a] A leguminous tree.
[b] A monocotyledon—the other higher plant species are dicotyledons.

B.   Cytochrome *c*

Cytochromes of the *c* type can be separated into at least three groups. Nearly all the sequence data have been obtained using cytochrome *c* of one of these, i.e. mitochondrial cytochrome *c*. These are heme-containing, low molecular weight proteins which act as an electron carrier in the respiratory chain of the mitochondria of eukaryotic organisms. A second group consists of those cytochromes which function in anaerobic energy-yielding reactions and which occur in chemosynthetic bacteria, as well as in some other organisms[69]. A third group of similar cytochromes *c*, functions in the photoreduction processes of photosynthetic organisms, and examples of this group are the cytochromes *f* of algae and higher plants and the cytochromes c_2 of *Rhodospirillum rubrum*. Probably cytochromes of these three groups are ancestrally related[32], but few sequences have been determined apart from the mitochondrial cytochromes *c*.

Table 3 gives the sixteen sequences of higher plant cytochromes *c* which have been determined so far. The results in Fig. 2 represent a phylogenetic tree generated from ten of these using a computer method based on the strategy of Dayhoff and Eck[31]. Sequence data from the six species not included in this tree have been excluded at this stage, because the location of these species cannot be established with certainty, i.e. there are several possible trees; it is hoped that additional sequences now being determined will soon clarify the situation. The evolutionary implications of these data have been discussed[14], although the number of species so far investigated is very small when considered against the total number of angiosperms—at least

Table 3. Higher plant cytochrome *c* sequences

										(10)										(20)							
Abutilon[118]	Acetyl	A	S	F	Q	E	A	P	P	G	N	A	K	A	G	E	K	I	F	K	T	K	C	A	Q	C	H
Arum[a]	Acetyl	A	S	F	A	E	A	P	P	G	N	P	K	A	G	E	K	I	F	K	T	K	C	A	Q	C	H
Buckwheat[119]	Acetyl	A	T	F	S	E	A	P	P	G	N	I	K	S	G	E	K	I	F	K	T	K	C	A	Q	C	H
Castor[121]	Acetyl	A	S	F	D	E	A	P	P	G	N	V	K	A	G	E	K	I	F	K	T	K	C	A	Q	C	H
Cauliflower[119]	Acetyl	A	S	F	D	E	A	P	P	G	N	S	K	A	G	E	K	I	F	K	T	K	C	A	Q	C	H
Cotton[118]	Acetyl	A	S	F	Q	E	A	P	P	G	N	A	K	A	G	E	K	I	F	K	T	K	C	A	Q	C	H
Ginkgo[97]	Acetyl	A	T	F	S	E	A	P	P	G	D	P	K	A	G	E	K	I	F	K	T	K	C	A	Q	C	H
Maize[a]	Acetyl	A	S	F	S	E	A	P	P	G	N	P	K	A	G	E	K	I	F	K	T	K	C	A	Q	C	H
Mungbean[117]	Acetyl	A	S	F	D	E	A	P	P	G	N	S	K	S	G	E	K	I	F	K	T	K	C	A	Q	C	H
Niger[b]	Acetyl	A	S	F	A	E	A	P	A	G	D	A	K	A	G	E	K	I	F	K	T	K	C	A	Q	C	H
Pumpkin[120]	Acetyl	A	S	F	D	E	A	P	P	G	N	S	K	A	G	E	K	I	F	K	T	K	C	A	Q	C	H
Rape[99]	Acetyl	A	S	F	D	E	A	P	P	G	N	S	K	A	G	E	K	I	F	K	T	K	C	A	Q	C	H
Sesame[121]	Acetyl	A	S	F	N	E	A	P	P	G	D	V	K	S	G	E	K	I	F	K	T	K	C	A	Q	C	H
Sunflower[96]	Acetyl	A	S	F	A	E	A	P	A	G	D	P	T	T	G	A	K	I	F	K	T	K	C	A	Q	C	H
Tomato[c]	Acetyl	A	S	F	D	E	A	P	P	G	N	P	K	A	G	E	K	I	F	K	T	K	C	A	Q	C	H
Wheat[113]	Acetyl	A	S	F	S	E	A	P	P	G	N	P	D	A	G	A	K	I	F	K	T	K	C	A	Q	C	H

				(30)										(40)										(50)					
Abutilon[118]	T	V	E	K	G	A	G	H	K	Q	G	P	N	L	N	G	L	F	G	R	Q	S	G	T	T	P	G	Y	S	Y
Arum[a]	T	V	E	K	G	A	G	H	K	Q	G	P	N	L	N	G	L	F	G	R	Q	S	G	T	T	A	G	Y	S	Y
Buckwheat[119]	T	V	E	K	G	A	G	H	K	Q	G	P	N	L	N	G	L	F	G	R	Q	S	G	T	T	A	G	Y	S	Y
Castor[121]	T	V	E	K	G	A	G	H	K	Q	G	P	N	L	N	G	L	F	G	R	Q	S	G	T	T	A	G	Y	S	Y
Cauliflower[119]	T	V	D	K	G	A	G	H	K	Q	G	P	N	L	N	G	L	F	G	R	Q	S	G	T	T	A	G	Y	S	Y
Cotton[118]	T	V	D	K	G	A	G	H	K	Q	G	P	N	L	N	G	L	F	G	R	Q	S	G	T	T	A	G	Y	S	Y
Ginkgo[97]	T	V	E	K	G	A	G	H	K	Q	G	P	N	L	H	G	L	F	G	R	Q	S	G	T	T	A	G	Y	S	Y
Maize[a]	T	V	E	K	G	A	G	H	K	Q	G	P	N	L	N	G	L	F	G	R	Q	S	G	T	T	A	G	Y	S	Y
Mungbean[117]	T	V	D	K	G	A	G	H	K	Q	G	P	N	L	N	G	L	F	G	R	Q	S	G	T	T	A	G	Y	S	Y
Niger[b]	T	V	E	K	G	A	G	H	K	Q	G	P	N	L	N	G	L	F	G	R	Q	S	G	T	T	A	G	Y	S	Y
Pumpkin[120]	T	V	D	K	G	A	G	H	K	Q	G	P	N	L	N	G	L	F	G	R	Q	S	G	T	T	P	G	Y	S	Y
Rape[99]	T	V	D	K	G	A	G	H	K	Q	G	P	N	L	N	G	L	F	G	R	Q	S	G	T	T	A	G	Y	S	Y
Sesame[121]	T	V	D	K	G	A	G	H	K	Q	G	P	N	L	N	G	L	F	G	R	Q	S	G	T	T	P	G	Y	S	Y
Sunflower[96]	T	V	E	K	G	A	G	H	K	Q	G	P	N	L	N	G	L	F	G	R	Q	S	G	T	T	A	G	Y	S	Y
Tomato[c]	T	V	E	K	G	A	G	H	K	Q	G	P	N	L	N	G	L	F	G	R	Q	S	G	T	T	A	G	Y	S	Y
Wheat[113]	T	V	D	A	G	A	G	H	K	Q	G	P	N	L	H	G	L	F	G	R	Q	S	G	T	T	A	G	Y	S	Y

Table 3.—*continued*

					(60)										(70)										(80)					
Abutilon[118]	S	A	A	N	K	N	M	A	V	N	W	G	E	N	T	L	Y	D	Y	L	L	N	P	J	K	Y	I	P	G	T
Arum[a]	S	A	A	N	K	N	M	A	V	I	W	E	E	S	T	L	Y	D	Y	L	L	N	P	J	K	Y	I	P	G	T
Buckwheat[119]	S	A	A	N	K	N	K	A	V	T	W	G	E	D	T	L	Y	E	Y	L	L	N	P	J	K	Y	I	P	G	T
Castor[121]	S	A	A	N	K	N	M	A	V	Q	W	G	E	N	T	L	Y	D	Y	L	L	N	P	J	K	Y	I	P	G	T
Cauliflower[119]	S	A	A	N	K	N	K	A	V	E	W	E	E	K	T	L	Y	D	Y	L	L	N	P	J	K	Y	I	P	G	T
Cotton[118]	S	A	A	N	K	N	K	A	V	E	W	E	E	K	T	L	Y	D	Y	L	L	N	P	J	K	Y	I	P	G	T
Ginkgo[97]	S	T	G	N	K	N	K	A	V	N	W	G	E	Q	T	L	Y	E	Y	L	L	N	P	J	K	Y	I	P	G	T
Maize[a]	S	A	A	N	K	N	K	A	V	V	W	E	E	N	T	L	Y	D	Y	L	L	N	P	J	K	Y	I	P	G	T
Mungbean[117]	S	T	A	N	K	N	M	A	V	I	W	E	E	K	T	L	Y	D	Y	L	L	N	P	J	K	Y	I	P	G	T
Niger[b]	S	A	A	N	K	N	K	A	V	A	W	E	E	N	S	L	Y	D	Y	L	L	N	P	J	K	Y	I	P	G	T
Pumpkin[120]	S	A	A	N	K	N	R	A	V	I	W	E	E	K	T	L	Y	D	Y	L	L	N	P	J	K	Y	I	P	G	T
Rape[99]	S	A	A	N	K	N	K	A	V	E	W	E	E	K	T	L	Y	D	Y	L	L	N	P	J	K	Y	I	P	G	T
Sesame[121]	S	A	A	N	K	N	M	A	V	I	W	G	E	N	T	L	Y	D	Y	L	L	N	P	J	K	Y	I	P	G	T
Sunflower[96]	S	A	A	N	K	N	M	A	V	I	W	E	E	N	T	L	Y	D	Y	L	L	N	P	J	K	Y	I	P	G	T
Tomato[c]	S	A	A	N	K	N	M	A	V	N	W	G	E	N	T	L	Y	D	Y	L	L	N	P	J	K	Y	I	P	G	T
Wheat[113]	S	A	A	N	K	N	K	A	V	E	W	E	E	N	T	L	Y	D	Y	L	L	N	P	J	K	Y	I	P	G	T

	(90)										(100)										(110)					
Abutilon[118]	K	M	V	F	P	G	L	J	K	P	Q	D	R	A	D	L	I	A	Y	L	K	E	S	T	A	-	-
Arum[a]	K	M	V	F	P	G	L	J	K	P	Q	E	R	A	D	L	I	A	Y	L	K	E	S	T	A	-	-
Buckwheat[119]	K	M	V	F	P	G	L	J	K	P	Q	E	R	A	D	L	I	A	Y	L	K	D	S	T	Q	-	-
Castor[121]	K	M	V	F	P	G	L	J	K	P	Q	D	R	A	D	L	I	A	Y	L	K	Q	A	T	A	-	-
Cauliflower[119]	K	M	V	F	P	G	L	J	K	P	Q	D	R	A	D	L	I	A	Y	L	K	E	A	T	A	-	-
Cotton[118]	K	M	V	F	P	G	L	J	K	P	Q	D	R	A	D	L	I	A	Y	L	K	E	A	T	A	-	-
Ginkgo[97]	K	M	V	F	P	G	L	J	K	P	Q	E	R	A	D	L	I	S	Y	L	K	Q	A	T	S	Q	E
Maize[a]	K	M	V	F	P	G	L	J	K	P	Q	E	R	A	D	L	I	A	Y	L	K	E	A	T	A	-	-
Mungbean[117]	K	M	V	F	P	G	L	J	K	P	Q	D	R	A	D	L	I	A	Y	L	K	E	S	T	A	-	-
Niger[b]	K	M	V	F	P	G	L	J	K	P	Q	E	R	A	D	L	I	A	Y	L	K	A	S	T	A	-	-
Pumpkin[120]	K	M	V	F	P	G	L	J	K	P	Q	D	R	A	D	L	I	A	Y	L	K	E	A	T	A	-	-
Rape[99]	K	M	V	F	P	G	L	J	K	P	Q	D	R	A	D	L	I	A	Y	L	K	E	A	T	A	-	-
Sesame[121]	K	M	V	F	P	G	L	J	K	P	Q	E	R	A	D	L	I	A	Y	L	K	E	A	T	A	-	-
Sunflower[96]	K	M	V	F	P	G	L	J	K	P	Q	E	R	A	D	L	I	A	Y	L	K	T	S	T	A	-	-
Tomato[c]	K	M	V	F	P	G	L	J	K	P	Q	E	R	A	D	L	I	A	Y	L	K	E	A	T	A	-	-
Wheat[113]	K	M	V	F	P	G	L	J	K	P	Q	D	R	A	D	L	I	A	Y	L	K	K	A	T	S	S	-

The single-letter code used is that given in the tentative rules of the IUPAC–IUB Commission on Biochemical Nomenclature—see Appendix.

[a] D. L. Richardson—unpublished.   [b] Ramshaw—unpublished.   [c] R. Scogin—unpublished.

250,000 species—and few phylogenetic insights are apparent at this stage of the investigation. The following points can be made however: (1) Sequences

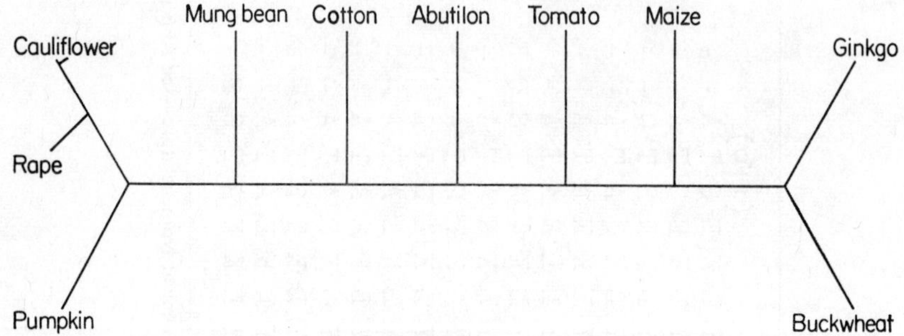

Figure 2. Ten-tree topology using a computer method based on that of Dayhoff and Eck[31].

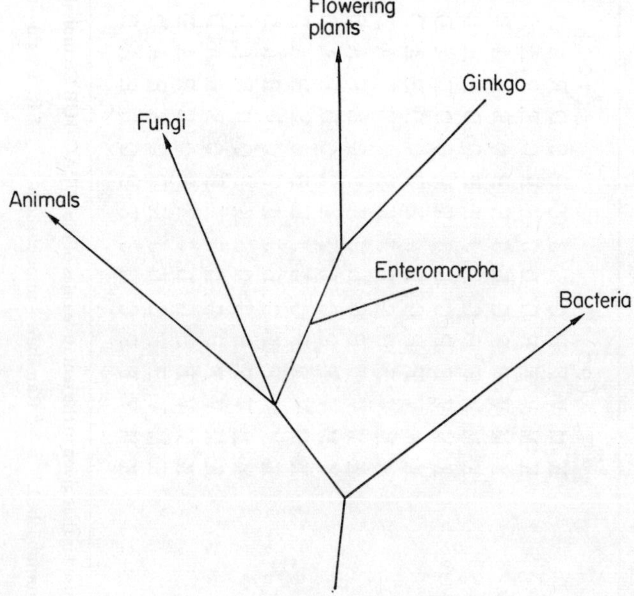

Figure 3. Relationship of major plant groups based on published data and unpublished sequence data of Boulter *et al*[14].

of cytochromes *c* of genera from the same family are more similar to each other than they are to that of any other genus, thus the cauliflower and rape sequences are identical (Brassicaceae)[99, 119]; niger is most similar to sunflower (Compositae)[96], and Abutilon to cotton (Malvaceae)[117];

(2) Families closely related on general biological grounds are also grouped together by this method, e.g. Brassicaceae and Cucurbitaceae[99, 120]; (3) It would appear that the ancestors of the large family Compositae diverged from the other dicotyledonous families quite early on in evolution; (4) Buckwheat, a member of the Polygonaceae, comes from an earlier offshoot than is generally accorded in modern phylogenetic schemes[74]. However, the Polygonaceae are generally accepted as being related to the Caryophyllales and several authors[28, 116] have noted the possible primitiveness of the latter order. The results given in Fig. 3 would also strongly support the placing of the fungi in a separate kingdom equivalent to those of the plant and animal kingdoms.

C.   Microheterogeneity

The work of Sanger[103] showed conclusively that insulin has a unique sequence, and it has been generally accepted that this is likely to be true for other proteins, although evidence of protein microheterogeneity has been indicated by ultracentrifuge studies[22]. Microheterogeneity in a determined sequence may be found because of: (1) heterozygosity of the allele specifying the polypeptide at one or more loci; (2) intergenic (intercistronic) differences of the gene specifying the protein either within a population (polymorphism), or between populations; (3) translational ambiguity of the messenger RNA template[20, 100]; or (4) as an artefact during the course of purification or sequencing of the protein[82].

Benson and Yasunobu[8] found that the sequence of the ferredoxin of *Leucaena*, which was obtained from a mixed population of plants, showed heterogeneity at four residues. Haemoglobin also shows considerable microheterogeneity, some of which is accounted for by polymorphism[78], and some possibly by translational ambiguity[39, 100].

Little evidence has been found for different forms of cytochrome *c* apart from isocytochromes *c* 1 and 2 of baker's yeast[82] (but see ref. 48) and direct attempts to demonstrate polymorphism for cytochrome *c* by Margoliash[80] were unsuccessful. He tested the cytochromes *c* from ~18 individual horses and found them all to be identical. A similar number of individual human cytochrome *c* preparations were also found to be the same. Heterogeneity of greater than 5% of the population in any one type of variant, i.e. at the same amino acid residue(s) would have been detected by his method. Cytochromes *c* obtained from different organs of the same species have been sequenced by Stewart and Margoliash[114], who showed that the primary structure of cytochrome *c* from hog liver, kidney, brain and skeletal muscle, was the same. In plants, some evidence was found[121] for the occurrence of two forms of pumpkin cytochrome *c*, whose sequences differed in three positions. However, attempts to separate two proteins by gradient elution on CM-cellulose proved unsuccessful.

V. CONCLUSIONS

In a paper written in 1969, Boulter *et al.*[13] suggested that the rate of accumulation of sequence data for plants was sufficiently rapid to hope that the relationships of the major groups of the higher plants could be established in this decade. Work subsequently[14] has strengthened our opinion that this will prove to be so.

The study of proteins in an evolutionary context, is not only important because of the light it throws on phylogeny, but also because of the structure/function information about an homologous set of proteins, which it may supply to the protein chemist. Further, this discipline has the important function of bringing together two types of scientists whose concepts, training and outlook differ; such interdisciplinary marriages tend to be scientifically fruitful. The major impetus for the development of macromolecular phylogenies must, however, rest with the taxonomists themselves. The rather haphazard development which has characterized micromolecular taxonomy, due to the different objectives of chemists and taxonomists[2], should be avoided. Choice of organisms and integration of sequence data with biological information is the task of the trained taxonomist, and the acquisition of sequence data is probably best done by a chemist or biochemist; clearly each must speak the other's language and be fully conversant with the biological aims of the project.

VI. APPENDIX

Aspartic	D	Cysteine	C
Asparagine	N	Methionine	M
Asp/or Asn	B	Isoleucine	I
Glutamic	E	Leucine	L
Glutamine	Q	Tyrosine	Y
Glu/or Gln	Z	Phenylalanine	F
Serine	S	Lysine	K
Threonine	T	Histidine	H
Glycine	G	Arginine	R
Alanine	A	Tryptophan	W
Proline	P	Trimethyllysine	J
Valine	V		

VII. ACKNOWLEDGEMENTS

I wish to thank Drs. R. E. Dickerson, T. H. Jukes, W. M. Fitch and A. C. Wilson, for allowing me to see manuscripts before publication, and Drs. M. E. Boulter and J. L. Crosby and Mr. J. A. M. Ramshaw, for reading the text and making several suggestions. The financial support of the Nuffield Foundation and Science Research Council, is gratefully acknowledged.

VIII. REFERENCES

1. P. Alexander and R. J. Block, *A Laboratory Manual of Analytical Methods of Protein Chemistry*, Vol. 1, Pergamon Press, London (1960).
2. R. E. Alston, in *Evolutionary Biology*, Vol. 1 (Eds. T. Dobzhansky, M. K. Hecht and W. C. Steere), Appleton-Century-Crofts, New York, 1967, p. 197.
3. N. Arnheim and A. C. Wilson, *J. Biol. Chem.*, **242**, 3951 (1967).
4. N. Arnheim and C. E. Taylor, *Nature*, **223**, 900 (1969).
5. N. Arnheim, E. M. Prager and A. C. Wilson, *J. Biol. Chem.*, **244**, 2085 (1969).
6. C. J. Bailey and D. Boulter, *Phytochemistry*, **11**, 59 (1972).
7. E. C. Bate Smith, in *Chemotaxonomy and Serotaxonomy* (Ed. J. G. Hawkes), Systematics Association Special Vol. 2, 1968, p. 193.
8. A. M. Benson and K. T. Yasunobu, *J. Biol. Chem.*, **244**, 955 (1969).
9. M. Birnstiel, *Ann. Rev. Pl. Phys.*, **18**, 25 (1967).
10. S. Blackburn, *Protein Sequence Determination: Methods and Techniques*, Marcel Dekker Ltd., 1970.
11. B. Blombäck and M. Blombäck, in *Chemotaxonomy and Serotaxonomy* (Ed. J. G. Hawkes), Systematics Association Special Vol. 2, 1968, p. 3.
12. D. Boulter, E. W. Thompson, J. A. M. Ramshaw and M. Richardson, *Nature*, **228**, No. 5271, 552 (1970).
13. D. Boulter, M. V. Laycock, J. A. M. Ramshaw and E. W. Thompson, in *Phytochemical Phylogeny* (Ed. J. B. Harborne), Academic Press, London and New York, 1970, p. 179.
14. D. Boulter, J. A. M. Ramshaw, E. W. Thompson, M. Richardson and R. H. Brown, *Proc. Roy. Soc. B* (1972), in press.
15. V. Bryson and H. J. Vogel (Eds.), *Evolving Genes and Proteins*, Academic Press, New York, 1965.
16. A. J. Cain and G. A. Harrison, *Proc. Zool. Soc. Lond.*, **135**, 1 (1960).
17. J. H. Camin and R. R. Sokal, *Evolution*, **19**, 311 (1965).
18. C. R. Cantor and T. H. Jukes, *Proc. U.S. Natl. Acad. Sci.*, **56**, 177 (1966).
19. J. D. Capra and H. G. Kunkel, *Proc. U.S. Natl. Acad. Sci.*, **67**, 87 (1970).
20. J. Carbon, P. Berg and C. Yanofsky, *Cold Spring Harbor Symp.*, **31**, 487 (1966).
21. G. T. Cocks and A. C. Wilson, *Science*, **164**, 188 (1969).
22. R. D. Cole, in *Proteins and Their Reactions* (Eds. H. W. Schultz and A. F. Anglemier), The Avi Publishing Co. Inc., Connecticut, 1964, p. 3.
23. S. P. Colowick and N. O. Kaplan (Eds.), several articles in *Methods in Enzymology*, Vol. 1, Academic Press, New York, 1955.
24. S. P. Colowick and N. O. Kaplan (Eds.), several articles in *Methods in Enzymology*, Vol. 4, Academic Press, New York, 1957.

25. S. P. Colowick and N. O. Kaplan (Eds.), several articles in *Methods in Enzymology*, Vol. 5, Academic Press, New York, 1962.

26. S. P. Colowick and N. O. Kaplan (Eds.), several articles in *Methods in Enzymology*, Vol. 6, Academic Press, New York, 1963.

27. F. H. C. Crick, in *The Biological Replication of Macromolecules*, Symp. 12, Soc. Exp. Biol., Cambridge University Press, 1958, p. 138.

28. A. Cronquist, *The Evolution and Classification of Flowering Plants*, Thomas Nelson (Printers) Ltd., London and Edinburgh, 1968.

29. P. D. Davis and V. H. Heywood, *Principles of Angiosperm Taxonomy*, Oliver and Boyd, Edinburgh and London, 1963.

30. M. O. Dayhoff, *Atlas of Protein Sequence and Structure*, Vol. 4, National Biomedical Research Foundation, Silver Spring, Md., 1969.

31. M. O. Dayhoff and E. Eck, *Atlas of Protein Sequence and Structure*, Vol. 2, National Biomedical Research Foundation, Silver Spring, Md., 1966.

32. R. E. Dickerson, *J. Mol. Biol.*, **57**, 1 (1971).

33. R. E. Dickerson, *J. Mol. Evolution* (1971), in press.

34. R. E. Dickerson and L. Geis, *The Structure and Action of Proteins*, Harper and Row, New York, 1969.

35. G. H. Dixon, *Essays in Biochemistry*, Vol. 2 (Eds. P. N. Campbell and G. D. Greville), Academic Press, 1966, p. 147.

36. K. Dus, K. Sletten and M. D. Kamen, *J. Biol. Chem.*, **243**, 5507 (1968).

37. P. Edman and G. Begg, *Eur. J. Biochem.*, **1**, 80 (1967).

38. A. W. F. Edwards and L. L. Cavalli-Sforza, *Biometrics*, **21**, 362 (1965).

39. G. von Ehrenstein, *Cold Spring Harbour Symp.*, **31**, 705 (1966).

40. M. C. W. Evans, D. O. Hall, H. Bothe and F. R. Whatley, *Biochem. J.*, **110**, 485 (1968).

41. D. E. Fairbrothers, *Serological Mus. Bull.*, **43**, 6 (1970).

42. W. M. Fitch, *J. Mol. Biol.*, **16**, 9 (1966).

43. W. M. Fitch, *Biochem. Gen.*, **3**, 99 (1969).

44. W. M. Fitch, *J. Mol. Biol.*, **49**, 1 (1970).

45. W. M. Fitch and E. Margoliash, *Science*, **155**, 279 (1967).

46. W. M. Fitch and E. Margoliash, *Biochem. Gen.* **1**, 65 (1967).

47. W. M. Fitch and E. Margoliash, in *Evolutionary Biology*, Vol. 4, 1971, in press.

48. T. Flatmark, in *Hemes and Hemoproteins* (Eds. B. Chance, R. W. Estabrook and T. Yonetani), Academic Press, New York, 1966, p. 411.

49. A. J. Gibbs, H. R. Kinns, M. B. Dale and H. G. MacKenzie, *Syst. Zool.* (1970), in press.

50. A. J. Gibbs and G. A. McIntyre, *Eur. J. Biochem.*, **16**, 1 (1970).

51. W. R. Gray and B. S. Hartley, *Biochem. J.*, **89**, 379 (1963).

52. J. B. Harborne, in *Progress in Phytochemistry*, Vol. 1 (Eds. L. Reinhold and Y. Liwschitz), Interscience Publishers, 1968, p. 545.

53. R. E. Harman, J. L. Patterson and W. J. A. Vandenheuvel, *Anal. Biochem.*, **25**, 452 (1968).

54. H. Harris, *Proc. Roy. Soc. B.*, **164**, 298 (1966).

55. C. H. W. Hirs (Ed.), Several articles in *Methods in Enzymology*, Vol. 11, Academic Press, New York, 1967.

56. C. H. W. Hirs, S. Moore and W. H. Stein, *J. Biol. Chem.*, **235**, No. 3, 633 (1960).

57. J. Hutchinson, *The Families of Flowering Plants*, 2nd ed. Oxford University Press, 1959.

58. D. Ilse and P. Edman, *Australian J. Chem.*, **16**, 411 (1963).

59. R. F. Inger, *Evolution*, **21**, 369 (1967).

60. N. Jardine, C. J. van Rijsbergen and C. J. Jardine, *Nature*, **224**, 185 (1969).

61. R. P. S. Jefferies, *Bull. Brit. Mus. Nat. Hist. (Geol.)*, **16**(6), (1968).

62. R. P. S. Jefferies, *Palaeontology*, **12**(3), 493 (1969).

63. T. H. Jukes, *Advan. Biol. Med. Phys.*, **9**, 1 (1963).

64. S. Keresztes-Nagy, F. Penini and E. Margoliash, *J. Biol. Chem.*, **244**, 981 (1969).

65. M. Kimura, *Nature*, **217**, 624 (1968).

66. M. Kimura, *Proc. U.S. Natl. Acad. Sci.*, **63**, 1181 (1969).

67. M. Kimura and T. Ohta, *Nature*, **229**, 467 (1971).

68. J. L. King and T. Jukes, *Science*, **164**, 788 (1969).

69. R. M. Klein and A. Cronquist, *Quart. Rev.*, **42**, 105 (1967).

70. R. K. Koehn, J. E. Perez and R. B. Merritt, *Amer. Nat.*, **105**, 51 (1971).

71. D. E. Kohne, in *Chemotaxonomy and Serotaxonomy* (Ed. J. G. Hawkes), Systematics Association Special Vol. 2, 1968, p. 117.

72. G. N. Lance and W. T. Williams, *Nature*, **212**, 218 (1966).

73. G. N. Lance and W. T. Williams, *Comp. J.*, **9**, 60 (1966).

74. G. H. M. Lawrence, *Taxonomy of Vascular Plants*, Macmillan, New York, 1951.

75. J. Leggett Bailey, *Techniques in Protein Chemistry*, 2nd ed., Elsevier Pub. Co., Amsterdam, 1967.

76. U. E. Loening, *Ann. Rev. Pl. Phys.*, **19**, 37 (1968).

77. R. C. Lowentin and J. L. Hubby, *Genetics*, **54**, 595 (1966).

78. C. Manwell and C. V. Schlesinger, *Comp. Biochem. Physiol.*, **18**, 627 (1966).

79. E. Margoliash, *Proc. U.S. Natl. Acad. Sci.*, **50**, 672 (1963).

80. E. Margoliash, *Discussion to paper by O. Smithies*, Brookhaven Symp. No. 21, **1**, 258 (1969).

81. E. Margoliash and E. L. Smith, in *Evolving Genes and Proteins* (Eds. V. Bryson and H. J. Vogel), Academic Press, New York, 1965, p. 221.

82. E. Margoliash and A. Schejter, *Advan. Prot. Chem.*, **21**, 113 (1966).

83. E. Margoliash, W. M. Fitch and R. E. Dickerson, *Brookhaven Symp. Biol.*, **21**, 259 (1968).

84. H. Matsubara, R. M. Sasaki and R. K. Chain, *J. Biol. Chem.*, **243**, 1725 (1968).

85. H. Matsubara, T. H. Jukes and C. R. Cantor, *Brookhaven Symp. Biol.* No. 21, **1**, 201 (1969).

86. J. Maynard Smith, *Amer. Nat.*, **104**, 231 (1970).

87. E. Mayr, *Fed. Proc.*, **23**, 1231 (1964).

88. O. Moritz, in *Mechanism of Mutation and Inducing Factors* (Ed. Z. Landa), Academia, Prague, Czech., 1966.

89. G. A. Mross and R. F. Doolittle, *Arch. Biochem. Biophys.*, **122**, 674 (1967).

90. S. B. Needleman (Ed.), *Protein Sequence Determination, Molecular Biology, Biochemistry and Biophysics*, Vol. 8, Springer-Verlag, Berlin–Heidelberg–New York, 1970.

91. S. B. Needleman and T. T. Blair, *Proc. U.S. Natl. Acad. Sci.*, **63**, 1227 (1969).

92. H. D. Niall and P. Edman, *Nature*, **216**, 262 (1967).

93. C. Nolan and E. Margoliash, *Ann. Rev. Biochem.*, **37**, 727 (1968).

94. R. E. Offord, *Nature*, **211**, 591 (1966).

95. S. Ohno, *Evolution by Gene Duplication*, George Allen and Unwin, England, 1970.

96. J. A. M. Ramshaw, E. W. Thompson and D. Boulter, *Biochem. J.*, **119**, 535 (1970).

97. J. A. M. Ramshaw, M. Richardson and D. Boulter, *Eur. J. Biochem.*, **23**, 475 (1971).

98. K. K. Rao and H. Matsubara, *Biochem. Biophys. Res. Commun.*, **38**, 500 (1970).

99. M. Richardson, J. A. M. Ramshaw and D. Boulter, *Biochimica et Biophysica Acta*, **251**, 331 (1971).

100. D. B. Rifkin, D. I. Hirsh, M. R. Rifkin and W. Konigsberg, *Cold Spring Harbor Symp.*, **31**, 715 (1966).

101. D. J. Rogers, H. S. Fleming and G. Estabrook, in *Evolutionary Biology*, Vol. 1 (Eds. T. Dobznasky, M. K. Hecht and W. C. Steere), Appleton-Century-Crofts, New York, 1967.

102. M. J. Sackin, in *Numerical Taxonomy* (Ed. A. J. Cole), Academic Press, London and New York, 1969, p. 315.

103. F. Sanger, *Advan. Prot. Chem.*, **7**, 1 (1952).

104. V. M. Sarich and A. C. Wilson, *Science*, **154**, 1563 (1966).

105. L. M. Shannon, *Ann. Rev. Pl. Phys.*, **19**, 187 (1968).

106. G. G. Simpson, *The Major Features of Evolution*, Columbia University Press, New York, 1953.

107. E. L. Smith, *Harvey Lect.*, **62**, 231 (1966–1967).

108. E. L. Smith, R. J. Delange and J. Bonner, *Fedn. Proc.* (1969), in press.

109. R. R. Sokal and J. H. Camin, *Syst. Zool.*, **14**, 176 (1965).

110. R. R. Sokal and C. D. Michener, *Univ. Kansas Sci. Bull.*, **38**, 1409 (1958).

111. R. R. Sokal and P. H. A. Sneath, *Principles of Numerical Taxonomy*, Freeman and Co., San Francisco and London, 1963.

112. G. L. Stebbins, *Variation and Evolution in Plants*, Columbia University Press, New York, 1950.

113. F. C. Stevens, A. N. Glazer and E. L. Smith, *J. Biol. Chem.*, **242**, 2762 (1967).

114. J. W. Stewart and E. Margoliash, *Can. J. Biochem.*, **43**, 1187 (1965).

115. K. Sugeno and H. Matsubara, *J. Biol. Chem.*, **244**, 2979 (1969).

116. A. Takhtajan, *Flowering Plants—Origin and Dispersal*, Oliver & Boyd, Edinburgh, 1969.

117. E. W. Thompson, M. V. Laycock, J. A. M. Ramshaw and D. Boulter, *Biochem. J.*, **117**, 183 (1970).

118. E. W. Thompson, B. A. Notton, M. Richardson and D. Boulter, *Biochem. J.* **124**, 787 (1971).

119. E. W. Thompson, M. Richardson and D. Boulter, *Biochem. J.*, **124**, 783 (1971).

120. E. W. Thompson, M. Richardson and D. Boulter, *Biochem. J.*, **124**, 779 (1971).

121. E. W. Thompson, M. Richardson and D. Boulter, *Biochem. J.*, **121**, 439 (1971).

122. B. L. Turner, *J. Pure Appl. Chem.*, **14**, 189 (1967).

123. B. L. Turner, *Taxon*, **18**, 134 (1969).

124. M. V. Volkenstein, *Nature*, **207**, 294 (1965).

125. S. M. Walters, in *Chemical Plant Taxonomy* (Ed. T. Swain), Academic Press, New York, 1963, p. 1.

126. M. Waterfield and E. Haber, *Biochemistry*, **9**, No. 4, 832 (1970).

127. M. Waterfield, C. Corbett and E. Haber, *Anal. Biochem.*, **38**(2), 475 (1970).

128. R. L. Watts and D. C. Watts, *J. Theor. Biol.*, **20**, 227 (1968).

129. R. L. Watts and D. C. Watts, *Nature*, **217**, 1125 (1968).

130. H. L. K. Whitehouse, *Towards an Understanding of the Mechanism of Heredity*, 2nd ed., Edward Arnold (Publishers) Ltd., London, 1969.

131. A. C. Wilson, R. S. Cahn and N. O. Kaplan, *Nature*, **197**, 331 (1963).

132. A. C. Wilson and V. M. Sarich, *Proc. U.S. Natl. Acad. Sci.*, **63**, 1088 (1969).

133. B. Witkop, *Advan. Prot. Chem.*, **16**, 221 (1961).
134. H. W. Woolhouse, in *Phytochemical Phylogeny* (Ed. J. B. Harborne), Academic Press, London and New York, 1970, p. 207.
135. K. Yarbrough and K. Kojima, *Genetics*, **57**, 677 (1967).
136. E. Zuckerkandl and L. Pauling, *Horizons in Biochemistry*, 1962, p. 189.
137. E. Zuckerkandl and L. Pauling, *J. Theor. Biol.*, **8**, 357 (1965).

The Bicyclic Diterpenes

J. R. HANSON SCHOOL OF MOLECULAR SCIENCES,
UNIVERSITY OF SUSSEX, BRIGHTON,
BN1 9QJ, ENGLAND

I. INTRODUCTION

The diterpenes form a large group of C_{20} substances derived from geranyl-geraniol pyrophosphate[1]. They are mainly of plant or fungal origin. Despite their apparent structural variety, it is possible to classify diterpenoid substances on clear biogenetic principles[2]. The majority of diterpenoid substances are cyclic. In the lower terpenes the characteristic mode of cyclization is based on the terminal pyrophosphate behaving as a leaving group and generating a carbonium ion which may then attack the starter isopropylidene unit leading, for example, to the monoterpenes based on the limonene skeleton. However in the diterpene series there are relatively few examples of this mode of cyclization. The major mode of cyclization involves protonation of the double bond of the starter isopropylidene unit of the geranylgeraniol pyrophosphate (**1**) leading to the bicyclic labdadienol pyrophosphate (**2**).

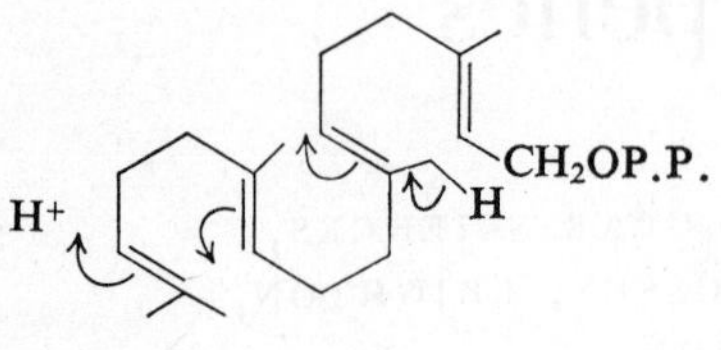

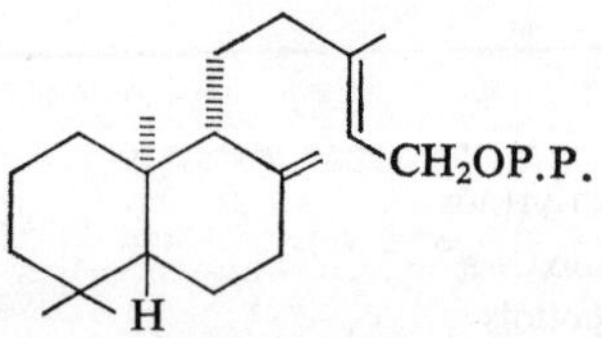

P.P. = Pyrophosphate

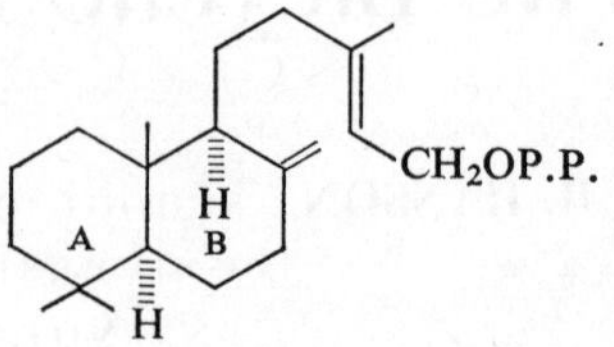

In contrast to the triterpenes and steroids, cyclization is initiated by protonation of a double bond rather than by attack on an epoxide. Consequently the characteristic C-3 oxygen function of the triterpenes and steroids is not a feature of the diterpenes. Another feature of the diterpenes that arises from the cyclization is the formation of normal (**2**) and antipodal (**3**) A/B ring junctions. The former belong to the same optical series as the steroids whilst the latter belong to the opposite series. The nature of the ring junction is a feature of the relative orientation of the double bonds involved in cyclization and may be visualized as arising through the different modes of coiling the open-chain precursor on an enzyme surface. Examples of both series are widespread and there is even a report of both ring fusions occurring alongside each other in the same plant species. Subsequent cyclization utilizing the pyrophosphate as a leaving group leads to the tricyclic and tetracyclic diterpenes. It is with the bicyclic products of the initial cyclization that this review is concerned.

Within the bicyclic series there are a number of structural variations. The side-chain may remain with an oxygen function at C-15 or an allylic rearrangement may occur to give compounds bearing an oxygen function at C-13 (**4**). A new asymmetric centre is generated when this rearrangement occurs or in those cases where C-13 is a saturated centre. Both stereoisomers are found. A number of side-chain dienes are known arising by dehydration of a C-13 or a C-15 alcohol. Another fairly common modification of the side-chain involves the formation of a furan ring (**5**) between C-15 and C-16.

(5)   (4)

In some instances each of the oxidative stages leading to this system have been found in the same plant.

Skeletal variations that are found involve the migration of the methyl group from C-10 to C-9 possibly initiated by protonation of the C-8 double bond and accompanied by a hydride shift from C-9 to C-8. The methyl migration may form part of a backbone or 'friedo' rearrangement in which there is a hydride shift from C-5 to C-10 and a further methyl migration from C-4 to C-5 completed by discharge of the carbonium ion in ring A (6 → 7). For a concerted process within the molecule this would lead to a

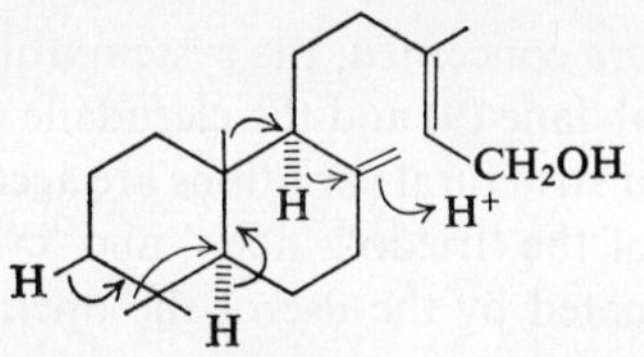

(6) 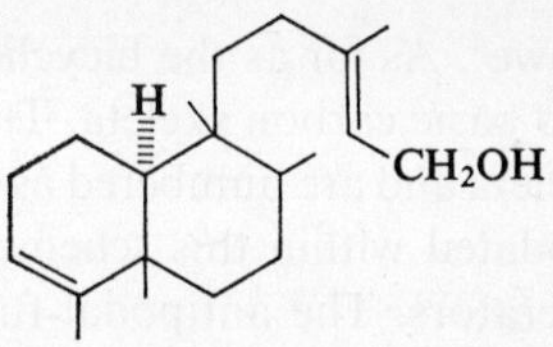   (7)

trans A/B fusion. However there are examples of rearranged *cis* A/B compounds. Possibly these arise through the intervention of a C-10 enzyme-bound intermediate—discharge from the enzyme then involving a further series of rearrangements.

In the organization of this review, we will follow the biogenetic pattern discussing, firstly the compounds of the labdane: manöol series and their side-chain variants followed secondly, by a discussion of the clerodane-derived systems.

II.   NOMENCLATURE

There are several numbering systems in use. In 1953 Klyne proposed[3] the numbering system (8) based on a *seco*-pimarane. Variations on this system are mainly associated with the numbering of the pendant groups. The establishment of the structures of many of the cyclic diterpenes led to the realization of biogenetic relationships and a subsequent effort to rationalize the numbering system. The following tentative rules have been proposed by

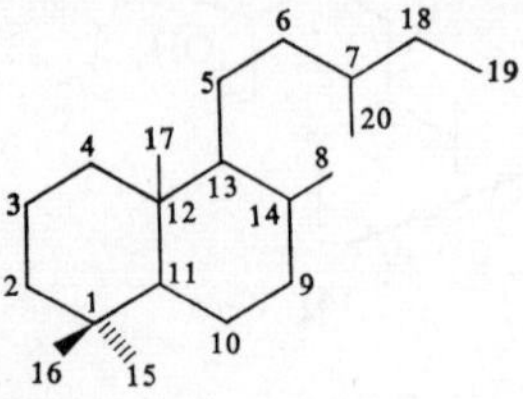

(8)

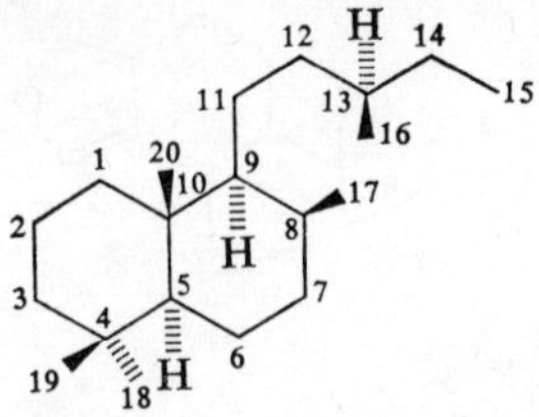

(9)

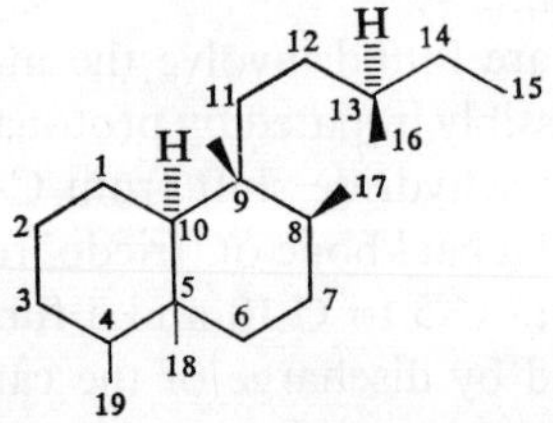

(10)

Rowe[4]. As far as the bicyclic diterpenes are concerned, the system utilizes two basic carbon skeleta. These are the labdane (9) and the clerodane (10) skeleta and are numbered as shown. Minor structural variations are accommodated within this scheme by the use of the 'friedo', 'abeo' and 'cyclo' operators. The antipodal fusion is designated by the use of the operator 'ent-'.

III.   THE LABDANES

A.   Alcohols

The bicyclic diterpenes of the labdanolic acid group may be classified firstly on the oxidation level of C-15 (alcohol, part of a diene, aldehyde or acid) and secondly on whether they are unsaturated, saturated or hydroxylated at C-13. In some instances it is a matter of argument as to the classification of a compound in the labdane or manöol series.

Surprisingly the parent alcohol, labda-8,13-dien-15-ol (11), has not been reported on many occasions. The acetate is amongst the products of the reaction[5] of sclareol with acetic anhydride. Its pyrophosphate, which has been isolated[6] in the antipodal series as copalol pyrophosphate from enzyme systems involved in gibberellin biosynthesis, has been prepared in both enantiomeric series and shown on the one hand, to be a precursor of the tricyclic rosane diterpenes[7] and on the other hand, to act as a precursor of the tetracyclic kauranoid and gibbane diterpenes[8]. The *in vitro* cyclization of this labdadienol is discussed later.

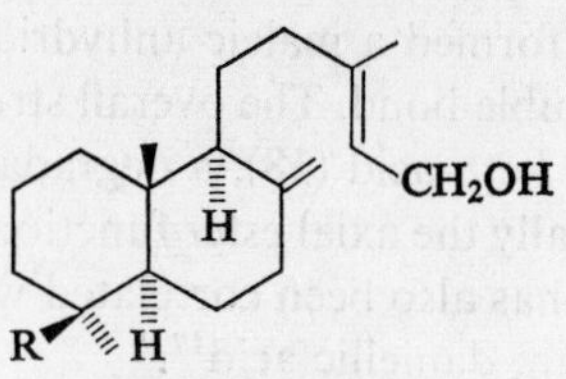

(11) R = CH$_3$
(12) R = CH$_2$OH
(13) R = CHO
(14) R = CO$_2$H

A number of hydroxylated derivatives of the labdadienol have been isolated. Agathadiol (contortadiol) (**12**) and agatholal (contortolal) (**13**) have been isolated from the bark of the lodgepole pine (*Pinus contorta*) and the jackpine (*P. banksiana*)[9]. Agathadiol has been isolated from the resin of *P. sibirica*[10]. The corresponding C-19 carboxylic acid, isocupressic acid (**14**), has been isolated[11] from *Cupressus sempervirens*. The resin also contains[12] manöol, torolusal, isoagatholal, agathadiol and torolusol. Reduction of the methyl ester of isocupressic acid with lithium aluminium hydride furnished agathadiol. The presence of an allylic primary alcohol was demonstrated by oxidation with manganese dioxide to an $\alpha\beta$-unsaturated aldehyde. Another diol, peregrinol has been isolated[13] from *Marrubium peregrinum* and shown to possess the structure **15**. It thus represents an early stage in marrubiin biogenesis. *Aplysin kurodai* has yielded some bromosesquiterpenoids and an unusual bromoditerpene, aplysin-20. This was shown[14] by X-ray measurements to have the structure **16** with the unusual axial hydroxyl group at C-8. It is interesting to speculate on whether the C-3 bromine atom represents the residue of the cyclizing agent in the biosynthesis of the diterpene.

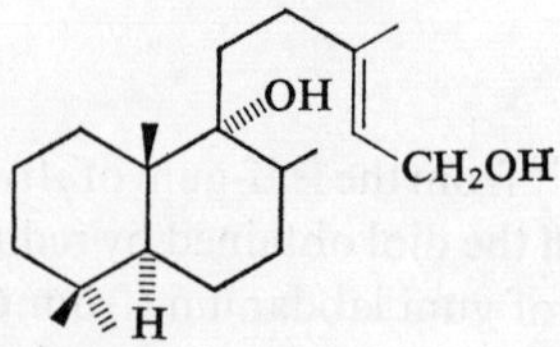

(15) 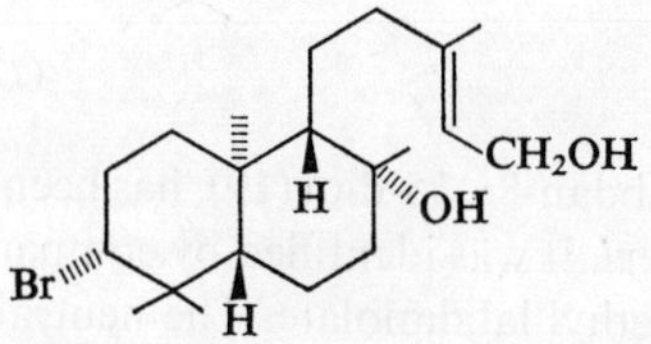   (16)

The heartwood of *Sciadopitys verticillata* contains[15] an interesting group of bicyclic diterpenes which form a biogenetic sequence. Methyl sciadopate (**17**) was characterized[16] as a dihydroxy monomethyl ester possessing two double bonds. One of these was shown to be a terminal methylene group and the other a trisubstituted double bond. Oxidation of the diol gave a dicarboxylic acid which formed a maleic anhydride thus relating the diol and the trisubstituted double bond. The overall structure followed from an ozonolysis, in which the keto acid (**18**), a degradation product of agathic acid, was isolated. Typically the axial ester function was resistant to hydrolysis. Methyl sciadopate has also been correlated with the diene acid, communic acid, and the furan, daniellic acid[17].

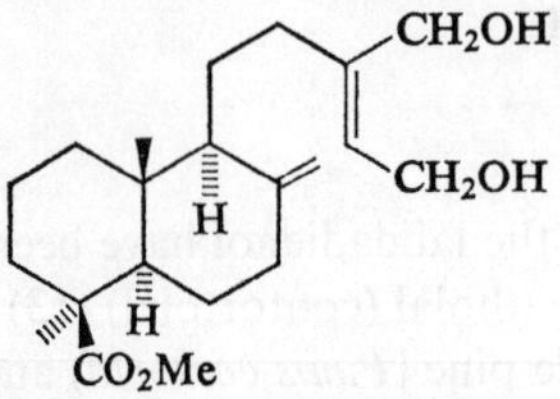

(17)

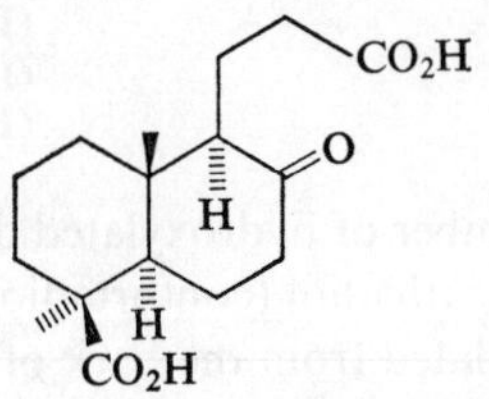

(18)

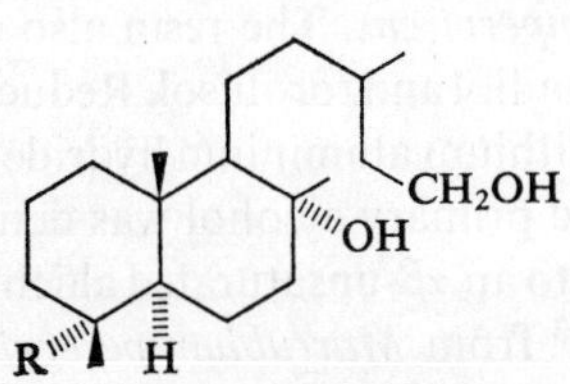

(19) R = CH$_3$
(20) R = CH$_2$OH

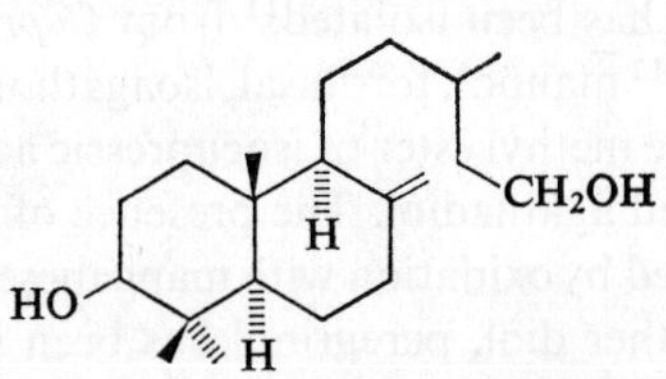

(21)

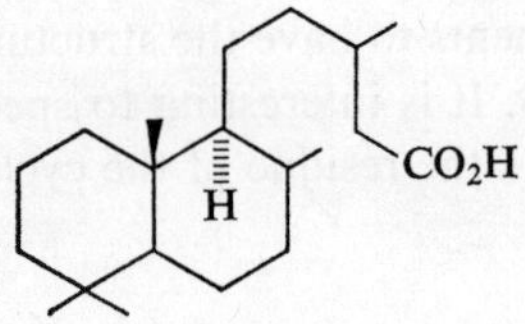

(22)

Labdan-8α,15-diol (**19**), has been isolated[18] from the leaf-gum of *Aeonium lindleyi*. It was identified by comparison with the diol obtained by reduction of methyl labdanolate. The neutral extract of gum labdanum (from *Cistus labdaniferus*) gave[19] the same diol together with labdan-8α,15,18-triol (**20**). This triol was converted to a labdan-15,18-diol, isomeric with the diol obtained by reduction of tetrahydroagathic acid.

Extraction of the 'monkey-puzzle tree' (*Araucaria imbricata*) gave[20] labda-8(17)-en-3β,15-diol (**21**), its corresponding C-15 aldehyde and C-15 acid. The spectral characteristics of the diol confirmed the presence of the hydroxyl groups, the terminal methylene grouping and the *gem*-dimethyl groups. On selenium dehydrogenation, 1,2,5-trimethylnaphthalene was obtained, indicative of the carbon skeleton. This was confirmed by conversion of the diol to dihydrocativic acid (22). Optical rotatory dispersion measurements on the 17-nor-8-ketone related the compound to the labdane stereochemistry. The presence of a hydroxyl group at C-3 was confirmed by ring contraction, analogous to the reactions of the 3β-hydroxytriterpenes. The resin of this tree contains[21] a further series of labdane derivatives. These are labdan-15,19-diol (imbricatadiol) (23), the corresponding C-19 aldehyde (imbricatolal) and C-19 acid (imbricatolic acid) and their corresponding acetates. The structures of these compounds were established by correlation with agathic acid. The stereochemistry at C-13 was shown[22] to be identical to that of labdanolic acid by conversion of imbricatolal to 17-nor-8-oxolabdan-15-oic acid which was also obtained from labdanolic acid (24). Studies on labdanolic acid (*vide infra*) have shown this to be the 13S configuration.

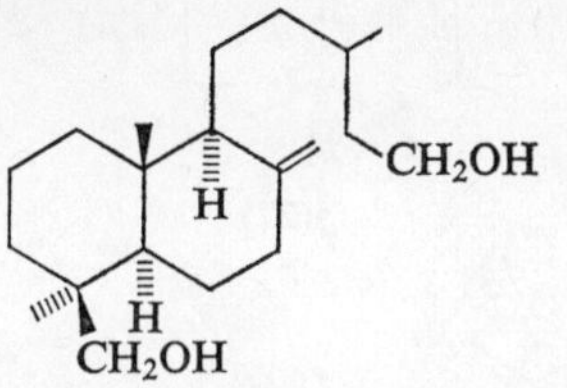

(23)

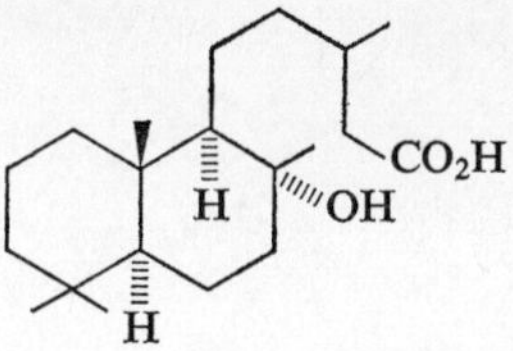

(24)

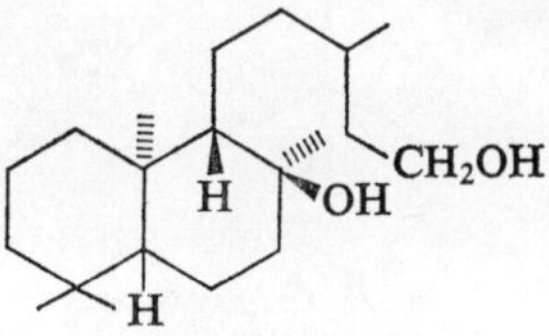

(25)

There are a further group of diterpenes which possess the opposite stereochemistry at this centre. In the enantiomeric series, i.e. with the antipodal A/B fusion and thus once again the 13S configuration, they are represented by the eperuane diterpenes. Typical of these are eperuane-8β,15-diol (25), eperu-13-en-8β,15-diol and the 8β,15,18-triol which together with methyl *ent*-13-epilabdanolate, have been isolated[23] from *Ricinocarpus muricatus*. The evidence for their structures is discussed later.

B. Dienes

A number of side-chain dienes occur naturally. Thus the heartwood extractives of *Dacrydium biforme* contain[24] manöol and a diterpene hydrocarbon, biformene **(26)**. The ultraviolet spectrum of biformene revealed the presence of a conjugated diene whilst the infrared spectrum showed the presence of vinyl and methyl groups. Selenium dehydrogenation gave a mixture of naphthalene and phenanthrene hydrocarbons comparable to that obtained by dehydrogenation of manöol. The identity of the carbon skeleton with that of manöol was revealed by the formation of a common trihydrochloride. Both the *cis* and *trans* isomers have been detected[25]. Sclarene, which is one of the dehydration products of sclareol, is the isomeric 13(16),14-diene.

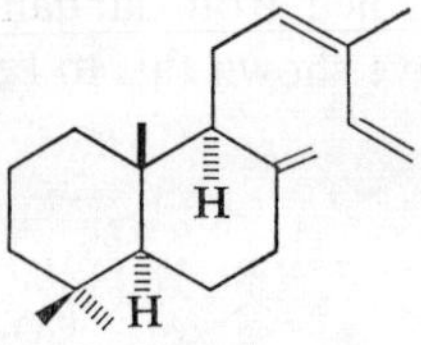

(26)

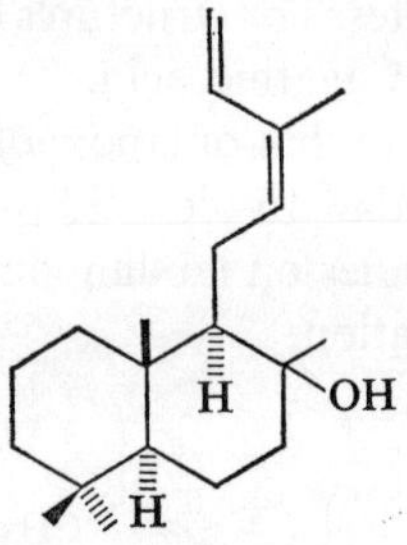

(27)

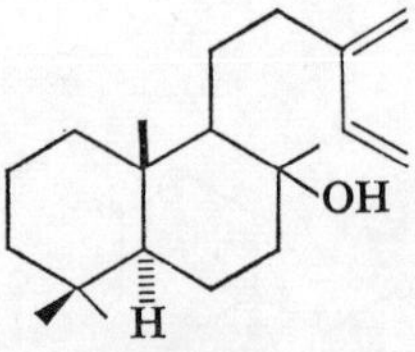

(28)

cis-Abienol **(27)** occurs widely in *Abies* oleoresins[26, 27], as for example, in *A. balsamea*[28]. However the *trans* isomer is present to only a small extent[29]. Abienol hydrate (isolated from *A. pectinata*) was described[30] in 1942 and it is probably the 'abietin' of the earlier literature. It has also been isolated from the resin of *A. sibirica*. Abienol was shown to be a bicyclic tertiary alcohol containing a conjugated diene. The structure **27** was established as a result of work in several laboratories. The absolute stereochemistry of abienol was elucidated[31] through its conversion into (+)-norambreinolide. *cis*-Abienol was isomerized[32] to the *trans* isomer by the action of mercuric acetate. On dehydration with phosphorus oxychloride in pyridine it gave *cis*-biformene. A double bond isomer, isoabienol **(28)** has been isolated[33] from *A. sibirica*.

Communic acid **(29)** was isolated[34] as its sodium salt from *Juniperus*

(29)

communis and as the free acid, from *J. horizontalis*[35]. It has also been isolated from a number of *Cupressus* species[36], *Callitris columellaris*[37] (in
which it co-occurs with $\Delta^{13(16)}$-communic acid) and from *Agathis robusta*
resin[38]. Two of the three olefinic linkages were accounted for in a transoid
diene and, since two moles of formaldehyde were isolated on ozonolysis, the
other double bond must be present as a terminal methylene group. The
larger carbon fragment which was isolated from this degradation, had also
been obtained from agathic acid. Consequently communic acid was
assigned the structure **(29)**. Reduction of the methyl ester of communic
acid gave communol. On the other hand reduction with sodium in isopropanol involved 1:4 addition to the diene to give isodihydrocommunol.
The same compound was also obtained from torolusol by dehydration and
reduction with sodium in isopropanol. The n.m.r. and u.v. spectra of the
diene of methyl communate were comparable to the *trans*-ocimenes and
consequently the *trans* configuration was assigned[39] to the side-chain of
communic acid. The correlation of communic acid with other labdanes has
been examined and a number of 17-nor-8-ketones have been prepared[40].
These ketones show a negative Cotton effect in their optical rotatory dispersion curve. Elliotinoic acid and elliotinol, isolated from the oleoresin of
the slash pine, *Pinus elliottii*, are identical with communic acid and
communol respectively[41].

Petroleum extraction of the wood of *Daniellia ogea* gives ozic acid[42]. The
spectral characteristics of the acid were consistent with the formulation **30**.
Reduction with sodium in isopropanol gave an isodihydro compound.
Ozonolysis of the methyl ester of this compound gave a diketone which was
then converted by an internal aldol condensation to the $\alpha\beta$-unsaturated
ketone **(32)** enantiomeric with a compound that had been obtained from
neoabietic acid. The corresponding alcohol to ozic acid, ozol **(31)**, has also
been isolated from the wood. The *cis* configuration about the 12(13) double

bond of these compounds was arrived at by a comparison of the ultraviolet and nuclear magnetic resonance spectra of the *cis*- and *trans*-ocimenes with the spectra of ozic acid.

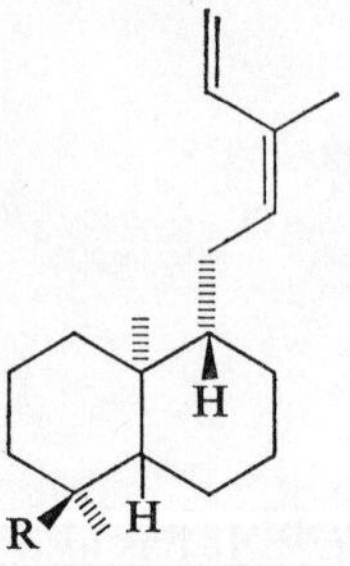

(30) R = CO_2H
(31) R = CH_2OH

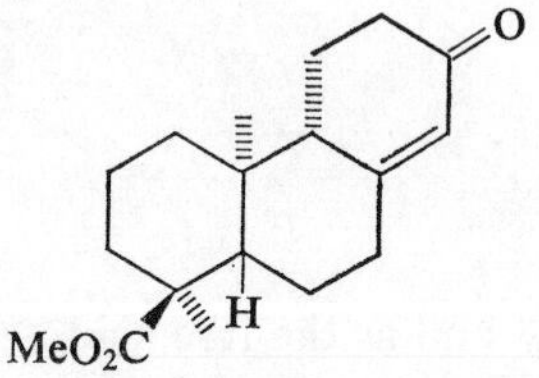

(32)

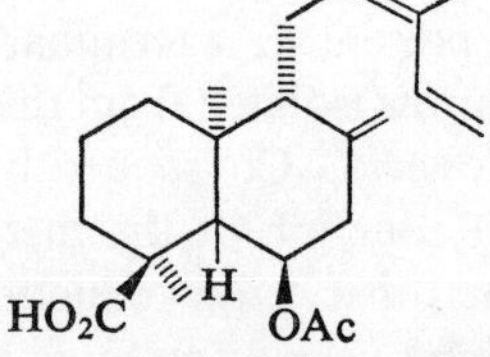

(33)

Zanzibaric acid (33) was isolated[43] from Madagascan *Trachylobium verrucosum*. Its structure was based on its spectral characteristics and on the formation of a dienone antipodal to a compound which was obtained from neoabietic acid. The location of the acetoxyl group at C-6 was established by the formation of a γ-lactone involving the C-18 carboxyl group.

C. Acids

The bicyclic diterpenes of the labdanolic acid group fall into two classes; those possessing a Δ^{13} double bond and those in which C-13 is a saturated centre and in which an additional asymmetric centre is therefore created. To simplify the discussion we will treat the unsaturated compounds first. Agathic acid (34) is the best known of these. Relatively pure agathic acid (agathene dicarboxylic acid) was first isolated by Ruzicka and Hosking in 1929 from Kauri resin obtained from *Agathis australis*[44]. It has been found in Manila copal and in other *Agathis* resins[45], e.g. *A. microstachya*, in which it co-occurs with its C-15 monomethyl ester. Dehydrogenation of this dibasic acid gave 1,2,5-trimethylnaphthalene, 1,7-dimethylphenanthrene and a C_{17}-hydrocarbon subsequently shown to be 1,1,4,7-tetramethyl phenalan (35)[46]. Hydrogenation[47] which resulted in the uptake of two moles of hydrogen, revealed the presence of two double bonds, one of which was shown to be part of an $\alpha\beta$-unsaturated acid. The relationship of the two double bonds was deduced from the results of ozonolysis in which formaldehyde and oxalic acid were isolated[48]. Ozonolysis of dimethyl agathate in glacial acetic acid led to the formation of a peroxide which, on reduction

with zinc dust, formed a diketoester (36). Similar peroxides and a hydroperoxide have been recorded[49] in the ozonolysis of abienol. In basic conditions the latter readily underwent an internal aldol condensation to form a β-hydroxy ketone (37) and thence an $\alpha\beta$-unsaturated ketone (38). This reaction finds a parallel in manöol chemistry and illustrates the interaction between C-8 and C-13 which is a characteristic feature of the chemistry of the bicyclic diterpenes. Treatment of the β-hydroxy ketone (37) with methyl magnesium iodide and dehydrogenation of the product gave pimanthrene thus establishing the location of the side-chain functionality.

On treatment with formic acid, agathic acid formed a tricyclic compound, isoagathic acid (**39**). This acid, like its parent, was readily decarboxylated giving in this case, isonoragathic acid. Although the acid was originally assigned the Δ^{13} structure, n.m.r. studies have shown that isoagathic acid possesses the Δ^{12} structure[50]. Both dimethyl agathate and dimethyl isoagathate possess a hindered ester grouping which was resistant to hydrolysis[51]. Its position on the carbon skeleton was demonstrated in a manner analogous to that used in the tricyclic diterpenes. Methyl isonoragathate was reduced to isonoragathenol which on dehydration underwent rearrangement to a hydrocarbon which was in turn dehydrogenated to 1-ethyl-7-methyl-phenanthrene[52].

The stereochemistry of agathic acid was established as follows. Correlation with manöol[53] clarified the A/B ring junction stereochemistry. The β-hydroxy ketone (**37**) obtained from agathic acid, was converted to an aromatic ester and the ester group was then reduced to a methyl group to give the hydrocarbon (**40**) which was also prepared from manöol. The stereochemistry of the carboxyl group at C-4 was established by the prolonged oxidation of agathic acid with alkaline potassium permanganate[54]. An optically active monomethyl ester acid anhydride derived from ring A was isolated[55], and shown to possess the structure **41**. Hence the carboxyl at C-4 was an axial grouping. Molecular rotation arguments led[56] to the conclusion that the side-chain was β-oriented whilst a study of the position of the side-chain methyl group resonances in the n.m.r. spectrum indicated[50] that the methyl group and carboxyl group were *cis* to one another.

Treatment of agathic acid with sulphuric acid in methanol led[57] to migration of the exocyclic 8(17) double bond into the endocyclic 8(9) position. This reaction also led to a C-15 monomethyl ester.

Isoagathic acid (**39**) has been the subject of a number of studies[58]. On pyrolysis, it lost the ring C carboxyl group. Ozonolysis of the nor-acid led to the location of the ring C double bond which had migrated during the

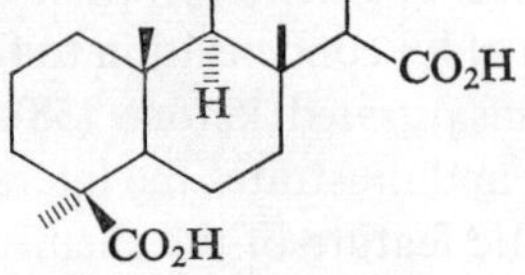

(39) 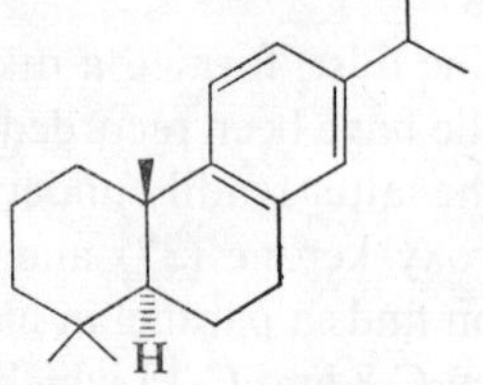   (40)

pyrolysis into the Δ^{13} position. The position of the carboxyl group on ring C was determined by reduction to a methyl group and dehydrogenation to give 1,7,8-trimethylphenanthrene. The *trans* stereochemistry of the B/C ring fusion in isoagathic acid was revealed by an examination of the optical rotatory dispersion curves of ketones at C-12 and C-14[50].

The soft grades of Manila copal also yield[59] agatholic acid (**41**) which contains a C-19 alcohol rather than a carboxylic acid grouping. The oleoresin of black Kauri from North Queensland contains[60] the C-19 aldehyde, agathalic acid, as well as the communic acids, abietic acid and agathic acid.

Copalic acid (**42**) which has been isolated[61] as a gummy mixture of double bond isomers from Brazilean copal (*Hymenea courbaril*) has been shown to possess the structure **42** in which the A/B ring junction possesses the antipodal fusion.

Whilst agathic acid and its relatives retain the Δ^{13} double bond of their labdadienol progenitors, there are many acids in which this centre has been reduced. They fall into two series, one related to labdanolic acid and the other related to eperuic acid.

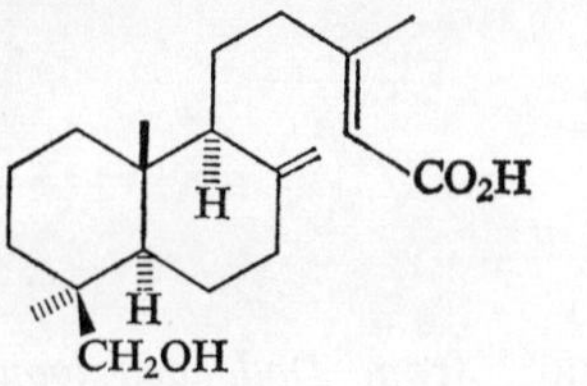

(41) 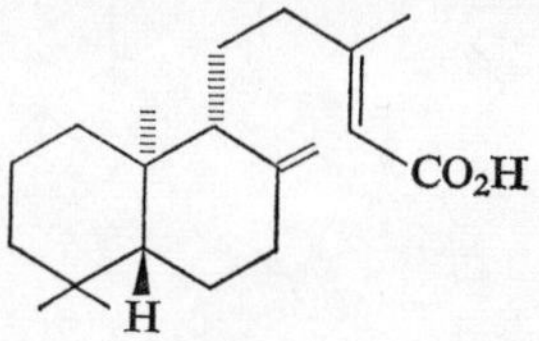　　　　　　　　　　　　　　　(42)

The resinous extract of *Prioria copaifera* (Cativa resin) contains a mono-carboxylic acid, cativic acid (**43**)[62]. Its bicyclic diterpenoid skeleton was apparent[63] from dehydrogenation to, rather unexpectedly, 1,2,5,6-tetra-methylnaphthalene by palladised charcoal and to 1,1,4,7-tetramethyl-phenalan (**35**) with selenium. Barbier–Wieland degradation of the side-chain gave a compound **44** which had also been obtained from manöol, thus establishing the nuclear structure and stereochemistry of the molecule. The position of the double bond was determined by the addition of nitrosyl chloride to form the oxime of an $\alpha\beta$-unsaturated ketone (**45**). The substitution pattern of this unsaturated ketone was deduced from the position of the ultraviolet absorption maximum. The side-chain stereochemistry followed from the relationship with labdanolic acid.

The resin of *Cistus labdaniferus* which is known as gum labdanum, contains a number of diterpenes of which the gummy acid, labdanolic acid (**46**) was the first to be described[64]. The acid was purified as its methyl ester.

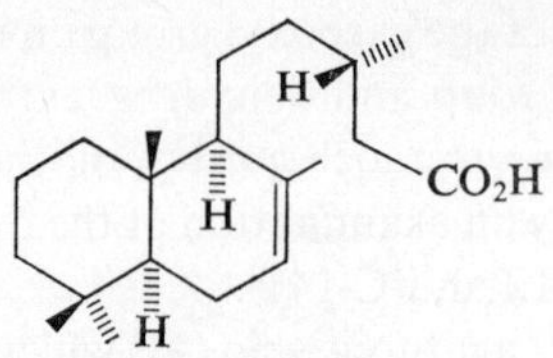

(43)

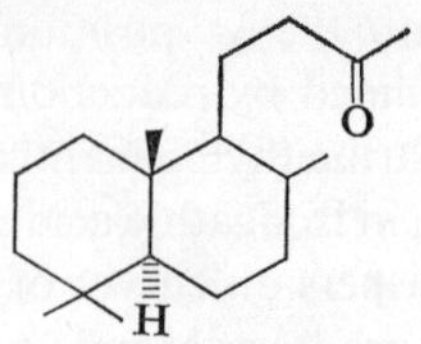

(44)

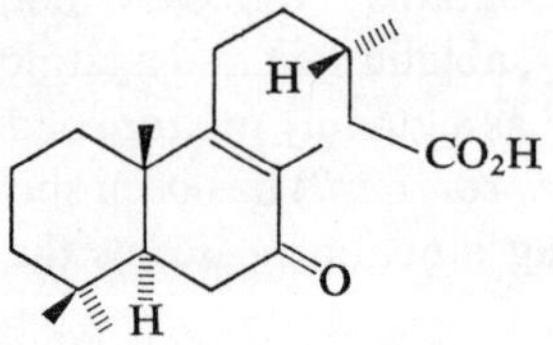

(45)

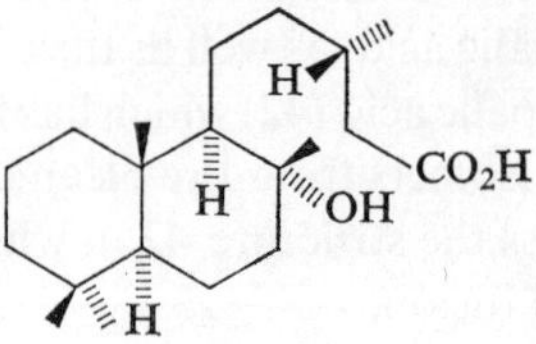

(46)

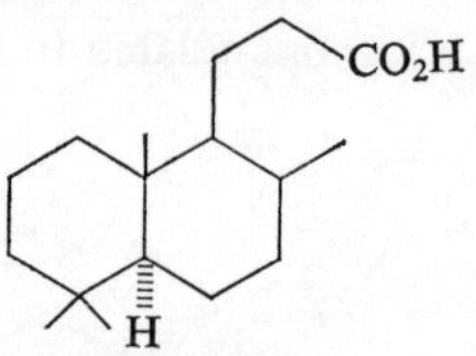

(47)

Methyl *ent*-labdanolate has been obtained[65] from *Dodonaea lobulata*. 1,2,5-Trimethylnaphthalene was obtained on dehydrogenation[66]. Dehydration of the inert hydroxyl group with phosphorus oxychloride gave an exocyclic methylene group, a feature which is typical of a C-8 equatorial hydroxyl group. Hydrogenation of the olefin gave an ester which was similar to the methyl dihydrocativate. Isomerization of the double bond and hydrogenation followed by Barbier–Wieland degradation of the side-chain, gave an acid **(47)**. This acid had also been obtained from marrubiin thus establishing a relationship with other bicyclic diterpenes. Ozonolysis of the exocyclic methylene gave a ketone which showed[67] a negative Cotton effect similar to a compound derived from manöol. Hence labdanolic acid possessed the normal A/B ring fusion. This keto acid was recovered unchanged from alkali in accord with the equatorial conformation for the side-chain. Evidence for the stereochemistry at C-13 and the relationship of labdanolic acid with eperuic acid, is presented later.

Several partial syntheses of labdanolic acid have been recorded[68]. Thus allylic rearrangement of the side-chain of sclareol, oxidation to an unsaturated acid and reduction of the double bond affords labdanolic acid.

6-Oxocativic acid (**48**) is a further constituent of gum labdanum[69]. Its ultraviolet spectrum revealed the presence of an $\alpha\beta$-unsaturated ketone although the carbonyl group was very inert towards the usual characterizing agents. Even lithium aluminium hydride reduced the double bond rather than the carbonyl group. However Wolff–Kishner reduction of the dihydro acid led to dihydrocativic acid. On the other hand oxidation of methyl cativate with selenium dioxide gave a diene (**49**) and an $\alpha\beta$-unsaturated aldehyde (**50**) whilst potassium dichromate oxidation led to the ketone (**51**). Dihydro-6-oxocativic acid possesses a positive Cotton effect in the optical rotatory dispersion curve similar to that of a 6-ketone derived from marrubiin (q.v.).

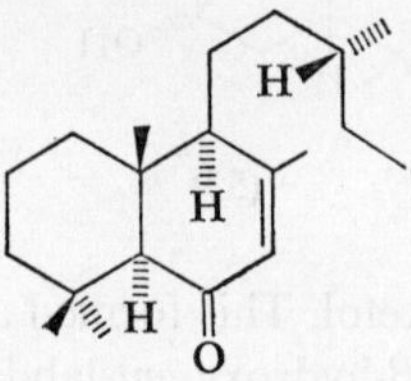

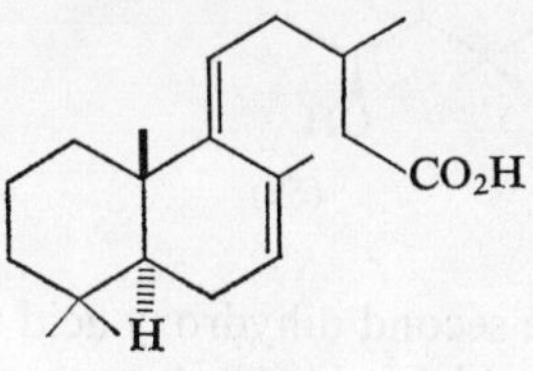

The acidic fraction of the light petroleum extract of the bark of *Araucaria imbricata* contains[70] 3β-hydroxy-labd-8(17)-en-15-oic acid. The corresponding diol is present in the neutral fraction. The structure of the acid rests on its interrelationship with this diol which was in turn related to methyl dihydrocativate.

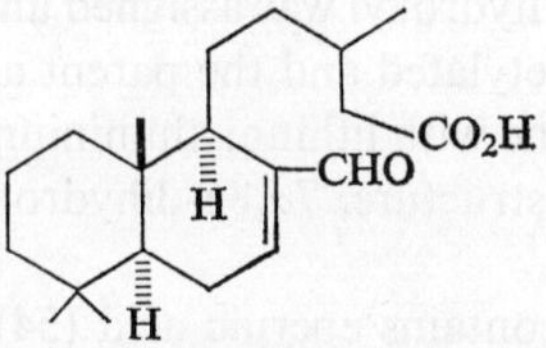

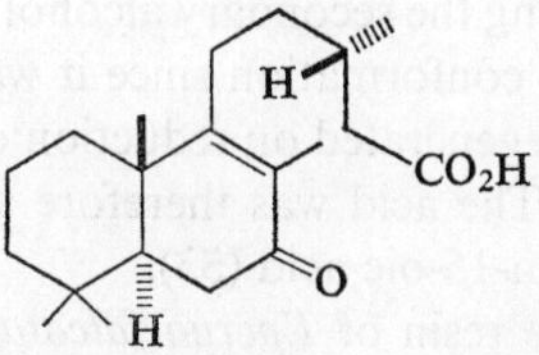

Many Western Australian shrubs including members of the *Beyeria*, *Dodonaea* and *Ricinocarpus* species, possess a resinous coating on the leaves. This has been a valuable source of diterpenes. Examination of *Dodonaea lobulata* led[71] to the isolation of *ent*-labdanolic acid and two dihydroxy acids (**52** and **53**). Dehydration of the *ent*-labdanolic acid afforded predominantly (70%) the exocyclic isomer and 20% of the 7(8)-trisubstituted isomer. Ozonolysis of the exocyclic isomer gave a nor-keto acid which formed an oxime. The specific rotations of these compounds, when compared to the corresponding compounds derived from labdanolic acid[64, 66],

9

clearly established that they arose from an 8β-hydroxy-*ent*-labdan-15-oic acid.

The first of the dihydroxy acids was oxidized to a β-hydroxy ketone which could be dehydrated to an $\alpha\beta$-unsaturated ketone. Hydrogenation and Wolff–Kishner reduction afforded methyl *ent*-labdan-15-oate. The spectral data of these compounds were consistent with the structure 6β,8β-di-hydroxy-*ent*-labdan-15-oic acid (**52**) for the dihydroxy acid.

(52)   (53)

The second dihydroxy acid was oxidized to a ketol. This formed a thio-ketal which was desulphurized to give methyl 8β-hydroxy-*ent*-labdan-15-oate. Oxidation of the diol with periodate gave a keto aldehyde thus locating the secondary alcohol at C-7. This hydroxyl was assigned an equatorial conformation since it was readily acetylated and the parent alcohol was regenerated on reduction of the 7-ketone with lithium aluminium hydride. The acid was therefore assigned the structure, 7α,8β-dihydroxy-*ent*-labdan-15-oic acid (**53**).

The resin of *Eperua falcata* (Wallaba) contains eperuic acid (**54**)[72]. Its enantiomer has been obtained[73] from *Oxystigma oxyphyllum*. The presence of a typical bicyclic diterpenoid carbon skeleton was revealed by dehydrogenation experiments which afforded 1,2,5-trimethylnaphthalene and 1,1,4,7-tetramethylphenalan. Ozonolysis of the double bond gave formaldehyde, typical of an exocyclic methylene and a nor-ketone. Dehydrogenation of the corresponding ester of this nor-ketone gave pimanthrene in which the tricyclic skeleton was generated by a base-catalysed cyclization. The positive Cotton effect in the optical rotatory dispersion curve of the 17-nor-8-ketone was nearly the mirror image of the corresponding keto ester derived[67] from labdanolic acid. Hence the A/B ring fusion was in the antipodal series. However molecular rotation differences suggested[65] that the molecule was not completely enantiomeric to labdanolic acid. It has now been shown[74] that eperuic acid is related to labdanolic acid in the sense that its carbon skeleton is enantiomeric to it except at C-13 where it possesses the same absolute stereochemistry. This was established[75] by converting the acids to tricyclic compounds. Labdanolic acid reacted with methyllithium to form a methyl ketone. Dehydration of the tertiary hydroxyl group and cleavage of the C-8 olefin gave a ketone (55) which readily cyclized to 56 under basic conditions. However the corresponding dione prepared from eperuic acid did not cyclize. Under acidic conditions both formed tricyclic unsaturated ketones (57). The difference in the ease of cyclization can be rationalized if the compound derived from eperuic acid possesses an axial methyl group at C-13 introducing strong diaxial interactions in the product. Thus labdanolic acid was assigned the stereochemistry 46 and eperuic acid (54).

(54)

(55)

(56)

(57)

(58)

The needles of *Pinus sylvestris* contain[76] a dibasic acid, pinifolic acid (58). On dehydrogenation it gave 1,2,5-trimethylnaphthalene characteristic of the bicyclic carbon skeleton, whilst S-(+)-3-methylpentanoic acid was obtained by cleavage of the side-chain. The double bond present in pinifolic acid was shown to be an exocyclic methylene. Pinifolic acid was related to labdanolic acid thus establishing a common C-13 stereochemistry. Furthermore it differed from dihydroagathic acid thus revealing the stereochemistry of the second carboxyl group at C-4.

Ricinocarpus muricatus contains[77] a number of diterpenes possessing the eperuane skeleton. The major diterpenoid constituents were a diol, a triol and a dihydroxy acid. The diol (59) was oxidized with chromium trioxide and the product methylated to give a hydroxy ester enantiomeric with methyl 13-epilabdanolate. Both C-13 epimers of this compound were prepared from 8β,15-dihydroxy-*ent*-labd-13-ene. The hydroxy ester was dehydrated to give a mixture of double bond isomers. The predominant exocyclic isomer was hydrolysed and shown to be identical to eperuic acid. This provided evidence for the relationship between eperuic acid and labdanolic acid. The triol (60) was related to the diol whilst the n.m.r. spectrum

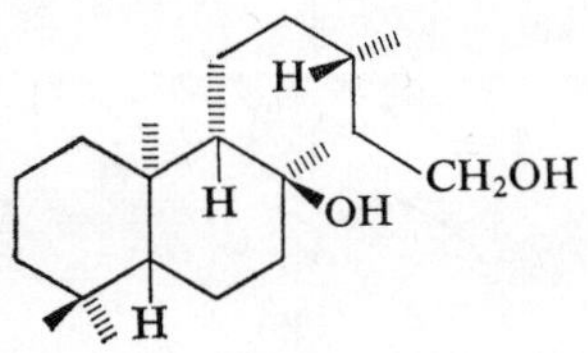

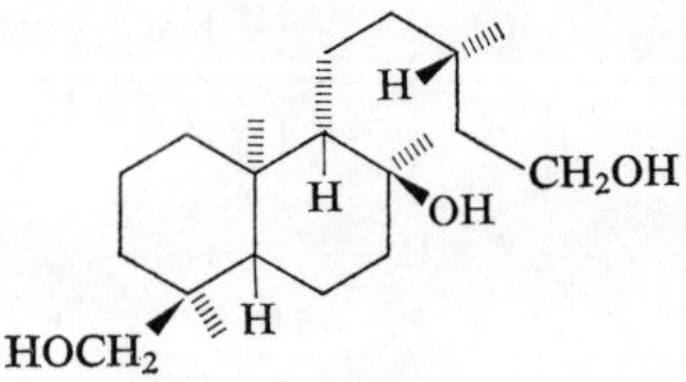

showed that the additional hydroxyl group was part of an equatorial alcohol at C-4. The dihydroxy acid (61) was shown by the formation of an ethylidene derivative, to contain a 1,4-diol. Dehydration with toluene-*p*-sulphonic acid gave a crystalline ether identical to hexahydropolyalthic acid. The stereochemistry at C-13 was established by reduction of the dihydroxy acid to a triol and hydrogenolysis of the toluene-*p*-sulphonate to give eperu-8(17)-ene which was characterized as its diol. Oxidation of the dihydroxy acid afforded a γ-lactone which was assigned the 16-15 structure. The acid had a pK^*_{MCS} consistent with an equatorial 4-carboxyl group[78].

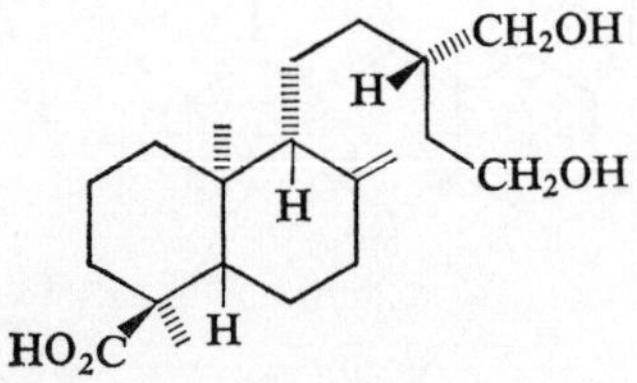

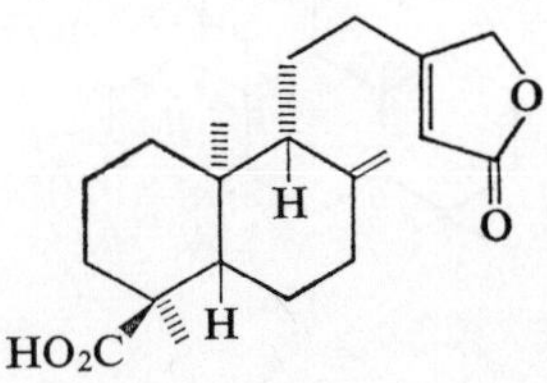

The minor diterpenoids of *Ricinocarpus muricatus* include[79] *ent*-13-epilabda-8(17)-en-15,18-dioic acid, 15-hydroxy-*ent*-13-epilabda-8(17)-en-18-oic acid and the $\Delta^{13(14)}$ butenolide (62). Their structures rest on spectral data and correlation with the major constituents.

IV. THE MANÖOL GROUP

A. Alcohols

The neutral fraction of the heartwood extract of *Dacrydium biforme* contains up to 90% manöol[80]. Manöol is of widespread occurrence and is found in a number of other *Dacrydium*, *Cupressus* and *Pinus* species[81-84]. Its 13-epimer has been isolated from *P. contorta*[85], *Larix europaea*[86] and the bark of the Sitka spruce[87]. The structure of manöol has been firmly established as **63**. On hydrogenation it forms di- and tetrahydro derivatives in which the vinyl group is reduced first. Manöol, 13-epimanöol, sclareol, 13-episclareol, manoyl oxide, 13-epimanoyl oxide and the biformenes give the same trihydrochloride[88] which has been shown to possess the structure **64** arising by allylic rearrangement during the addition of hydrogen chloride. Earlier proposals[89] had suggested that this compound possesses a C-8 equatorial chlorine atom. However in a paper clarifying[90] the structure of various chlorolabdanes, the monohydrochloride derived from tetrahydroabienol was assigned the 8-axial stereochemistry. This led to a reassignment of the structure of the other chlorolabdanes. Dehydrochlorination of the axial isomer yields all three expected olefins whilst the equatorial epimer affords only the $\Delta^{7(8)}$-and $\Delta^{8(17)}$-olefins.

The carbon skeleton of manöol was identified through dehydrogenation and oxidation experiments[91,92]. Amongst the ozonolysis and potassium permanganate oxidation products was a C_{17}-diketone (**65**) which readily underwent an internal condensation through a β-hydroxy ketone to an $\alpha\beta$-unsaturated ketone (**66**)[93,94]. Reaction of the hydroxy ketone with methyl

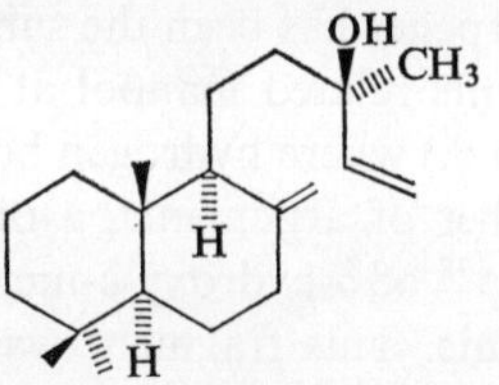

(63)

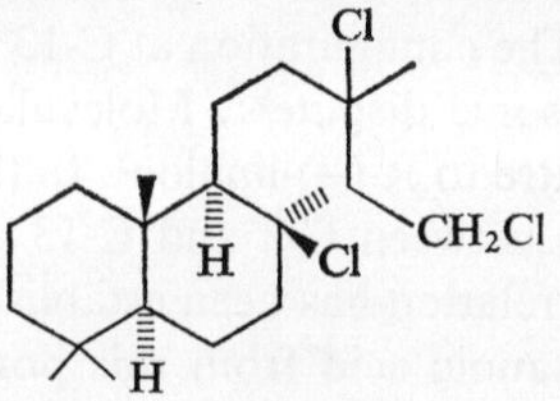

(64)

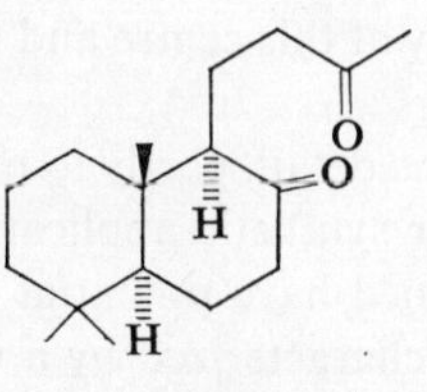

(65)

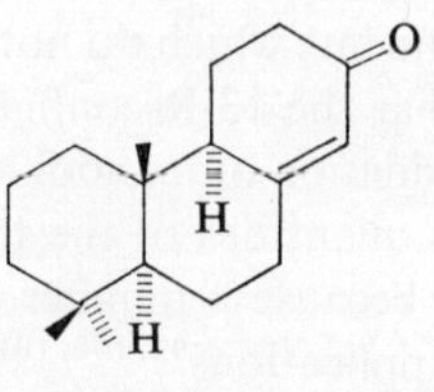

(66)

magnesium iodide followed by dehydrogenation afforded 1,7-dimethyl-phenanthrene (pimanthrene) thus locating the carbonyl function on the tricyclic nucleus. Use of isopropyl magnesium bromide followed by stepwise dehydrogenation experiments gave dehydroabietane (**40**) providing evidence for the identity of the A/B ring fusion between manöol and the tricyclic abietic acid[95]. A similar relationship has been established with neoabietic acid. The tricyclic $\alpha\beta$-unsaturated ketone (**66**) has been synthesized and has formed a useful intermediate for diterpene total synthesis in the bi-, tri-, and tetracyclic series.

The stereochemistry of manöol rests on a number of pieces of evidence. The interrelationship with dehydroabietic acid, which was of known stereochemistry, indicated that manöol possessed the normal *trans* A/B fusion. The stereochemistry at C-9 also follows the normal pattern with an α-hydrogen atom. There were four main pieces of evidence for this. Firstly manöol was degraded[92] to a lactone (**67**) related to ambreinolide. This is an oxidation product of the triterpene ambrein in which the stereochemistry is known. Secondly the tricyclic $\alpha\beta$-unsaturated ketone (**66**) has an optical rotatory dispersion curve comparable to that of cholest-4-en-3-one[93]. This evidence supplemented an earlier series of molecular rotation arguments[56]. Thirdly catalytic hydrogenation[66] of methyl $\Delta^{8(9)}$-labdanolate gave methyl dihydrocativate which had in turn been related to manöol. Here it was assumed that reduction took place from the less hindered α-face of the molecule to give an equatorial side-chain. Finally ozonolysis of methyl labdanolate under mild conditions gave a nor-ketone which was stable to alkali, a proof which was strengthened by the subsequent isolation[96] of the less stable isomer during a synthetic sequence.

The configuration at C-13 of the bicyclic diterpenes has been the subject of some dispute[97]. Molecular rotation arguments related manöol at this centre to R-(−)-linalool. In the sclareol series (q.v.) where hydrogen bonding between C-8 and C-13 invalidated a number of arguments, a direct correlation has been established by the isolation[98] of δ-hydroxy-δ-methyl-hexanoic acid from this portion of the molecule. This fragment was of known absolute configuration. Sclareol was thus shown to possess the 13-R-configuration at this centre. Manöol and sclareol have been related[99] by experiments which do not disturb the asymmetry at this centre and thus manöol has the 13-R-configuration.

The oxidation of manöol has been studied in some detail, partly to make available quantities of the tricyclic ketone (**66**) for synthetic applications and also because a number of ketals that are formed have potential perfumery applications[92, 100, 101]. These reactions are characterized by a very ready ether formation between C-8 and C-13. The initial ketonic product

(68) of the potassium permanganate oxidation readily undergoes further hydroxylation with the formation of a stable ketal (69). The C-13 epimer of this ketal was produced by the selective epoxidation of manöol at C-8 followed by oxidation of the vinyl grouping and cleavage of the epoxide. The stereochemistry of these ketals was assigned on the basis of a careful examination of their n.m.r. spectra and of the chemistry of the C-12 ketones

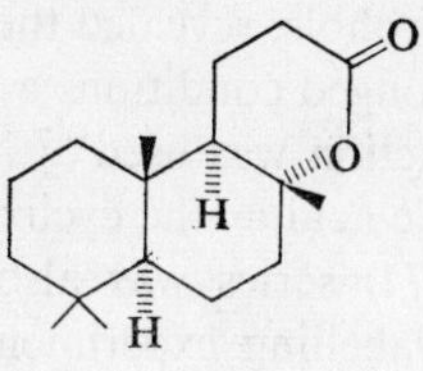

(67)

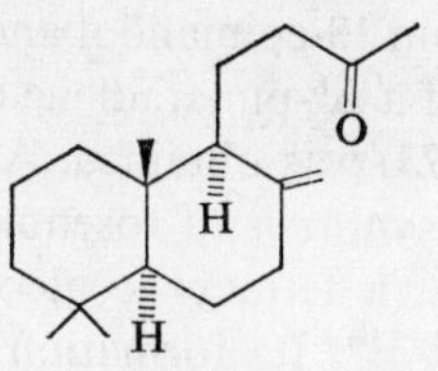

(68)

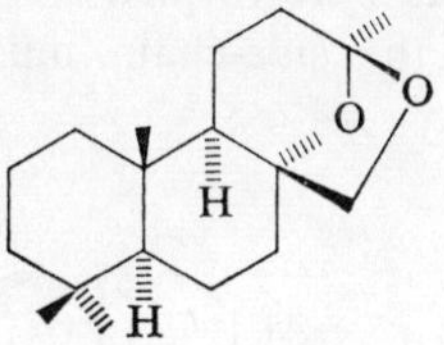

(69)

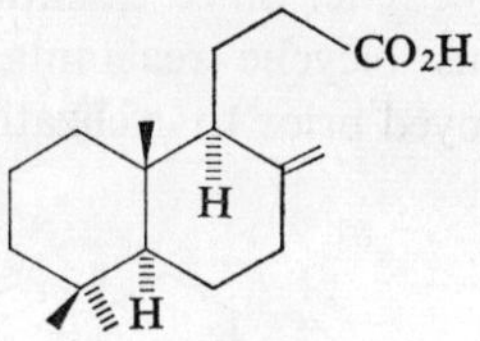

(70)

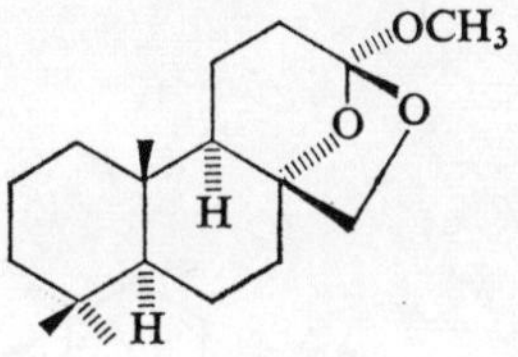

(71)

formed on oxidation of the ketals with chromium trioxide. Further oxidation of the ketone (68) with sodium hypoiodite led to an acid (70). Lactonization of this acid gave a C-8 epimer of ambreinolide. The corresponding methyl ester was converted to a C-13 epimeric pair of methoxy ketals (71). The reaction of the C-8 methylene with osmium tetroxide led to attack which was predominantly from the α-face of the molecule and to the formation of the ketal (69). Epoxidation and cyclization led to its isomer. Selective epoxidation of manöol itself leads[99] to attack on the C-8 methylene first. Again attack is mainly, but not entirely, from the α-face of the molecule. Reduction of these C-8 epimeric epoxides affords sclareol and 8-episclareol[102]. Examination of the effect of pH on the oxidation of manöol with osmium tetroxide–sodium periodate afforded a pentaol at high pH, the tricyclic ketol around neutrality and the enol-ether at low pH[103].

A second aspect of manöol chemistry which has been the subject of study, has been its acid-catalysed cyclization. Cyclization of manöol with acetic acid–sulphuric acid mixtures gave a hydrocarbon mixture from which Δ^8-isopimaradiene was isolated[104]. Further examination of this mixture, which could be obtained[105, 106] with essentially the same composition from 13-epimanöol and the allylic primary alcohols, revealed the presence of a Δ^8-pimaradiene (**72**). Under more prolonged conditions a rosadiene (**73**) was obtained. An extension of this reaction was used[107] in the partial synthesis of rosenonolactone. Using formic acid as the cyclization catalyst, a tetracyclic alcohol of the hibaene (**74**) series was also isolated[108–110]. Its formation has been shown by labelling experiments to involve an eight-membered ring (**75**) rather than a pimaradiene type of intermediate. Since essentially the same products were formed from the various bicyclic reactants, the asymmetry in the side-chain must be destroyed prior to cyclization.

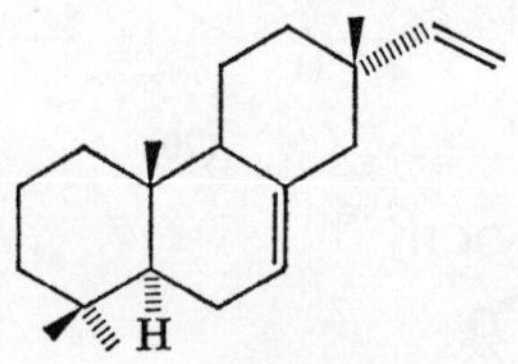

(72)

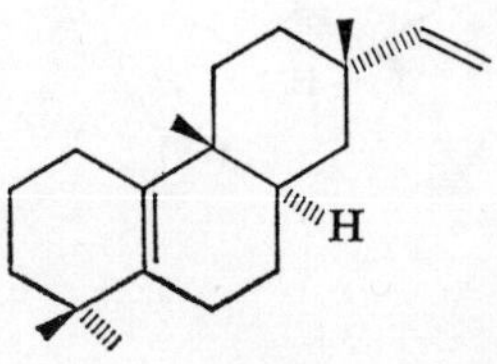

(73)

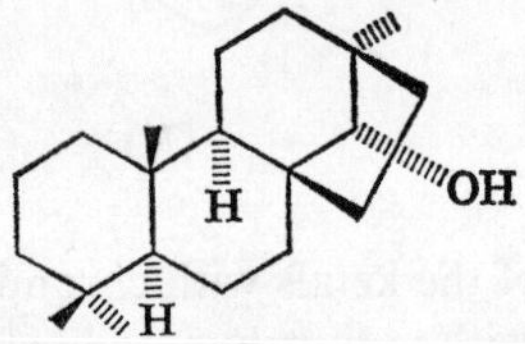

(74)

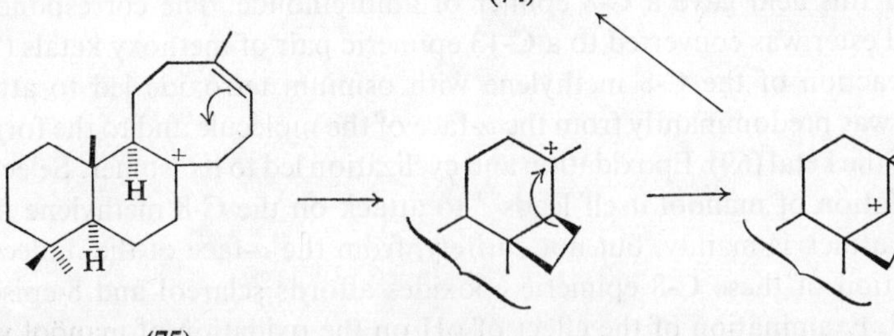

(75)

The essential oil of *Salvia sclarea* contains[111] a diterpene, sclareol (**76**) which has been studied for many years. It was shown[112] to contain one double bond and two hydroxyl groups which were difficult to acetylate. The underlying carbon skeleton was revealed[113] by dehydrogenation to 1,2,5-trimethylnaphthalene. Evidence for the structure of the side-chain of sclareol stems from oxidative degradation[114]. Mild oxidation with potassium permanganate led to formaldehyde and a C_{19}-acid, sclareotic acid. Further oxidation of this afforded a hydroxy ketone or the corresponding enol-ether (**77**). This enol-ether was degraded to ambreinolide (**67**)[115]. The reactions of the enol-ether have been studied in the search for compounds with a musk-like odour[116,117]. The products are comparable to those found in the manöol series and are dominated by the ready ether and lactone formation between C-8 and C-12 or C-13.

Sclareol undergoes a number of allylic rearrangements. For example oxidation with chromium trioxide and methylation of the acidic material led to a separable mixture of *cis* and *trans* esters (**78**) together with an oxide (**79**) related to manoyl oxide[117]. Digestion of sclareol with acetic anhydride and acetic acid affords the biformene:sclarene mixture of hydrocarbons together with manoyl oxide, 13-epimanoyl oxide, and the labda-8,13-dien-15-ol acetate[118].

Sclareol has an additional asymmetric centre at C-8 the stereochemistry of which was defined as follows. Firstly oxidation of the enol-ether (**77**) gave a *trans* fused γ-lactone which underwent acid-catalysed isomerization to a more stable *cis* γ-lactone[116]. Secondly manöol underwent epoxidation predominantly from the α-face of the molecule to give an epoxide which was reduced with lithium aluminium hydride to sclareol. Thirdly pyrolysis of sclareol diacetate regenerated the exocyclic methylene of manöol, indicative of an equatorial C-8 acetoxyl group.

A number of other diols related to manöol have been isolated. Thus 13-episclareol has been found in *S. sclarea*[119]. Torolusol (**80**), was isolated[120] along with the corresponding aldehyde, torolusal, from *Cupressus torolusa*. The two were related[121] by mild oxidation whilst Wolff–Kishner reduction of torolusal gave manöol thus defining the underlying carbon skeleton, the position of one oxygen atom and the C-13 stereochemistry. The position of the second oxygen atom was established by the allylic rearrangement of torolusol to agathadiol (**12**).

Extraction of the heartwood of *Dacrydium kirkii* led[122] to the isolation of a number of bi- and tricyclic diterpenes. These include isopimaradiene, sclarene, *cis*- and *trans*-biformene, manoyl oxide, 14,15-bisnorlabda-8(17)-en-13-one, labda-8(17)-13-dien-15-ol, manöol, isopimaradiene, torolusol, sandaracopimaradien-3β,19-diol and isopimaric acid. Two new diterpene alcohols were isolated. One was shown by mass spectral and n.m.r. studies to be 7α-hydroxymanöol and the other to be 2β,3β-dihydroxymanoyl oxide.

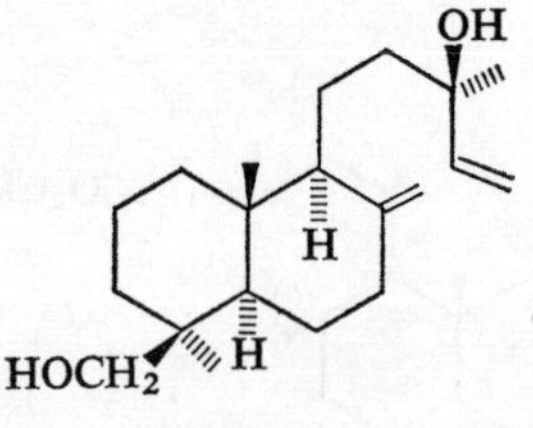

(80) 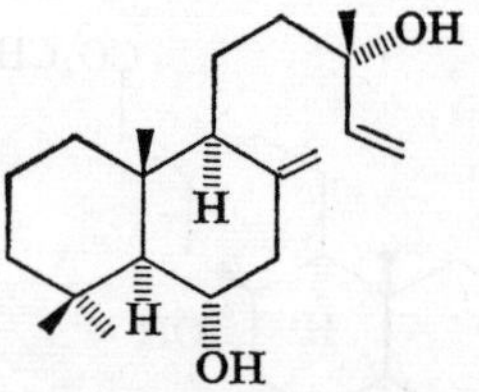   (81)

The heartwood of the European larch, *Larix europaea*, which contains 13-epimanöol[86], gave larixyl acetate[123,124] which could be hydrolysed to the diol, larixol (**81**). This diol has also been isolated[125] from *Larix sibirica*. The bicyclic skeleton was revealed by dehydrogenation which gave 1,7-dimethylphenanthrene and 1,2,5-trimethylnaphthalene. The nature and position of the oxygen functions was revealed by oxidation experiments. Thus the secondary alcohol was oxidized to a ketone which was isomerized to an $\alpha\beta$-unsaturated ketone (λ_{max} 240 nm) thus locating the hydroxyl group at position 6. Oxidation with potassium permanganate gave products characteristic of the cleavage of the side-chain. One of these, an unsaturated ketone, was dehydrogenated to 1-methyl-7-phenanthrol. The C-13 stereochemistry of larixol has been defined[126] by a relationship with 13-epimanöol, whilst the hydroxyl group at C-6 has been shown to possess the equatorial (α) configuration.

B.   The C-8–C-13 Ethers

The ready ether formation between C-8 and C-13 is reflected in the occurrence of manoyl oxide, 13-epimanoyl oxide and a number of oxygenated derivatives of these ethers. Manoyl oxide (**82**) and ketomanoyl oxide (2-oxomanoyl oxide) (**83**) were isolated from the silver pine (*Dacrydium colensoi*)[127]. The underlying diterpenoid carbon skeleton was revealed by dehydrogenation experiments[128]. Dehydrogenation with selenium afforded 1,2,5-trimethylnaphthalene and 1,7,8-trimethylphenanthrene. Although manoyl oxide gave only a dihydro derivative on hydrogenation, addition of hydrogen chloride led to the formation of a trihydrochloride identical to that obtained from manöol and sclareol, thus establishing a link with these diterpenes. The presence of a vinyl group in manoyl oxide was demonstrated by its ozonolysis to give formaldehyde and by oxidation with potassium permanganate, when an acid, $C_{19}H_{32}O_3$ was obtained. Treatment of this acid with hydrogen chloride gave a dihydrochloride by cleavage of the ether ring. Consequently manoyl oxide was assigned the structure **82**. 13-Epimanoyl oxide (**84**) was isolated[129] as its antipodal enantiomer, olearyl oxide, from the bush, *Olearia paniculata* and from the mould, *Gibberella fujikuroi*[130]. The enantiomer with the normal A/B ring fusion, occurs amongst the dehydration products of sclareol[131]. The relationship with manoyl oxide was established[132] by hydrogenolysis of both oxides to the same 8α-hydroxylabda-13-ene and by the formation of the same trihydrochloride. It therefore

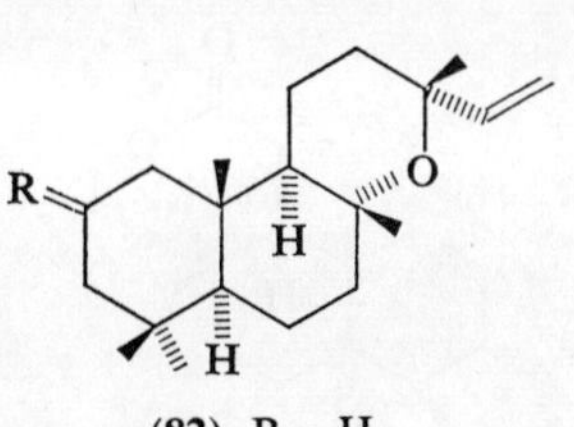

(82)   R = H₂
(83)   R = O

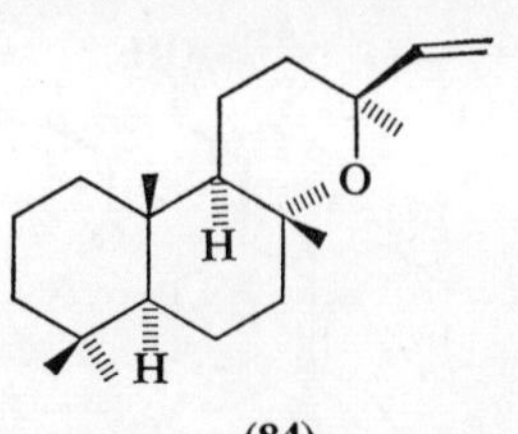

(84)

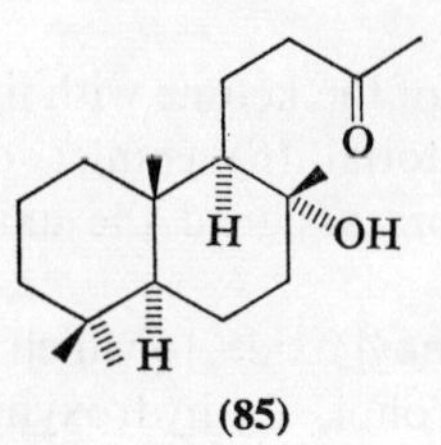

(85)

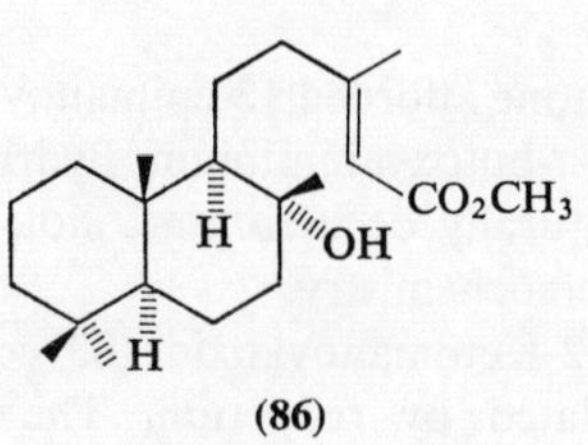

(86)

remained to assign the stereochemistry at C-13. Three pieces of evidence have been described for this. Each rests on the assumption that ring c remains in a chair form in which the vinyl and methyl substituents take up axial or equatorial conformations. Thus in a study of mass spectral intensities, it was assumed[132] that an axial group would be expelled more readily than an equatorial one. Secondly[133] the methyl- and vinyl-proton resonances will be deshielded to different extents by the oxygen ring. Thirdly cleavage of the vinyl group afforded the corresponding carboxylic acids. The relative rates of hydrolysis of the methyl esters were studied[134] in order to define the axial epimer. In each case an axial (β) vinyl group was assigned to 13-epimanoyl oxide. The ether ring of manoyl oxide and its 13-epimer has been reconstructed[135] starting with the ketol (85). Treatment with the phosphonate derived from methyl bromoacetate, under Arbusov conditions afforded a separable mixture of unsaturated esters (86). On reaction with sodium hydride the ether ring was formed and the respective 13-epimers were converted to manoyl oxide and 13-epimanoyl oxide.

The remaining manoyl oxide isomers, 8-epimanoyl oxide and 8,13-epimanoyl oxide, have been isolated[136] from *Chaemacyparis nootkatensis*. Their C-13 stereochemistry was based on an interesting and potentially useful cyclization catalysed by Amberlyst XN-1005 resin. The oxides gave isopimara-8,15-diene and pimara-8,15-diene, in which the C-13 stereochemistry was retained in the tricyclic product.

12α-Hydroxy-13-epimanoyl oxide (87) occurs[134] along with a number of other interesting diterpenes in Turkish tobacco. Ozonolysis led to the known lactone (88) whilst Wolff–Kishner reduction of the corresponding

(87)　　　　　　　　　　　　　　　　(88)

ketone afforded 13-epimanoyl oxide. Reduction of the ketone with lithium tri-*t*-butoxyaluminium hydride gave the equatorial (β) epimer of the naturally occurring alcohol, which was therefore assigned the axial (α) stereochemistry.

2-Ketomanoyl oxide (83) co-occurs[137] with manoyl oxide, to which it was related by reduction. The corresponding alcohol, 2α-hydroxymanoyl oxide, has also been isolated[138] from *Dacrydium colensoi*. The unusual position of the oxygen function was established in several ways. Reaction with methyl magnesium iodide and dehydrogenation of the product gave 1,2,5,7-tetramethylnaphthalene[137]. Deuteration studies indicated[139] the presence of four exchangeable protons whilst reduction of the ketone to the alcohol followed by elimination with phosphorus pentachloride gave[140] a cyclohexene rather than a product of ring contraction. Allylic oxidation of the olefin afforded[141] a 1-ketone whilst conversion of the olefin to an epoxide followed by reduction with lithium aluminium hydride gave a 3α-(axial) alcohol. As expected the corresponding 3β-(equatorial) alcohol undergoes ring contraction on treatment with phosphorus pentachloride. The heartwood of *Dacrydium colensoi* also contains[142] a nor-diterpene, colensenone (**89**) in which ring A is a cyclopentanone. Its relationship with 2-oxomanoyl oxide was established by partial synthesis. Thus autoxidation of dihydro-2-ketomanoyl oxide gave a diosphenol (**90**) which underwent benzilic acid rearrangement to a ring-contracted hydroxy acid. This was in turn oxidized to colensanone. The isomer, colensan-1-one, was prepared[143] from 1-ketomanoyl oxide by a similar route.

Cleavage of the 2,3-diosphenol afforded[138] a 2,3-dicarboxylic acid (**91**). This was identical to the dihydro derivative of 2,3-dicarboxy-2,3-seco-manoyl oxide which was isolated from *D. colensoi*. A δ-lactone was also obtained from this pine. Its n.m.r. spectrum showed that it contained a methylene group adjacent to the lactone carbonyl thus restricting the lactone to ring A. Reduction of the lactone ring gave a diol which formed a monoacetate. This was dehydrated to isomeric isopropenyl and isopropylidene olefins consistent with the 2-oxo-3-oxamanoyl oxide structure (**92**) for the lactone.

The neutral fraction of *D. colensoi* also contains[144] an unusual lactonic homoditerpene (**93**) which was correlated with colensan-2-one. Thus reduction of the vinyl group and the lactone ring afforded a triol in which the vicinal diol was then cleaved to form a nor-ketone. A base-catalysed retro-aldol cleavage of this then gave colensan-2-one and a C-1 methylene ketone as a β-elimination product. Nuclear magnetic resonance coupling constant studies were used to establish the stereochemistry of the lactone/ring A fusion. Similar measurements, taken together with chemical evidence, suggest that the five-membered ring A of the colensane derivatives takes up either the β-envelope or half-chair conformations. The structure of this unusual diterpene was confirmed by a partial synthesis[145] from 1α-methoxycarbonylcolensan-2-one. *Dacrydium colensoi* contains both homo- and A-nor-diterpenes and it is interesting to speculate on the origin of the extra carbon atom of the homo-diterpene. Another oxide isolated[146] from

D. colensoi is 19-hydroxy-2-oxomanoyl oxide (**94**). Conversion of the primary alcohol to its toluene-*p*-sulphonate, followed by treatment with base, led to the formation of a cyclopropane ring thus relating the carbonyl and alcohol functions. The n.m.r. spectrum of the compound indicated that the alcohol was an axial group at C-4.

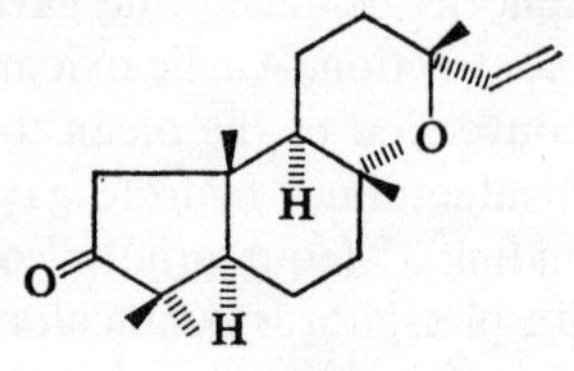

(**89**) 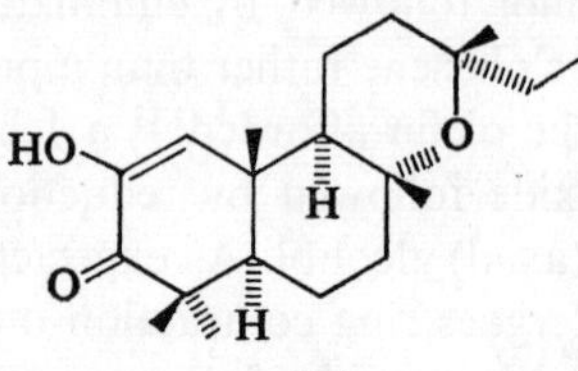   (**90**)

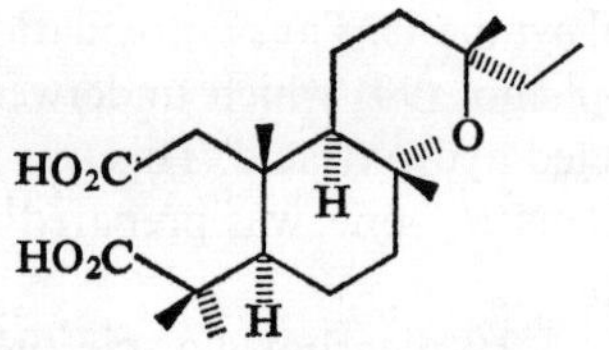

(**91**) 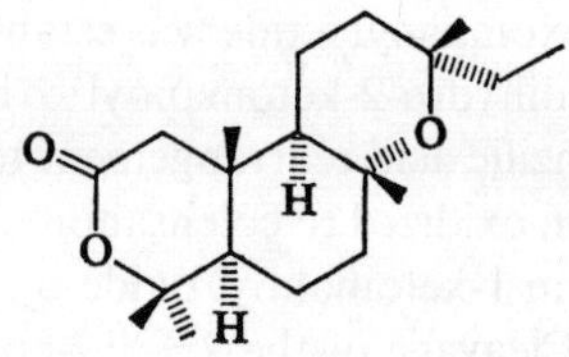   (**92**)

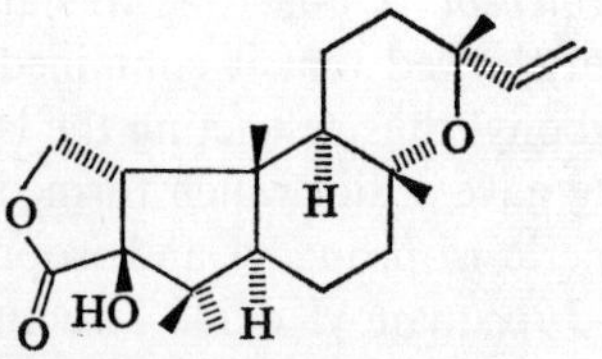

(**93**) 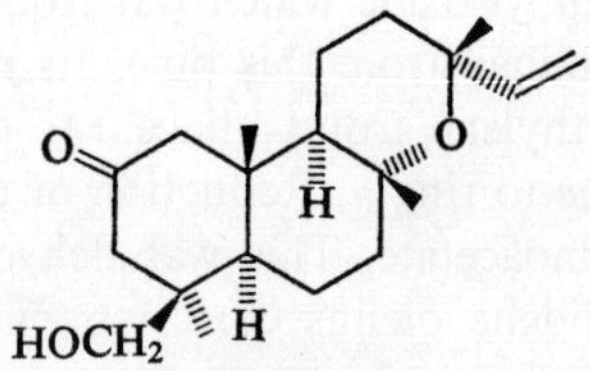   (**94**)

3-Ketomanoyl oxide occurs[147] in the wood of *Xylia dolabriformis*. Its oxygenation pattern was established through interrelation with manoyl oxide.

A number of diterpenes possessing the 13-epimanoyl oxide skeleton of olearyl oxide have been isolated from the resin of Australian plants. A Western Australian *Beyeria* species contains[148] 8β,15-dihydroxy-*ent*-labda-13-ene, the enantiomer of which had been prepared from sclareol. The same species also contained olearyl oxide, together with its 18-hydroxy and 18-carboxylic acid derivatives. The latter were interrelated by oxidation whilst Wolff–Kishner reduction of the corresponding aldehyde afforded olearyl oxide. The presence of an oxygen function at C-18 followed from the chemical shift of the C-18 protons in the n.m.r. spectrum. A triol, 14R,15,18-trihydroxyolearyl oxide, has been isolated from the same source.

C. The C-9–C-13 Ethers

Both C-8 and C-9 are involved in ether formation with C-13. The C-9–C-13 ethers include compounds that are the potential progenitors of the furanoid diterpenes such as marrubiin. Grindelic acid (**95**) was amongst the first of these ethers to be described[149]. Extraction of the resin of *Grindelia robusta* gave a group of diterpene acids. Typically[150] dehydrogenation of grindelic acid gave 1,2,5-trimethylnaphthalene. Hydrogenation gave a dihydro derivative which was accompanied by hydrogenolysis products of the ether. Reduction of the ester with lithium aluminium hydride gave a primary monohydroxylic alcohol. On further reduction with lithium in liquid ammonia, the ether was cleaved to a diol which contained a tetrasubstituted double bond. Oxidation and methylation of this product gave a γ-hydroxy ester which was hydrogenated and dehydrated with phosphorus oxychloride to form an $\alpha\beta$-unsaturated ester. Ozonolysis of this dehydration product gave the known 15,16-bisnor-labdan-13-one (**96**). The position of

(95)   (96)

the double bond in grindelic acid was established by hydroxylation and cleavage of the corresponding diol to form a keto aldehyde. Since the ether linkage underwent hydrogenolysis on reduction of the double bond, these groups must be allylic to one another. Hence grindelic acid was assigned the structure **95**.

6-Oxogrindelic acid and 6,7-epoxygrindelic acid have also been isolated[151] from the resin. Their structures rest upon interrelationships with grindelic acid and upon spectral data.

V. THE FURANOID BICYCLIC DITERPENES

A further oxidation level of the side-chain involves the formation of the furan ring. Several of these diterpenes occur as C-9–C-13 spiro-ethers which generate the furan ring system during the isolation procedure. For example marrubiin is regarded as a probable artefact rather than as a natural product. However the biosynthetic origin of the furan ring is not clear.

Indeed examination of the structures in which some compounds retain unsaturation at C-8 and a hydrogen atom at C-9, suggests that more than one pathway may be operating in this series.

Daniellic acid (**97**) has been isolated[152] from *Daniellia oliveri* whilst its enantiomer, lambertianic acid, was found in *Pinus lambertiana*[153]. On hydrogenation, daniellic acid formed an oily hexahydro compound. The presence of a terminal methylene group was indicated by the infrared spectrum and by the isolation of formaldehyde on ozonolysis. Spectral data indicated the presence of a furan ring. The major fragment from the ozonolysis was a keto acid. This was the optical antipode of a degradation product (**98**) of agathic acid. This established both the structure and the stereochemistry of daniellic acid.

The bark of *Polyalthia fragrans* contains[154] polyalthic acid (**99**) which is closely related to daniellic acid. Polyalthic acid is also amongst the diterpenoid constituents of *Ricinocarpus stylosus*[155]. 1,2,5-Trimethylnaphthalene was formed on dehydrogenation with selenium thus defining the carbon

skeleton. The nuclear magnetic resonances spectrum indicated the presence of a β-substituted furan ring as the second oxygen function. This was
confirmed by the pyrolysis of the adduct with acetylene dicarboxylic acid.
Furan-3,4-dicarboxylic acid was obtained by a retro-Diels–Alder reaction.
Ozonolysis of polyalthic acid gave a keto dibasic acid (**100**), the enantiomer
of which was produced by ozonolysis of neoabietic acid. Thus polyalthic
acid possessed an antipodal A/B ring fusion and was isomeric at C-4 with
daniellic acid.

Psiadiol (**101**) is the most abundant of the diterpenes of *Psiadia altissima*[156]. The spectral features of the diol and its diacetate, indicated the
presence of a β-substituted furan ring, two *C*-methyl groups, a terminal
methylene group, and a primary and secondary alcohol. Oxidation of the
diol afforded a keto aldehyde which was in turn reduced by sodium borohydride to an epimeric diol. This only formed a monoacetate and thus the
secondary alcohol in the latter was in a more hindered (axial) conformation.
The n.m.r. spectrum suggested that the secondary alcohol was at C-6 and
that the primary alcohol was an axial substituent at C-4. Confirmation of
this structure was provided by a correlation with daniellic acid.

Marrubiin (**102**) forms the bitter principle of Horehound, *Marrubium
vulgare*, from which it was first isolated[157] in 1842. It possesses a γ-lactone,
an inert oxygen atom and a tertiary alcohol[158]. Elimination of the latter
gave anhydromarrubiin[159]. Hydrogenation of marrubiin revealed the presence of two double bonds which, together with the inert oxygen atom,
formed a furan ring[160]. Dehydrogenation with selenium gave 1,2,5-tri-
methylnaphthalene, indicative of the carbon skeleton. The oxygen functions were interrelated in the following way. Ozonolysis or oxidation of
marrubiin with chromium trioxide, gave a saturated di-γ-lactone (**103**)
thus establishing the relationship between the furan ring and the tertiary
hydroxyl group. Anhydromarrubiin, on ozonolysis, gave a keto lactone
(**104**) which was also produced by the stepwise degradation of the lactone
(**103**). Hydrolysis of this lactone gave an acid which was oxidized to a diketone. These carbonyl groups were related by further oxidation with selenium
dioxide[161] to form an ene-dione (**105**). This related the tertiary hydroxyl
group and the original γ-lactone of marrubiin. The dilactone (**103**), on
partial hydrolysis and oxidation gave a keto acid which on relactonization
formed an enol-lactone (**106**). This is sterically possible only on the basis of
oxygenation in ring B[160]. Ozonolysis of the enol-lactone produced an
aldehyde acid anhydride, this serving to confirm the position of the
double bond. The lactone (**103**) was converted[161] to a degradation product (**107**) of ambrein thus putting the chemistry of marrubiin on a firm
basis.

(102)

(103)

(104)

(105)

(106)

(107)

The stereochemistry of marrubiin has been the subject of some controversy. The relationship with ambrein clarified the A/B stereochemistry. The presence of an oxygen function at C-6 vitiated the stereochemical interpretation of physical data such as n.m.r. and pK measurements, on the oxygen function at C-4[162]. Once the C-6 functional group was removed[163], n.m.r. data were clearly interpreted in terms of an axial functional group. The C-6 proton resonance indicated that it had the equatorial conformation and hence the lactone ring was *cis* and β-oriented[cf.164]. The ready ether and lactone formation between these centres is in accord with this stereochemistry. Solvent shift data in the n.m.r. spectrum led to the conclusion that the methyl group at C-8 was equatorial and α-oriented whilst the formation of a $\Delta^{8(9)}$-endocyclic olefin on dehydration of the C-9 alcohol with phosphoryl chloride in pyridine, was an indication of the *trans* diaxial relationship between the C-9 hydroxyl group and the C-8 proton. In the absence of pyridine, the $\Delta^{9(11)}$-olefin is formed. Alternative evidence[165] for the stereochemistry of marrubiin came from the partial synthesis of marrubiin and some of its degradation products from the keto-γ-lactone (**104**) under stereochemically defined conditions.

Reduction of the 6-ketones in the marrubiin series has been studied[166] in some detail. In the presence of a C-4 carboxyl group, reduction with sodium borohydride and with lithium in liquid ammonia furnished the axial alcohol. However in the presence of a C-4 ester, lithium aluminium hydride reduction gave the axial alcohol whilst lithium in liquid ammonia gave an equatorial alcohol. Strongly alkaline potassium permanganate oxidation of the keto acid (**108**) gives rise to a keto dilactone (**109**) formed by a ring:ring tautomerization reaction[167].

A spiro-ether, premarrubiin (**110**) containing a C-9–C-13 ether has been isolated[168] from *Marrubiium vulgare*. On heating this ether undergoes a fragmentation reaction leading to the formation of the furan ring of marrubiin. Indeed it has been suggested that marrubiin itself is an artefact arising during the isolation procedure. Marrubenol (**111**) and a hemi-acetal (**112**) have also been isolated[169] from this plant.

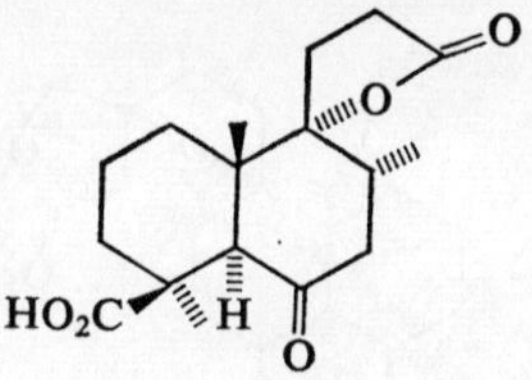

(**108**)

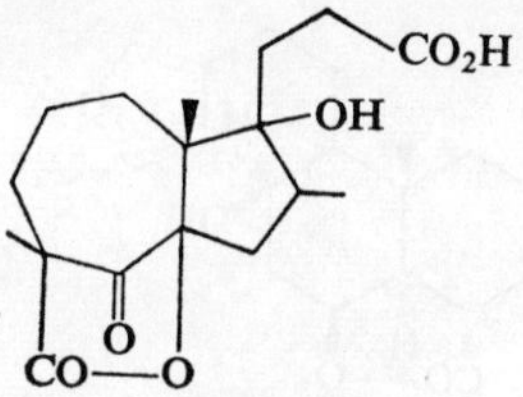

(**109**)

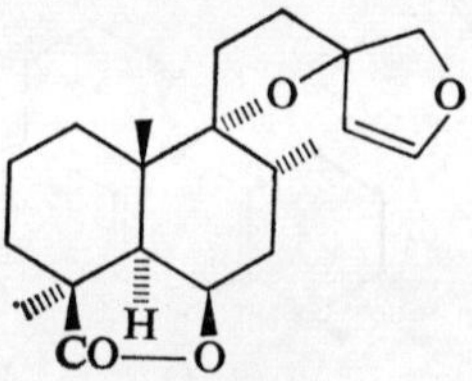

(**110**)

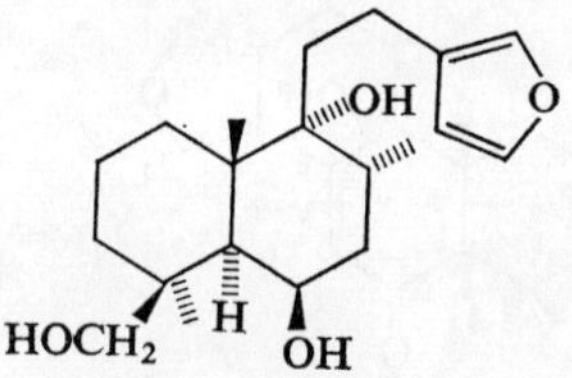

(**111**)

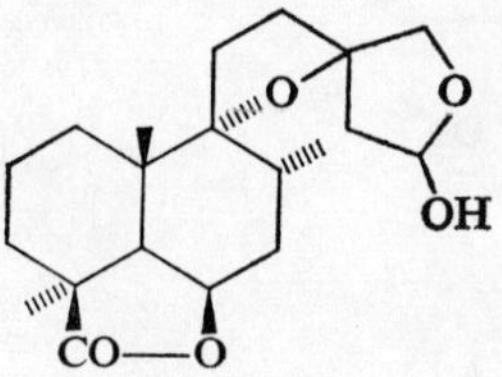

(**112**)

Two other *Marrubium* species have been investigated. The major diterpenoid isolated[170] from *M. icanum* is probably identical to peregrinine, which was isolated[171] from *M. peregrinum*. Spectral data implied a relationship with marrubiin which was also present in the plant. The compound was transformed into marrubiin thus locating four out of the five oxygen functions. The position of the fifth oxygen as a C-3 carbonyl was established by the decarboxylation of the lactone under basic conditions.

Leonotin was isolated from *Leonotis nepetaefolia* and shown[172] to be 8β-hydroxymarrubiin (**113**). It has also been isolated[173] from *L. dysophylla*. When 8β-hydroxymarrubiin was refluxed with phosphorus trichloride in pyridine, it was transformed into an epoxide. Reduction of this with lithium aluminium hydride gave marrubenol. The presence of a *trans* diaxial diol in 8β-hydroxymarrubiin was supported by its failure to react with periodate. Oxidation of leonotin with chromium trioxide led to destruction of the furan ring and the formation of 13 → 9 and 13 → 8 lactones. A 13-9 spiro-ether (**114**) reminiscent of premarrubiin, has been isolated from *Leonotis leonorus*[174].

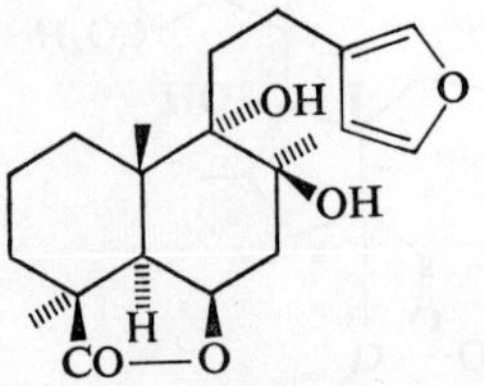

(113) 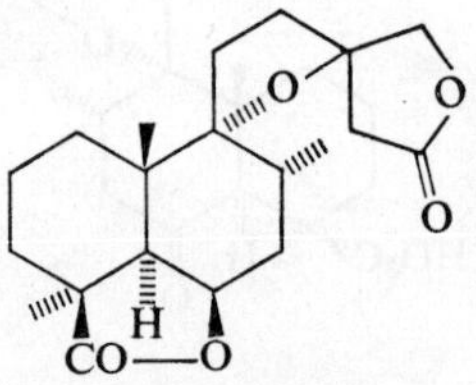   (114)

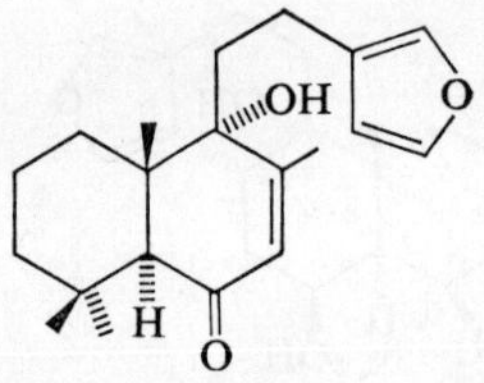

(115) 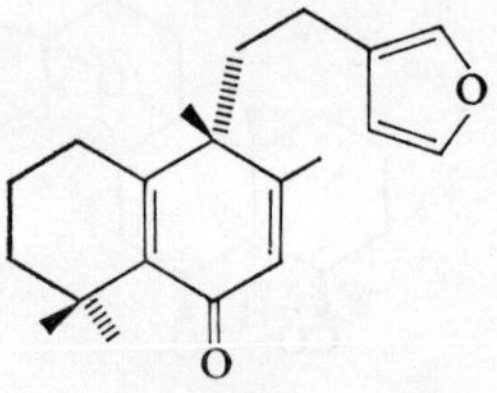   (116)

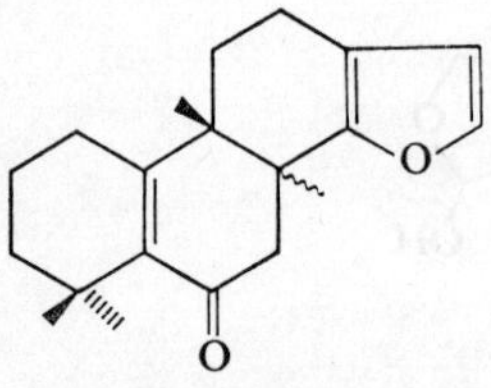

(117) 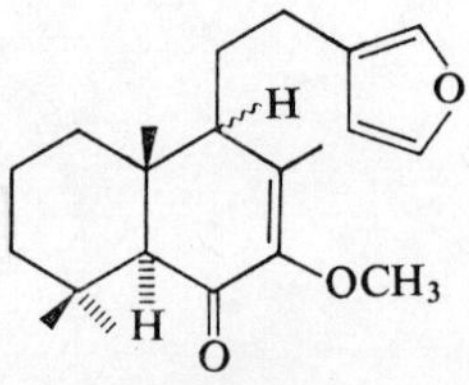   (118)

Solidagenone **(115)** and the related 9-13 ethers have been isolated[175] from *Solidago canadensis*. The structure was assigned on the basis of spectral data and on the hydrogenolysis of the hydroxyl group from the γ-hydroxy $\alpha\beta$-unsaturated ketone. The relationship of the tertiary hydroxyl group, furan ring and $\alpha\beta$-unsaturated ketone led[176] to a number of acid-catalysed rearrangements. Solidagenone first rearranges to give a dienone **(116)** which then cyclizes to **117**. A methoxy ketone **(118)** is a minor product of this rearrangement.

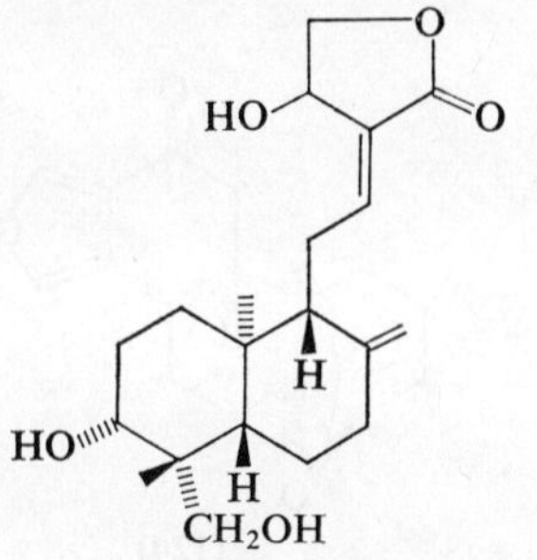

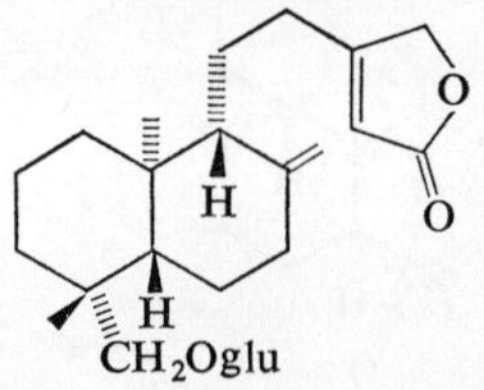

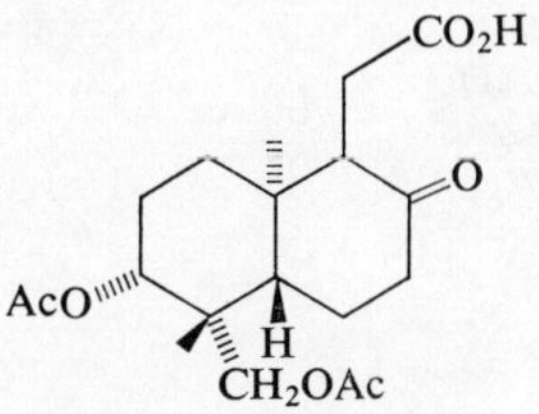

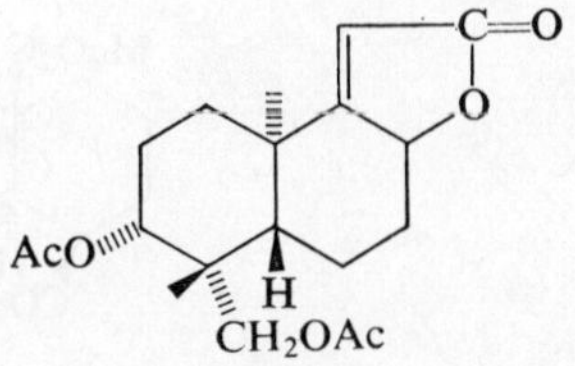

Andrographolide **(119)** and neoandrographolide **(120)** were isolated[177] as the bitter principles of *Andrographis paniculata*. Andrographolide contained three hydroxyl groups one of which was readily dehydrated, an $\alpha\beta$-unsaturated lactone and an exocyclic methylene. The nature of the carbocyclic ring system was demonstrated[178] as follows. Oxidation of triacetyl andrographolide gave a diacetylketo acid **(121)** which was shown by dehydrogenation to a naphthol to contain a bicyclic nucleus. The ready formation of a lactone on reduction of the keto group in this acid, established the relationship between the ketone and the carboxyl group. The further substitution of the bicyclic nucleus was revealed by oxidation of the anhydro-19-monotrityl ether of andrographolide to a ketone which, with acid, underwent a retro-aldol cleavage to give formaldehyde. Since andrographolide contained only two C-methyl groups and yet on dehydrogenation gave 1,2,5,6-tetramethyl-naphthalene, the secondary alcohol β to the primary alcohol was placed at position 3 where it might facilitate a Wagner–Meerwein rearrangement during the dehydrogenation. The position of the $\alpha\beta$-unsaturated lactone was established by a study of the n.m.r. spectrum. Furthermore one of the acetyl groups of triacetylandrographolide underwent ready hydrogenolysis indicative of its allylic position. The absolute stereochemistry of the molecule followed from its optical rotatory dispersion curve. The structure and stereochemistry of rings A and B have been confirmed[179] by the total synthesis of the fragment (122). This fragment had been obtained previously by ozonolysis of triacetylandrographolide followed by treatment with acetyl chloride.

Neoandrographolide is the glucoside of an aglycone (120). Spectral data indicated[180] that two of the oxygen atoms were present in a butenolide and the third in an axial primary alcohol to which the sugar residue was attached. The general similarity of the data to those obtained from andrographolide led to the proposed structure. Investigation of *Andrographis wightiana* has led to the isolation[181] of a compound, wightionolide, which is possibly related to andrographolide.

Sciadoptys verticillata contains methyl sciadopate[182], and a group of furanoid diterpenes including sciadin (123)[183], sciadinone (124) and dimethyl sciadinonate (125). The spectral properties of sciadin revealed the presence of a furan and a lactone ring whilst the inertness of the fourth oxygen atom indicated that it was present in an ether[184]. The underlying carbon skeleton

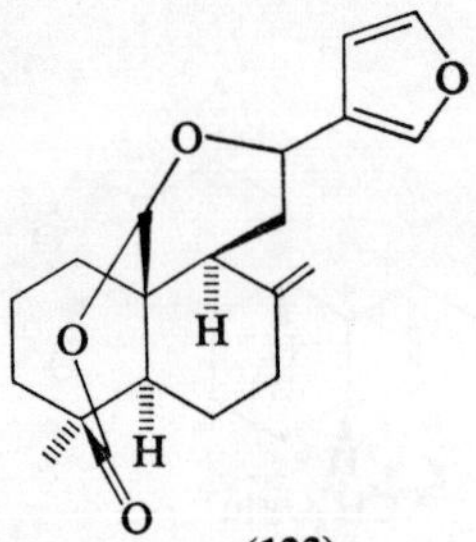

(123)

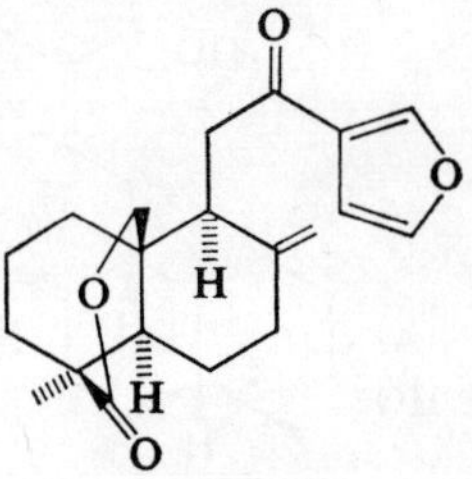

(124)

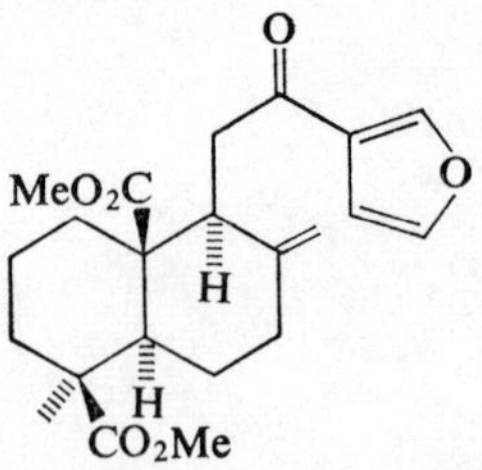

(125)

of sciadin was established by dehydrogenation experiments which afforded 1,2,5-trimethylnaphthalene. Hydrolysis of sciadin gave sciadinic acid which readily relactonized. Ozonolysis of sciadin gave formaldehyde from the terminal methylene group and a keto acid lacking the furan ring which thus formed part of a terminal function. Reduction of sciadin with lithium aluminium hydride gave a triol which was then oxidized to a δ-lactone containing a carbonyl group conjugated with the furan ring. Similarly oxidation of sciadinic acid gave a lactol which also contained this furano-ketone. Hydrolysis and methylation of the lactol gave an aldehyde ester which was in turn related to dimethyl sciadinonate. The *trans* A/B stereochemistry of the latter was established[185] by lactonization onto C-8 leading to the structures of these diterpenes.

Two closely related $\alpha\beta$-unsaturated γ-lactones, the α- and β-levanteno-lides (**126** and **127**) have been isolated[186] from Turkish tobacco. On reduction with lithium aluminium hydride, they both gave the same triol. Saponification of either lactone followed by relactonization, gave predominantly α-levantenolide, a relationship which led to the conclusion that the compounds were the epimeric lactones of a hemi-ketal. Ozonolysis of α-levantenolide gave oxalic acid and the known lactone (**128**). Bearing in mind the presence of the unsaturated lactone, this led to the structures for the

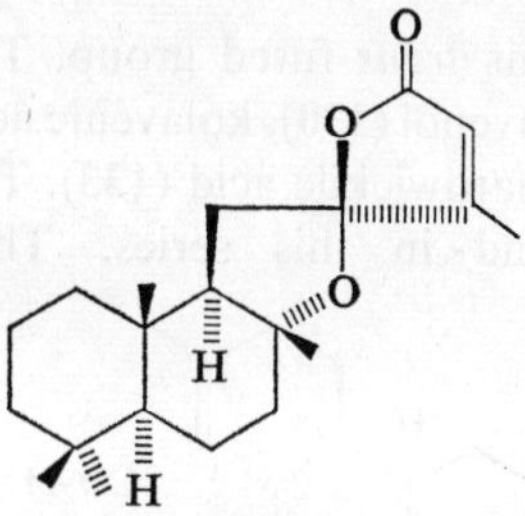

(126)

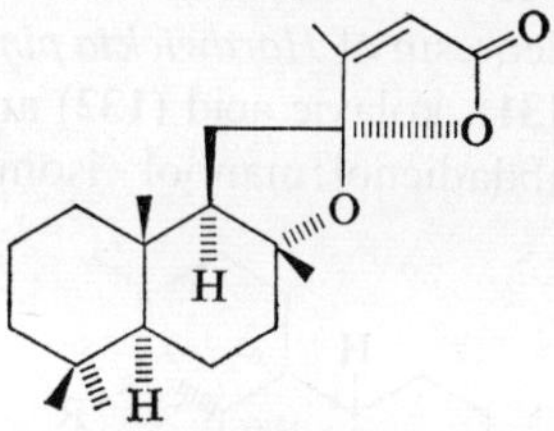

(127)

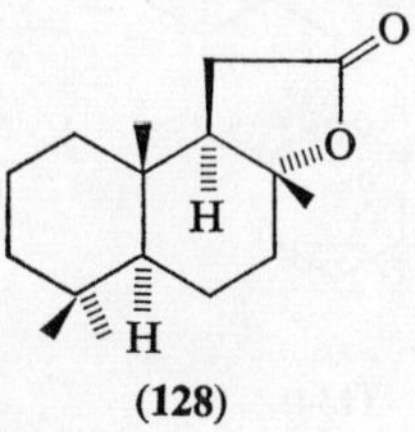

(128)

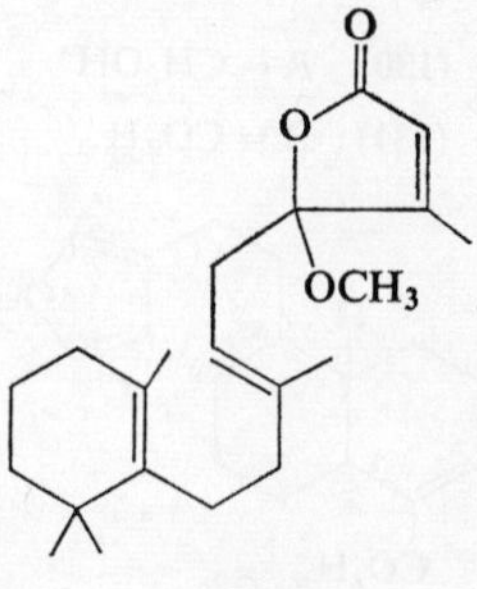

(129)

levantenolides. The more stable α-levantenolide was assigned the structure
126 in which the C-8 and C-13 methyl interactions were minimized. The
levantenolides have been synthesized[187] by two closely related routes. The
first cyclization of the butenolide (**129**) derived from monocyclofarnesyl
bromide afforded the lactones. In the second cyclization the corresponding
acyclic butenolide derived from farnesyl bromide was cyclized with stannic
chloride.

VI. THE REARRANGED LABDANES

A. *Trans* A/B Fused Compounds

There are an increasing number of bicyclic diterpenes that have been found
to possess a rearranged carbon skeleton of the clerodane type. In the majority of cases the C-8 methylene is protonated and possibly a C-9 → C-8
hydride shift occurs inducing a further backbone rearrangement in which the
C-10 methyl group migrates to C-9, the C-5 hydrogen migrates to C-10 and
a methyl group migrates from C-4 to C-5. The carbonium ion is then dis-
charged by olefin formation on ring A. This must represent an overall simpli-
fication because both *cis* and *trans* A/B ring junctions occur naturally. In
this review we shall treat both groups separately. An interesting feature is
the similarity of oxygenation pattern between the normal and the rearranged
labdanes.

Kolavenol (**130**) represents the parent of this *trans* fused group. The
oleoresin of *Hardwickia pinnata* contains[188] kolavenol (**130**), kolavenic acid
(**131**), kolavic acid (**132**) and a furanoid acid, hardwickiic acid (**133**). The
labdadienol:manöol isomerism is also found in this series. Thus

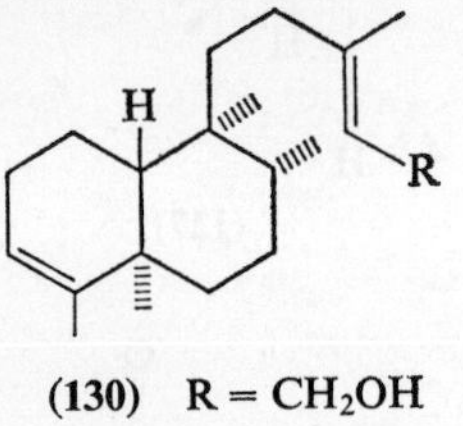

(130)   R = CH$_2$OH

(131)   R = CO$_2$H

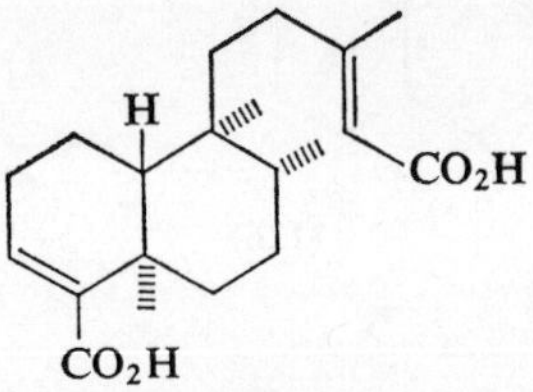

(132)

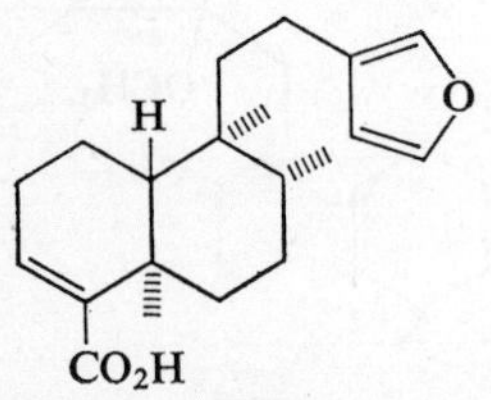

(133)

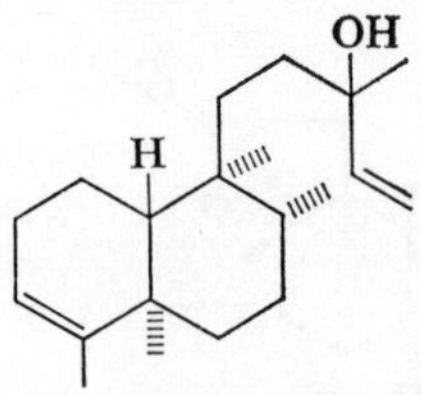

(134)

kolavelöol **(134)** was also isolated from this resin. The enantiomer of hardwickiic acid has been isolated[189] from *Copaifera officinalis*. The structure of kolavenol is based on dehydrogenation experiments and spectral measurements. Further evidence[190] comes from a relationship with hardwickiic acid **(133)**. The Wolff–Kishner reduction of 13,14-epoxykolavenol afforded kolavelöol (and presumably) its 13-epimer.

Solidago species have been the source of a number of diterpenes. *Solidago elongata* contains[191] methyl kolavenate, its 6β-acetate **(135)** and 6β-angelate together with kolavenol, kolavelöol and its 6β-angelate. The position of the hydroxyl group at C-6 in these diterpenes was established through oxidative studies. Since it was not possible to generate a conjugated system involving the olefin and the carbonyl group arising from oxidation of the hydroxyl group, these functions must occupy different rings. Spectral data led to their location at Δ^4 and C-6. *Solidago elongata* also contains an interesting group of $\alpha\beta$-unsaturated butenolides, the elongatolides, whose structures **(136)** were established mainly by spectroscopic studies. Solidagonic acid **(137)**

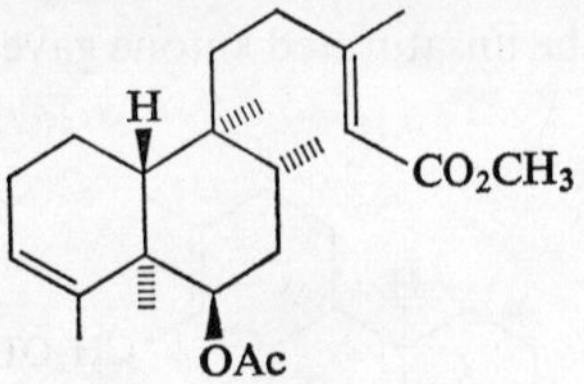

CO_2CH_3

OAc

(135)

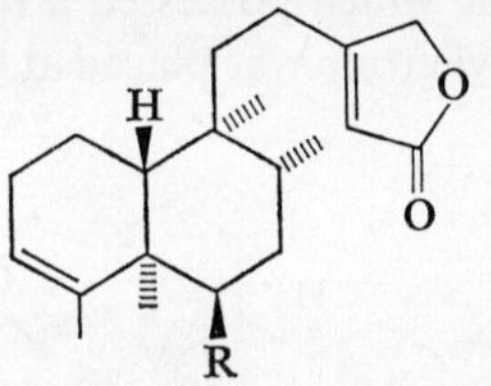

R

(136) R = OH, OAc, OAng

CO_2H

OAc

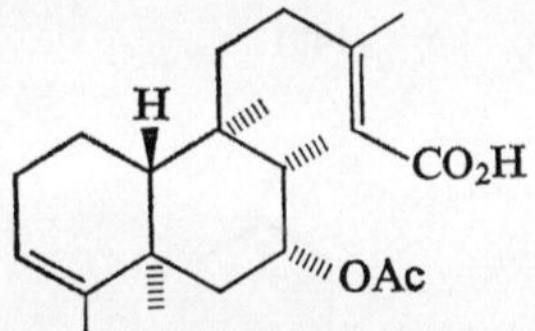

(137)

has been isolated[192] from the root of *S. altissima* and it has been assigned the structure of 7α-acetoxykolavenic acid. The n.m.r. spectrum revealed the presence of two tertiary methyl groups, a secondary methyl group and a methyl group attached to an olefin. The position of the nuclear functional groups was determined by methyl labelling. Thus dehydrogenation of the methyl ester of solidagonic acid gave 1,2,5-trimethylnaphthalene. Conversion of the olefin to an epoxide and reaction with methyl magnesium iodide, gave a triol which on dehydrogenation afforded 1,2,5,6-tetramethylnaphthalene thus locating the nuclear double bond. Hydrolysis of the acetate and oxidation of the resultant hydroxy acid, gave a keto acid which also reacted with methyl magnesium iodide. Dehydrogenation of this product afforded 1,2,3,5-tetramethylnaphthalene. The presence of an $\alpha\beta$-unsaturated acid in the side-chain was supported by spectral considerations and by its reduction to an allylic alcohol. Hence the overall structure of 7α-acetoxykolavenic acid was proposed for solidagonic acid.

The widespread occurrence of the clerodane skeleton is exemplified by the isolation[193] of (−)-hardwickiic acid, the C-15 monomethyl ester of kolavenic acid and two furanoid substances, agbaninol (**138**) and agbanindiol B (**139**) from *Gossweilerodendron balsamiferum*. Agbanindiol A, which was isolated from the same source, was assigned the structure **140**. These compounds have been correlated with kolavenic acid and hardwickiic acid.

The initial evidence for the structure of hardwickiic acid was based on spectral and dehydrogenation studies. Further evidence came from ozonolysis experiments in which two products were formed (**141** and **142**). Bromination and dehydrobromination of **142** gave an $\alpha\beta$-unsaturated ketone which possessed a fully substituted allylic carbon atom and thus a methyl group was placed at C-9. Oxidation of the unsaturated ketone gave a

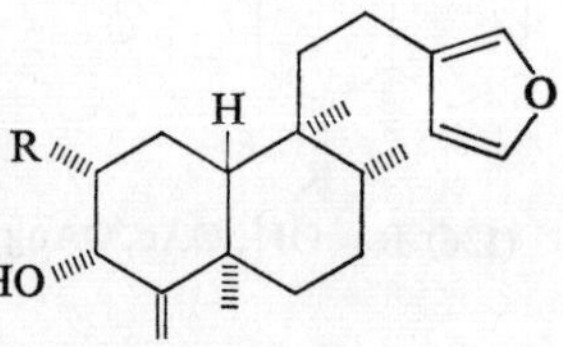

(**138**) R = H

(**139**) R = OH

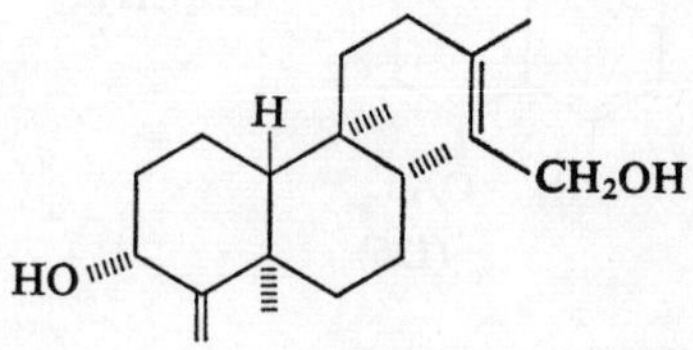

(**140**)

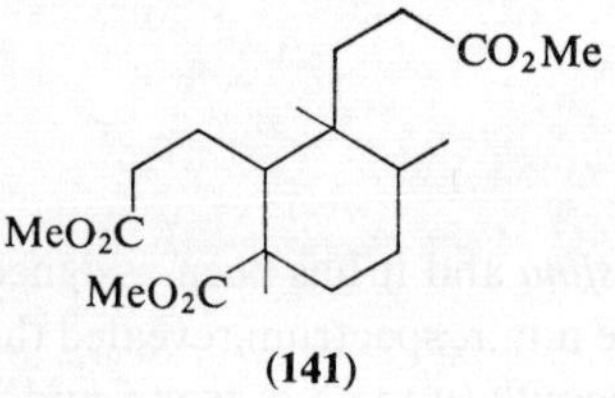

(**141**)

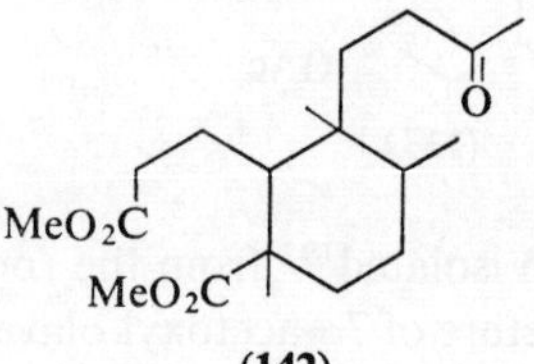

(**142**)

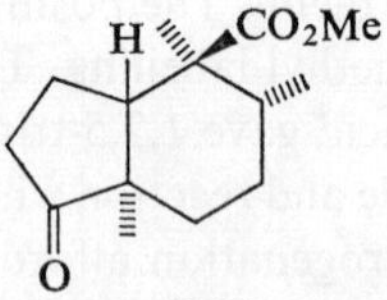

(**143**)

monocarboxylic acid. Hydrolysis of the primary ester derived from ring A then gave a dicarboxylic acid. These carboxyl groups were converted to a cyclopentanone thus establishing a relationship between the double bond in ring A and the side-chain. Dieckmann cyclization of the triester gave a cyclopentanone from ring A. The circular dichroism of this was consistent with the configuration **133** for hardwickiic acid. Conversion of this cyclopentanone to **143** gave a compound, the n.m.r. spectrum of which afforded information on the stereochemistry at C-9.

Dodonaea attenuata, an Australian species, contains[194] an acetoxy hydroxy acid **(144)** which possesses a carbon skeleton of the rearranged

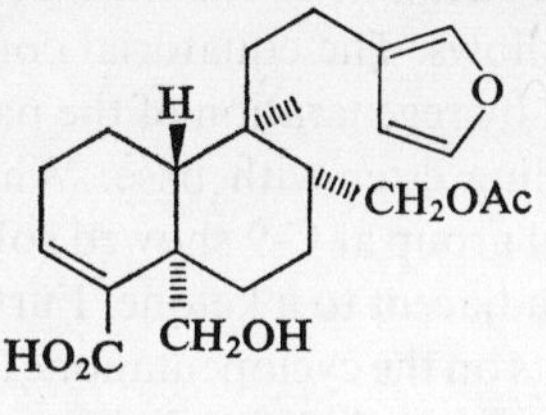

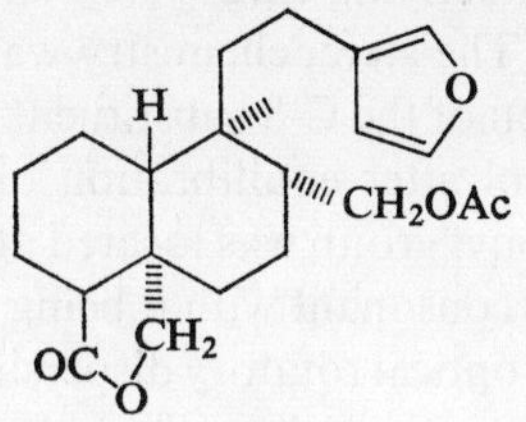

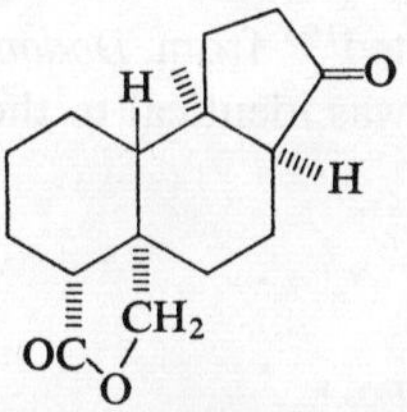

labdane type. The n.m.r. spectrum of the acid revealed a furan ring, an $\alpha\beta$-unsaturated carbonyl group a tertiary methyl and an acetate group. The acid could be converted into an $\alpha\beta$-unsaturated γ-lactone thus interrelating the carboxyl and hydroxyl functions. Reduction of the hydroxy acid with sodium in ethanol gave a saturated γ-lactone **(145)** and a saturated hydroxy acid. The latter lactonized only after epimerization with base and hence this epimer has the carboxyl and hydroxyl groups in a *trans* diaxial relationship. The relationship between the furan ring and the other oxygen function was demonstrated by ozonolysis of the furan to a tris-nor-carboxylic acid. The acetoxyl group was then hydrolysed and the resultant primary alcohol oxidized to a carboxylic acid. The carboxyl functions of this dicarboxylic acid were linked by anhydride formation. Pyrolysis of this anhydride gave a cyclopentanone thus revealing a four-carbon chain between the acetoxymethyl group and the furan ring. The acetoxymethyl group in the saturated γ-lactone (**145**) was converted through the corresponding aldehyde and enol-acetate to a nor-ketone. Bromination and dehydrobromination of this afforded an $\alpha\beta$-unsaturated ketone. The n.m.r. spectrum of this showed its olefinic protons as an AB quartet and thus the adjacent carbon atoms were fully substituted. Furthermore treatment with alkali led to the retro-aldol loss of formaldehyde thus locating the terminal alcohol of the lactone function.

The acetoxymethyl group could be converted through the corresponding alcohol and chloride to a methyl group. Dehydrogenation of the corresponding saturated lactone led to 1,2-dimethylnaphthalene thus indicating the underlying ring system and hence the constitution of the natural product. The stereochemistry was assigned as follows. The equatorial conformation of the C-8 substituent was established by regeneration of the parent alcohol after equilibration of its corresponding ester with base. When a carbonyl group was located at C-8, the methyl group at C-9 showed solvent shifts consonant with it being an axial group adjacent to a ketone. Furthermore optical rotatory dispersion measurements on the cyclopentanone (**146**) when compared to those from A-nor-lupan-3-one indicated the stereochemistry as in **144**.

Hautriwaic acid was isolated[195] from *Dodonaea viscosa* and also from *D. attenuata* var. *linearis*. It was identical to the lactone (**147**) prepared[194] from the above diterpenes.

(147)   (148)

Olearin was isolated[196] from *Olearia heterocarpa* and has been assigned the structure **148**. The spectral data showed that it contained two $\alpha\beta$-unsaturated lactones. After hydrogenation, these gave infrared absorption characteristic of γ-lactones. The fifth oxygen atom was present as a secondary alcohol. The relationship of this alcohol to one of the lactone rings was established by a retro-aldol cleavage of the corresponding ketone, when formaldehyde was lost. Tetrahydroolearin underwent an isomerization to a different γ-lactone in which a primary alcohol was generated and the secondary alcohol was blocked by lactone formation. The alcohol may be eliminated through its methane sulphonate with the formation of a *trans* double bond. The multiplicity of the olefinic protons in the n.m.r. spectrum indicated that the adjacent carbon atoms were fully substituted. The side-chain of the butenolide of dibromoolearin can be oxidatively removed. The C_{16}-acid then undergoes a fragmentation reaction in alkali with the elimination of bromine and formaldehyde and the formation of a dione which was readily oxidized to a dicarboxylic acid. In support of the rearranged labdane nucleus implied by this structure, dehydrogenation gave 1,2-dimethylnaphthalene. Thus olearin was assigned the structure **148**.

The bitter principle, clerodin (**149**), isolated from *Clerodendron infortunatum*, was the subject of two complementary studies, one using the X-ray method[197] and the other a classical chemical degradation[198] coupled with spectroscopic techniques. Clerodin was converted to a bromolactone which was shown by X-ray analysis to have the structure **150**. Analytical and spectroscopic data revealed the presence of two acetoxyl groups, one present as a CHOAc and the other as $\geq$CCH$_2$OAc, two *C*-methyl groups, one secondary and one tertiary, and a vinyl ether. Addition of acetic acid to the vinyl ether followed by hydrolysis gave a hemi-acetal. This also co-occurs with clerodin. Oxidation of the hemi-acetal gave a γ-lactone. On treatment with methanolic ammonia, this formed an amide which readily recyclized to a lactam. The substitution pattern of the vinyl ether was established by n.m.r. spectroscopy in one of the earlier applications of this technique to the diterpene area. Reduction of clerodin with lithium aluminium hydride generated a tertiary hydroxyl group and a new *C*-methyl group and hence the seventh oxygen atom was present as a terminal epoxide. The relationship of the other nuclear oxygen atoms was shown by the formation of a cyclic *cis* 1:3 carbonate from bisdeacetyl dihydroclerodin.

Cascarillin (**151**) from Cascarilla bark (*Croton eleuteria*) has been known for many years[199]. It is a furanoid diterpene acetate. Its spectral properties showed that the remaining oxygen atoms may be accounted for in two secondary hydroxyl groups, one tertiary hydroxyl group on a carbon atom bearing a methyl group and a tertiary aldehyde. The latter readily participated in hemi-acetal formation with the secondary hydroxyl group adjacent to the furan ring. The acetate group in cascarillin is readily hydrolysed and the free hydroxyl group then participates in acetal formation with the aldehyde. The remaining secondary hydroxyl group may be oxidized to a ketone. X-ray analysis of the iodoacetate showed[200] that cascarillin had the structure **151**.

Crotonin (**152**) is a nor-diterpene which has been isolated[201] from a related species, *C. lucidus*, and has a similar structure. The oxygen functions are accounted for in a γ-lactone, furan ring and a carbonyl group. Since

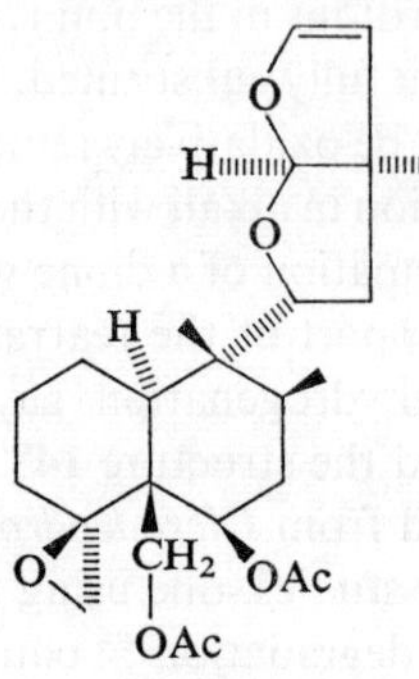

(149)

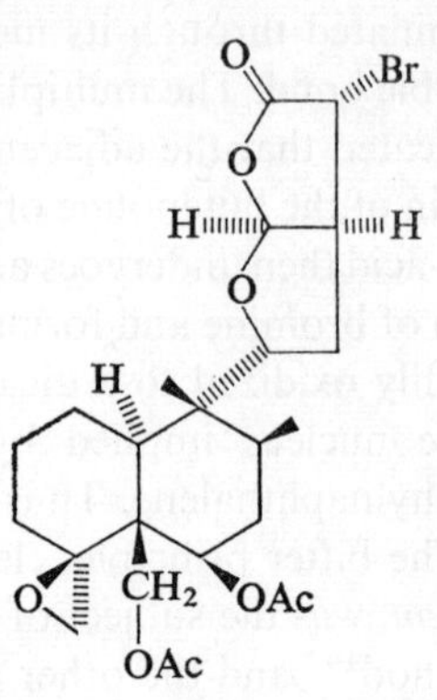

(150)

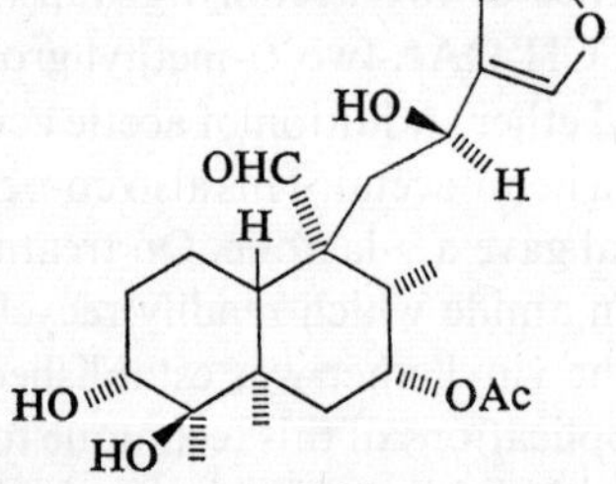

(151)

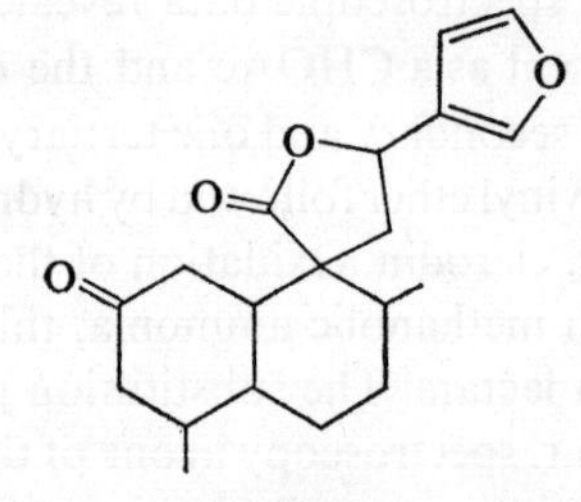

(152)

hydrogenation led not only to saturation of the furan ring but also to hydrogenolysis of the lactone, this must be attached to the allylic position of the furan ring. The lactone carbonyl was converted to an aldehyde in which the aldehydic proton appeared as a singlet and was therefore attached to a tertiary centre. The location of the carbonyl group at C-2 was demonstrated by aromatization of ring A with selenium dioxide to a phenol. This could be oxidized at C-6 to a ketone showing a u.v. characteristic of a *p*-methoxyacetophenone.

B.   *Cis* A/B Fused Compounds

The *cis* fused clerodanes contain an interesting group of diterpenoid bitter principles. However the simplest compound possessing this carbon skeleton is plathyterpol. This alcohol, which was isolated[202] from *Plathymenia reticulata*, has been formulated as **153**. The spectral properties of plathyterpol revealed the presence of a vinyl group and a trisubstituted double bond. The side-chain vinyl group and allylic hydroxyl group were cleaved with the formation of a bisnor-ketone. X-ray analysis[203] and optical rotatory dispersion measurements on the dibromo derivative of this nor-ketone defined the structure and absolute stereochemistry of plathyterpol as **153**.

Cistodiol (**154**) and cistodioic acid (**155**) have been isolated[204] from *Cistus monspeliensis* and assigned the *cis* fused clerodane structure. It is interesting to note that they co-occur with $8\alpha,15$-labdanediol.

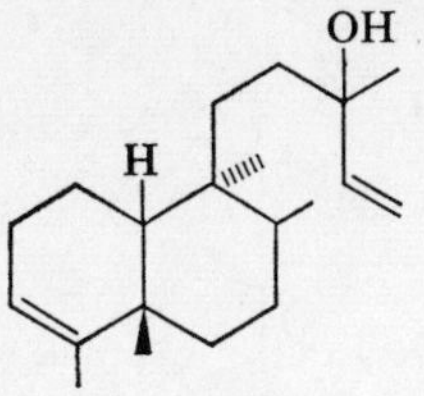

(153)

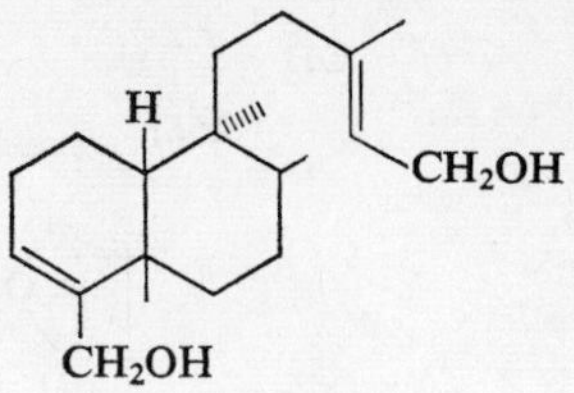

(154)

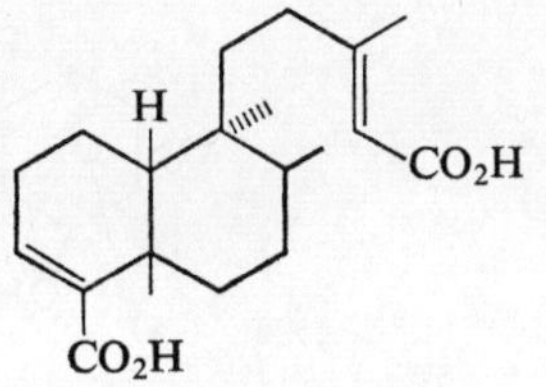

(155)

Columbin (**156**), chasmanthin (**157**), palmarin (**158**) and isojateorin (**159**) form the bitter principles of the Colombo root (*Jateorrhiza palmata*)[205]. The structure of columbin was the first to be elucidated[206]. Hydrogenation although not a simple reaction of columbin, revealed[207, 208] the presence of three double bonds and one hydrogenolysable lactone. The oxygen functions were located in one hindered hydroxyl group and two lactone rings whilst the sixth oxygen atom formed part of a furan ring. Dehydrogenation with zinc dust gave 1,2,5-trimethylnaphthalene, characteristic of the bicyclic diterpenes. 1-Methylnaphthalene and 1,5-dimethylnaphthoic acid were also formed. On heating, columbin readily decarboxylated, generating in the process a ketone, decarboxycolumbin (**160**). However the dihydro compound was resistant to decarboxylation and hence one double bond, was placed in the $\beta\gamma$-position to the δ-lactone ring. Further oxidative evidence also associated the hydroxyl group with the lactone ring. Hydrogenation of decarboxycolumbin gave decarboxyoctahydrocolumbinic acid (**161**) by hydrogenation and hydrogenolysis. This, on further perchloric acid hydrogenation followed by dehydrogenation, gave 1,5-dimethyl-2-naphthoic acid. However Wolff–Kishner reduction of the

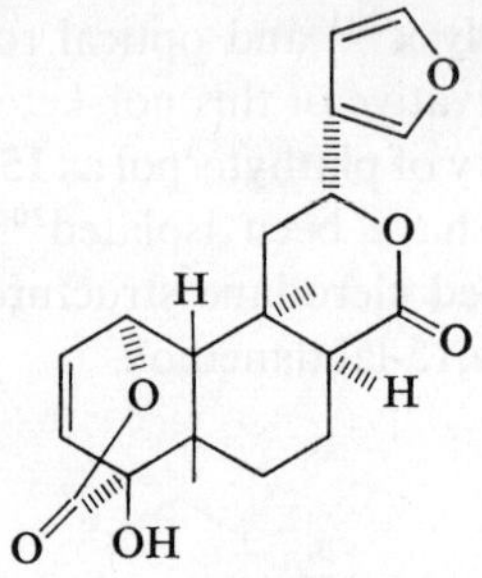

(156)

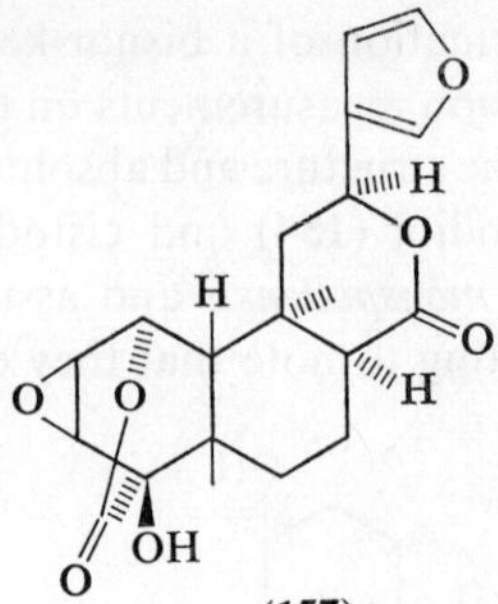

(157)

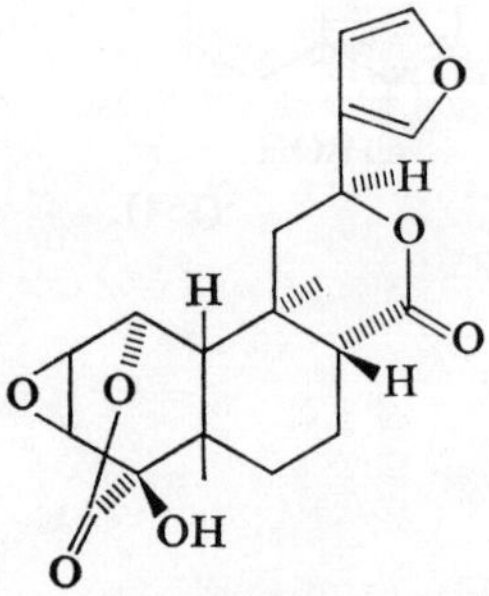

(158)

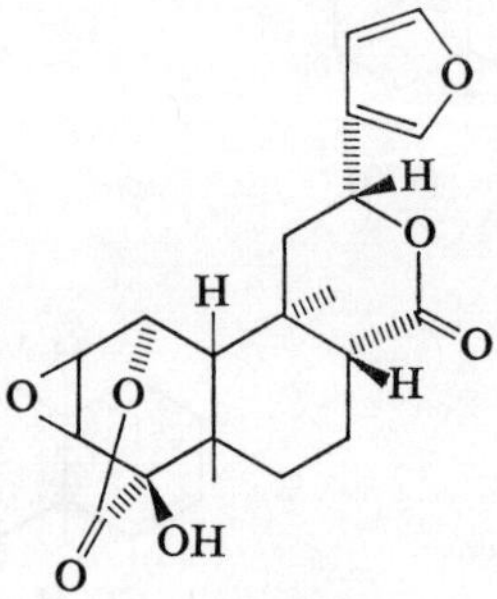

(159)

ketone and subsequent dehydrogenation gave only 1-methyl-2-naphthoic acid. Hence a tertiary methyl group was placed adjacent to the carbonyl group to account for a Wagner–Meerwein rearrangement during dehydrogenation. Combining these points, the nuclear substitution pattern of columbin becomes apparent. The hydrogenolysis of the second lactone ring suggested that it was attached to the allylic position of the furan ring thus leading to the overall structure **156** for columbin.

On treatment with alkali, columbin gave isocolumbin, an isomerization which was interpreted as the isomerization of a *cis* fused lactone to a *trans* fused lactone. The chemical evidence[208] for the stereochemistry of columbin rests on a comparison of the reactions of these substances. Octahydro-isocolumbinic acid was obtained from isocolumbin by hydrogenation of the nuclear double bond, the furan ring and hydrogenolysis of the side-chain lactone ring. On treatment with base, it gave an isomer which was formulated as **162**. On the other hand octahydrocolumbic acid showed no evidence for any interaction between C-17 and C-1. The formation of a bridged structure requires rings A and B to be *cis* fused. Furthermore the two substituents which interact to form the new lactone must be on the same side of the *cis* decalin whilst the methyl group at C-5 must be on the opposite face of the

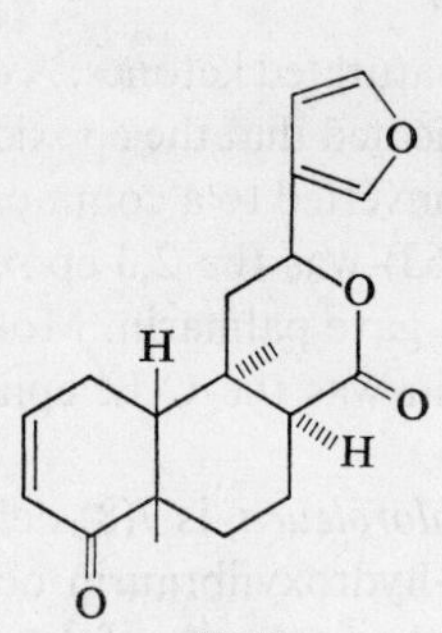

(160)

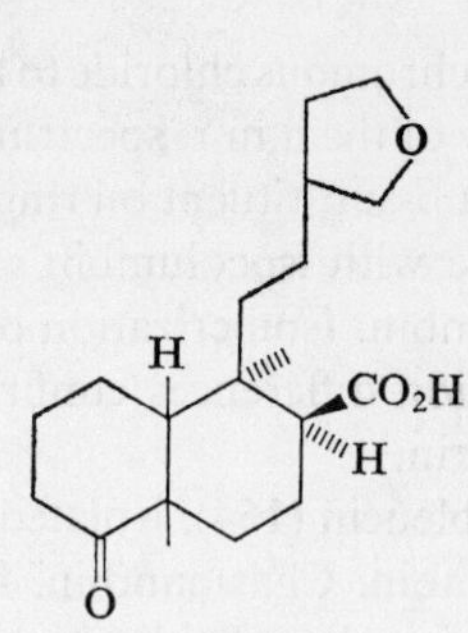

(161)

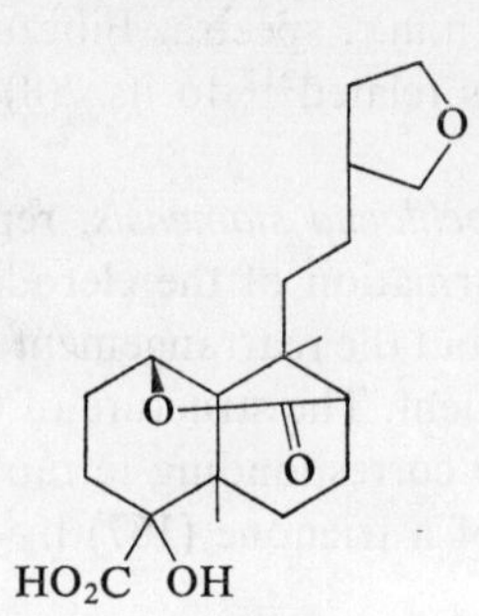

(162)

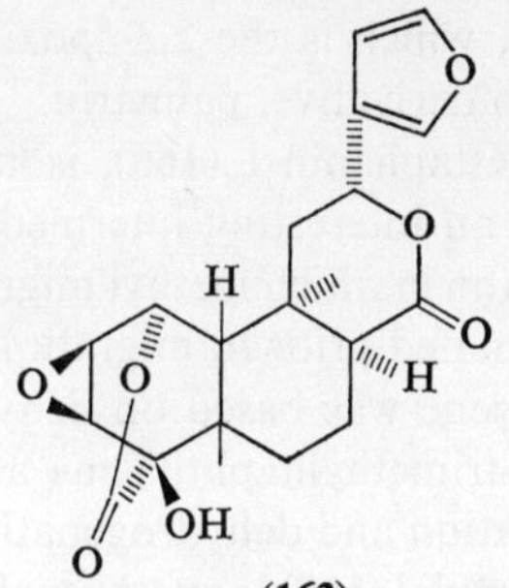

(163)

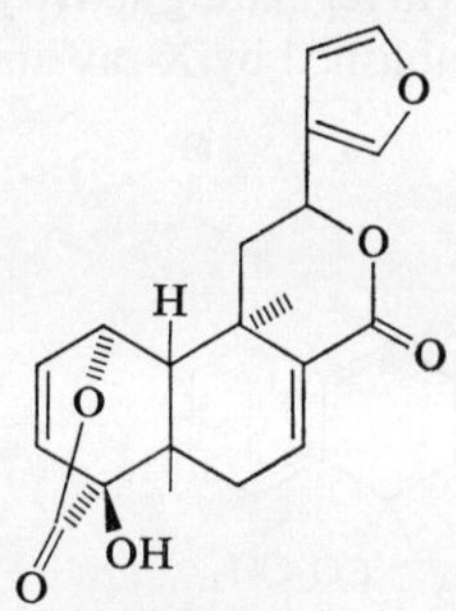

(164)

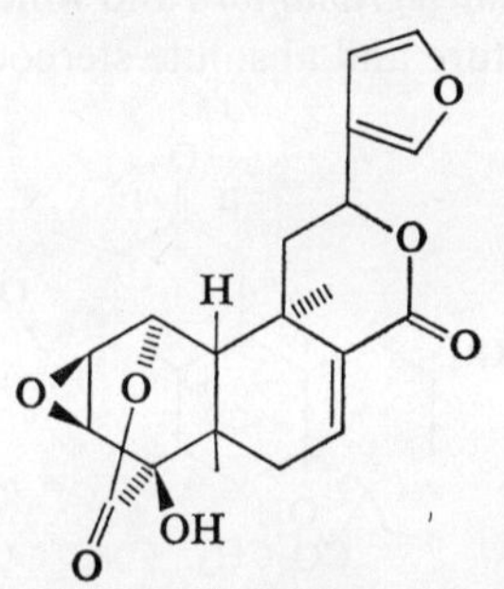

(165)

molecule. Optical rotatory dispersion measurements on a number of C-4 ketones clarified the absolute stereochemistry of this fragment. Application of the method of molecular rotation differences to the hydrolysis of the side-chain lactone indicated the C-12 stereochemistry. The overall structure for columbin has been confirmed[209] by the X-ray analysis of a derivative of isocolumbin.

Palmarin[210] and isojateorin were both converted to C-4 ketones lacking the ring A lactone. These were shown to be $\alpha\beta$-epoxyketones by reduction

10

with chromous chloride to the corresponding unsaturated ketones. A careful study of the n.m.r. spectrum of the epoxides indicated that the epoxide ring was a β-substituent on ring A. Isojateorin was converted to a common derivative with isocolumbin and hence jateorin (**163**) was the 2,3-epoxide of columbin. Isomerization of chasmanthin at C-8 gave palmarin. Molecular rotation differences confirmed that chasmanthin was the C-12 epimer of jateorin.

Fibleucin (**164**), isolated[211] from *Fibraurea chloroleuca*, is 7(8)-dehydrocolumbin. Chasmanthin, fibraurin (**165**) and 6-hydroxyfibraurin occur in the same plant. With a high level of oxygenation, the structures of these compounds were derived from an analysis of their n.m.r. spectra. Fibraurin (**165**), which is the 2,3-epoxide of fibleucin, was related[212] to its 7(8)-dihydro derivative, palmarin.

Chettaphanin-1 (**166**), isolated[213] from *Adenochlaena siamensis*, represents an interesting intermediate stage in the formation of the clerodane skeleton in which methyl migration has occurred but the rearrangement has terminated prior to migration of the C-4 substituent. The structure of this diterpene was based on dehydrogenation of the corresponding tetraol to 1,2,5-trimethylnaphthalene and the formation of a trienone (**167**) by dehydration and dehydrogenation.

Portulal (**168**) is an unusual bicyclic diterpene which was isolated[214] from *Portulaca grandiflora* and which shows plant growth regulating activity. The structure and absolute stereochemistry were established by X-ray analysis

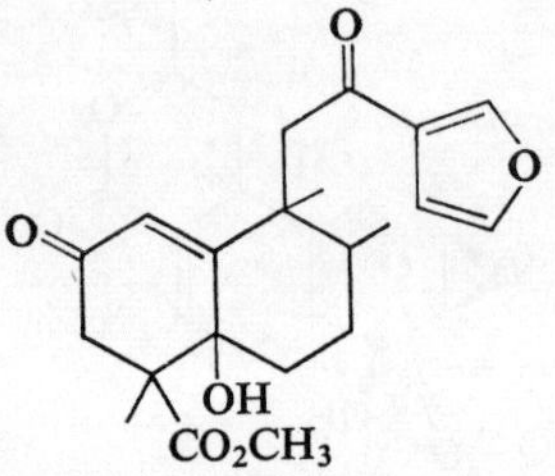

(166)

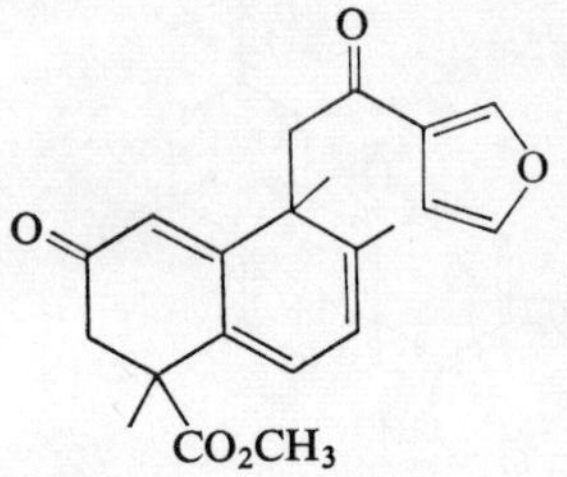

(167)

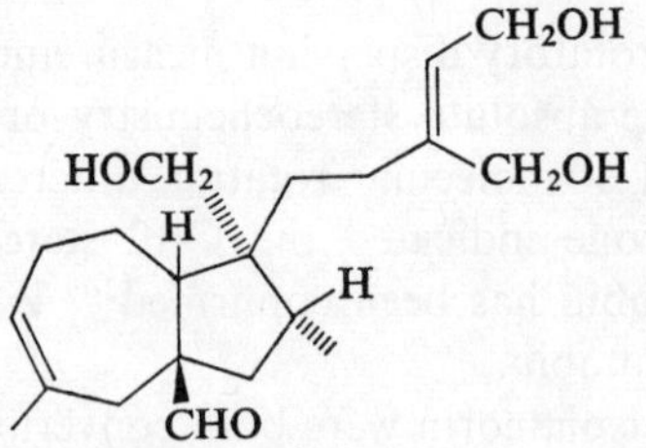

(168)

of its *p*-bromosulphonylhydrazone. The biogenesis of this compound may proceed through a labdadienol and could involve the collapse of a C-4,5,19 cyclopropane ring initiated by a C-3 leaving group. This would permit a synchronous collapse of ring B to a five-membered ring with the extrusion of an angular substituent.

VII. REFERENCES

1. For previous reviews see: Sir J. Simonsen and D. H. R. Barton, *The Terpenes*, Vol. 3, Cambridge University Press, London, 1951; R. McCrindle and K. H. Overton, The diterpenoids, sesterterpenoids and triterpenoids' in *Chemistry of Carbon Compounds*, Vol. 2C (Ed. S. Coffey), Elsevier, London, 1969.
2. For reviews of diterpenoid biosynthesis see: R. B. Clayton, *Quart. Rev. (London)*, **19**, 168, 201 (1965); B. Achilladelis and J. R. Hanson, *Perfum. Essent. Oil Record.*, 802 (1968).
3. W. Klyne, *J. Chem. Soc.*, 3072 (1953).
4. J. W. Rowe, *The Common and Systematic Nomenclature of Cyclic Diterpenes*, I.U.P.A.C., in press.
5. G. Ohloff, *Ann. Chem.*, **617**, 314 (1958).
6. D. R. Robinson and C. A. West, *Biochemistry*, **9**, 80 (1970) and refs. therein.
7. B. Achilladelis and J. R. Hanson, *Phytochemistry*, **7**, 589 (1968).
8. J. R. Hanson and A. F. White, *J. Chem. Soc. C*, 981 (1969).
9. J. W. Rowe and J. H. Scroggins, *J. Org. Chem.*, **29**, 1554 (1964); J. W. Rowe and G. W. Schaeffer, *Tetrahedron Letters*, 2633 (1965); cf. C. A. Henrick and P. R. Jefferies, *Tetrahedron*, **21**, 1175 (1965) for a comment on the n.m.r. data.
10. A. I. Lisina, N. K. Kashtanova, A. K. Dzizenko and V. A. Pentegova, *Chem. Abstr.*, **68**, 290 (1968).
11. L. Mangoni and M. Belardini, *Gazetta*, **94**, 1108 (1964).
12. L. Mangoni and R. Caputo, *Gazetta*, **97**, 908 (1967).
13. L. A. Salei, D. P. Popa and G. V. Lazur'veski, *Chem. Abstr.*, **68**, 3880 (1968).
14. H. Matsuda, Y. Tomiie, S. Yamamura and Y. Hirata, *Chem. Commun.*, 898 (1967).
15. M. Sumimoto, *Tetrahedron*, **19**, 643 (1963).
16. M. Sumimoto, Y. Tanaka and K. Matsufuji, *Chem. Ind. (London)*, 1928 (1963); *Tetrahedron*, **20**, 1427 (1964).
17. T. Miyasaka, *Chem. Pharm. Bull.*, **12**, 744 (1964).
18. A. J. Baker, G. Eglinton, A. G. Gonzalez, R. J. Hamilton and R. A. Raphael, *J. Chem. Soc.*, 4705 (1962).
19. M. Mme C. Tabacik-Wlotzka, M. Mousseron and Mme A. Chafai, *Bull. Soc. Chim. France*, 2299 (1963).
20. G. Chandra, J. Clark, J. McLean, P. L. Pauson and J. Watson, *J. Chem. Soc.*, 3684 (1964).
21. K. Bruns, G. Weissman and H. Fr. Grutzmacher, *Tetrahedron Letters*, 4623 (1965); 1901 (1966); *Naturwissenschaften*, **52**, 185 (1965).
22. K. Bruns, *Tetrahedron*, **24**, 3417 (1968); *Tetrahedron Letters*, 3263 (1970).
23. C. A. Henrick and P. R. Jefferies, *Tetrahedron*, **21**, 3219 (1965).
24. R. M. Carman and P. K. Grant, *J. Chem. Soc.*, 2187 (1961).
25. R. M. Carman and N. Dennis, *Australian J. Chem.*, **20**, 157 (1967).
26. G. V. Pigulevski, V. G. Kostenko and L. D. Kostenko, *Zhur. Obshch. Khim.*, **31**, 3143 (1961); **30**, 1057 (1960).

27. P. S. Gray and J. S. Mills, *J. Chem. Soc.*, 5822 (1964).
28. R. M. Carman, *Australian J. Chem.*, **19**, 629 (1966).
29. R. M. Carman and N. Dennis, *Australian J. Chem.*, **21**, 823 (1968).
30. H. Wienhaus and K. Mucke, *Chem. Ber.*, **75B**, 1830 (1942).
31. R. M. Carman, *Australian J. Chem.*, **19**, 1535 (1966).
32. J. S. Mills, *J. Chem. Soc. C*, 2514 (1967).
33. M. A. Chirkova, A. E. Gorbunova, A. I. Lisina and V. A. Pentegova, *Chem. Abstr.*, **65**, 13772 (1966).
34. V. P. Arya, C. Enzell, H. Erdtmann and T. Kubota, *Acta Chem. Scand.*, **15**, 225 1303, 1313 (1961).
35. N. Narasimhachan and E. von Rudloff, *Canad. J. Chem.*, **39**, 2572 (1961).
36. L. J. Gough and J. S. Mills, *Phytochemistry*, **9**, 1093 (1970).
37. P. W. Atkinson and W. D. Crow, *Tetrahedron*, **26**, 1935 (1970).
38. R. M. Carman and N. Dennis, *Australian J. Chem.*, **17**, 390 (1964).
39. T. Norin, *Acta Chem. Scand.*, **19**, 1020 (1965).
40. R. M. Carman, D. E. Cowley and R. A. Marty, *Australian J. Chem.*, **22**, 1681 (1969).
41. N. M. Joye, E. M. Roberts, R. V. Lawrence, L. J. Gough, M. D. Soffer and O. Korman, *J. Org. Chem.*, **30**, 429 (1965).
42. C. W. L. Bevan, D. E. U. Ekong and J. I. Okugon, *Chem. Commun.*, 44 (1966); *J. Chem. Soc. C*, 1063 (1968).
43. G. Hugel, L. Lods, J. M. Mellor, D. W. Theobald and G. Ourisson, *Bull. Soc. Chim. France*, 2882 (1965); G. Hugel and G. Ourisson, *Bull. Soc. Chim. France*, 2903 (1965).
44. L. Ruzicka, R. Steiger and H. Schinz, *Helv. Chim. Acta*, **9**, 962 (1926); L. Ruzicka and J. R. Hosking, *Ann. Chem.*, **469**, 147 (1929); For a recent study of the resin see B. R. Thomas, *Acta Chem. Scand.*, **20**, 1074 (1966).
45. R. M. Carman, *Australian J. Chem.*, **17**, 393 (1964); **19**, 2403 (1966).
46. L. Ruzicka and E. Rey, *Helv. Chim. Acta*, **26**, 2136 (1943); G. Buchi and J. J. Pappas, *J. Am. Chem. Soc.*, **56**, 2963 (1954).
47. L. Ruzicka and J. R. Hosking, *Helv. Chim. Acta*, **13**, 1402 (1930).
48. L. Ruzicka, E. Bernold and A. Tallichet, *Helv. Chim. Acta*, **24**, 223 (1941).
49. R. M. Carman and D. E. Cowley, *Tetrahedron Letters*, 2723 (1968).
50. S. Bory, M. Fetizon and P. Laszlo, *Bull. Soc. Chim. France*, 2310 (1963).
51. L. Ruzicka and J. R. Hosking, *Helv. Chim. Acta*, **14**, 203 (1931).
52. L. Ruzicka and H. Jacobs, *Rec. Trav. Chim.*, **57**, 509 (1938).
53. L. Ruzicka, R. Zwicky and O. Jeger, *Helv. Chim. Acta*, **31**, 2143 (1948).
54. L. Ruzicka and E. Bernold, *Helv. Chim. Acta*, **24**, 931 (1941).
55. D. H. R. Barton and G. Schmiedler, *J. Chem. Soc.*, 1197 (1948); S 232 (1949).
56. W. Klyne, *J. Chem. Soc.*, 3072 (1953).
57. R. A. Marty and R. M. Carman, *Australian J. Chem.*, **22**, 491 (1969); V. P. Arya, H. Erdtmann and T. Kubota, *Tetrahedron*, **14**, 255 (1961).
58. L. Ruzicka and E. Bernold, *Helv. Chim. Acta*, **24**, 1167 (1941).
59. C. Enzell, *Acta Chem. Scand.*, **15**, 1303 (1961).
60. R. M. Carman and R. A. Marty, *Australian J. Chem.*, **21**, 1923 (1968).
61. T. Nakano and C. Djerassi, *J. Org. Chem.*, **26**, 167 (1961).
62. N. V. Kalman, *J. Am. Chem. Soc.*, **60**, 1423 (1938).
63. F. W. Grant and H. H. Zeiss, *J. Am. Chem. Soc.*, **76**, 5001 (1954); **79**, 1201 (1957).
64. J. D. Cocker, T. G. Halsall and A. Bowers, *J. Chem. Soc.*, 4259 (1956).
65. C. A. Henrick, P. R. Jefferies and R. S. Rosich, *Tetrahedron Letters*, 3475 (1964).

66. J. D. Cocker and T. G. Halsall, *J. Chem. Soc.*, 4262 (1956); 4401 (1957).
67. C. Djerassi and D. Marshall, *Tetrahedron*, **1**, 238 (1957).
68. S. Bory and E. Lederer, *Croatica Chim. Acta*, **29**, 157 (1957).
69. T. G. Halsall and M. Moyle, *J. Chem. Soc.*, 1324 (1960).
70. G. Chandra, J. Clark, J. McLean, P. L. Pauson, J. Watson, R. I. Reed and F. M. Tabrizi, *J. Chem. Soc.*, 3648 (1964).
71. R. M. Dawson, M. W. Jarvis, P. R. Jefferies, T. G. Payne and R. S. Rosich, *Australian J. Chem.*, **19**, 2133 (1966).
72. F. E. King and G. Jones, *Chem. Ind.* (*London*), 1325 (1953); *J. Chem. Soc.*, 658 (1955).
73. W. Sandermann, K. Bruns and W. Reichhelm, *Tetrahedron Letters*, 2685 (1967).
74. E. M. Graham and K. H. Overton, *J. Chem. Soc.*, 126 (1965).
75. K. H. Overton and A. J. Renfrew, *J. Chem. Soc. C*, 931 (1967).
76. C. Enzell and O. Theander, *Acta Chem. Scand.*, **16**, 607 (1962).
77. C. A. Henrick and P. R. Jefferies, *Tetrahedron*, **21**, 1175 (1965).
78. P. F. Sommer, C. Pascaul, V. P. Arya and W. Simon, *Helv. Chim. Acta*, **46**, 1734 (1963).
79. C. A. Henrick and P. R. Jefferies, *Tetrahedron*, **21**, 3219 (1965).
80. J. R. Hosking and C. W. Brandt, *Chem. Ber.*, **68**, 1311 (1935); *New Zealand J. Sci. Tech.*, **17**, 755 (1936).
81. H. S. Baretto and C. Enzell, *Acta Chem. Scand.*, **15**, 1313 (1961).
82. R. C. Cambie and R. J. Weston, *Chemistry in New Zealand*, **32**, 105 (1968).
83. P. K. Grant, C. Huntrakul and D. R. J. Sheppard, *Australian J. Chem.*, **20**, 969 (1967).
84. L. Mangoni and R. Caputo, *Gazzetta*, **97**, 908 (1967).
85. J. W. Rowe and J. H. Scroggins, *J. Org. Chem.*, **29**, 1554 (1964).
86. K. Bruns, *Tetrahedron*, **25**, 1771 (1969).
87. I. H. Rogers and L. R. Rozon, *Canad. J. Chem.*, **48**, 1021 (1970).
88. R. M. Carman, *Tetrahedron*, **18**, 285 (1962); cf. G. Ohloff, *Ann. Chem.*, **617**, 134 (1958).
89. R. M. Carman and N. Dennis, *Australian J. Chem.*, **20**, 163 (1967).
90. R. M. Carman and H. C. Deeth, *Australian J. Chem.*, **22**, 2161 (1969).
91. J. R. Hosking, *Chem. Ber.*, **69**, 780 (1936).
92. H. R. Schenk, H. Gutmann, O. Jeger and L. Ruzicka, *Helv. Chim. Acta*, **35**, 817 (1952); **37**, 543 (1954).
93. P. K. Grant and R. Hodges, *J. Chem. Soc.*, 5274 (1960).
94. J. A. Barltrop, D. Giles, J. R. Hanson and N. A. J. Rogers, *J. Chem. Soc.*, 2534 (1962).
95. O. Jeger, O. Durst and G. Buchi, *Helv. Chim. Acta*, **30**, 1853 (1947).
96. R. Church, R. Ireland and J. Marshall, *Tetrahedron Letters*, 34 (1961).
97. G. Buchi and K. Biemann, *Croatica Chim. Acta*, **29**, 163 (1957); J. A. Barltrop and D. B. Bigley, *Chem. Ind.* (*London*), 1378 (1959); D. B. Bigley, N. A. J. Rogers and J. A. Barltrop, *J. Chem. Soc.*, 6413 (1960).
98. M. Souček and P. Vlad, *Chem. Ind.* (*London*), 1946 (1962).
99. M. Mousseron Canet, M. Mousseron Millet and J. C. Mani, *Israel J. Chem.*, **1**, 468 (1963).
100. U. Scheidegger, K. Schaffner and O. Jeger, *Helv. Chim. Acta*, **45**, 400 (1962).
101. E. Demole, *Experientia*, **20**, 609 (1964); *Helv. Chim. Acta*, **50**, 1314 (1967).
102. M. Mousseron Canet and J. C. Mani, *Bull. Soc. Chim. France*, 481 (1965).

103. E. Wenkert, J. R. Mahajan, M. Mussin and F. Schenker, *Canad. J. Chem.*, **44**, 2575 (1966).
104. S. Bory and C. Asselineau, *Bull. Soc. Chim. France*, 1355 (1961).
105. T. McCreadie and K. H. Overton, *Chem. Commun.*, 288 (1968).
106. E. Wenkert and Z. Kumazawa, *Chem. Commun.*, 140 (1968).
107. T. McCreadie, K. H. Overton and A. J. Allison, *Chem. Commun.*, 959 (1969).
108. J. L. Fourrey, J. Polonsky and E. Wenkert, *Chem. Commun.*, 714 (1969).
109. O. E. Edwards and R. S. Rosich, *Canad. J. Chem.*, **46**, 1113 (1968); O. E. Edwards and B. S. Mootoo, *Canad. J. Chem.*, **47**, 1189 (1969).
110. S. F. Hall and A. C. Oehlschlager, *Chem. Commun.*, 1157 (1969).
111. Y. Volmer and A. Jermstad, *Compt. Rend.*, **186**, 517, 783 (1928); M. M. Janot, *Compt. Rend.*, **191**, 847 (1930); **192**, 845 (1931).
112. L. Ruzicka and M. M. Janot, *Helv. Chim. Acta*, **14**, 645 (1931).
113. L. Ruzicka, L. L. Engel and W. H. Fischer, *Helv. Chim. Acta*, **21**, 364 (1938).
114. L. Ruzicka, C. F. Siedal and L. L. Engel, *Helv. Chim. Acta*, **25**, 621 (1942).
115. E. Lederer and D. Mercier, *Experientia*, **3**, 188 (1947).
116. M. Stoll and M. Hinder, *Helv. Chim. Acta*, **33**, 1251, 1308, 1345 (1950); **36**, 1995 (1953); **37**, 1856, 1859, 1866 (1954).
117. C. Asselineau, S. Bory, M. Fetizon and P. Laszlo, *Bull. Soc. Chim. France*, 1429 (1961); 19 (1963); H. Audier, S. Bory and M. Fetizon, *Bull. Soc. Chim. France*, 1381 (1964).
118. G. Ohloff, *Helv. Chim. Acta*, **41**, 845 (1958); A. Ahond and B. Gastambide, *Bull. Soc. Chim. France*, 4533 (1967).
119. D. P. Popa and G. V. Lazurveski, *Zhur. Obshch. Khim.*, **33**, 303 (1963).
120. V. P. Arya, C. Enzell, H. Erdtmann and T. Kubota, *Acta Chem. Scand.*, **15**, 225 (1961); H. S. Baretto and C. Enzell, *Acta Chem. Scand.*, **15**, 1313 (1961).
121. C. Enzell, *Acta Chem. Scand.*, **15**, 1303 (1961).
122. R. C. Cambie, P. K. Grant, C. Huntrakul and R. J. Weston, *Australian J. Chem.*, **22**, 1691 (1969).
123. H. Weinhaus, *Angew. Chem.*, **59**, 248 (1947); H. Weinhaus, W. Pilz, H. Seibt and H. G. Dessler, *Chem. Ber.*, **93**, 2625 (1960).
124. J. Heuser, *Bull. Soc. Chim. France*, 1490 (1961).
125. E. N. Schmidt, A. I. Lisina and V. A. Pentegova, *Chem. Abstr.*, **61**, 12042 (1964).
126. T. Norin, G. Ohloff and B. Willhalm, *Tetrahedron Letters*, 3523 (1965); W. Sandermann and K. Bruns, *Chem. Ber.*, **99**, 2835 (1966).
127. J. R. Hosking and C. W. Brandt, *Chem. Ber.*, **67**, 1173 (1934); *New Zealand J. Sci. Tech.*, **17**, 750 (1936).
128. J. R. Hosking and C. W. Brandt, *Chem. Ber.*, **68**, 37 (1935); **69**, 780 (1936).
129. J. McLean and S. N. Slater, *J. Soc. Chem. Ind.*, **64**, 28 (1945).
130. B. E. Cross, R. H. B. Galt and J. R. Hanson, *J. Chem. Soc.*, 2937 (1963).
131. G. Ohloff, *Ann. Chem.*, **617**, 134 (1958).
132. R. Hodges and R. I. Reed, *Tetrahedron*, **10**, 71 (1960).
133. E. Wenkert and P. Beak, *Chem. Ind.* (*London*), 1174 (1961); E. Wenkert, P. Beak, and P. K. Grant, *Chem. Ind.* (*London*), 1574 (1961).
134. J. A. Giles, J. N. Schumacher, S. S. Mims and E. Bernasek, *Tetrahedron*, **18**, 169 (1962).
135. M. Belardini, G. Scuderi and L. Mangoni, *Gazetta*, **92**, 1379 (1962); **94**, 829 (1964).
136. Y. S. Cheng and E. von Rudloff, *Tetrahedron Letters*, 1131 (1970).
137. J. R. Hosking and C. W. Brandt, *Chem. Ber.*, **68B**, 286 (1935).

138. P. K. Grant, M. H. G. Munro and N. R. Hill, *J. Chem. Soc.*, 3846 (1965).
139. P. K. Grant and R. Hodges, *Chem. Ind.* (*London*), 1300 (1960); C. Enzell, *Acta Chem. Scand.*, **14**, 2053 (1960).
140. P. K. Grant, *J. Chem. Soc.*, 860 (1959).
141. R. Hodges, *Tetrahedron*, **12**, 215 (1961).
142. P. K. Grant and R. M. Carman, *J. Chem. Soc.*, 3740 (1962).
143. P. K. Grant and N. R. Hill, *Australian J. Chem.*, **17**, 66 (1964).
144. P. K. Grant and M. J. A. McGrath, *Tetrahedron*, **26**, 1619 (1970).
145. P. K. Grant, L. N. Nixon and J. M. Robertson, *Tetrahedron*, **26**, 1631 (1970).
146. P. K. Grant and M. H. G. Munro, *Tetrahedron*, **21**, 3599 (1965).
147. R. A. Laidlaw and J. W. W. Morgan, *J. Chem. Soc.*, 644 (1963).
148. P. R. Jefferies and T. G. Payne, *Australian J. Chem.*, **18**, 1441 (1965).
149. F. B. Power and F. Tutin, *Chem. Zentr.*, **1**, 1401 (1908).
150. L. Panizzi, L. Mangoni and M. Belardini, *Tetrahedron Letters*, 376 (1961).
151. L. Mangoni and M. Belardini, *Gazetta*, **92**, 983, 995 (1962).
152. J. Heuser, R. Lombard, F. Lederer and G. Ourisson, *Tetrahedron*, **12**, 205 (1961).
153. W. G. Dauben and V. F. German, *Tetrahedron*, **22**, 679 (1966).
154. K. W. Gopinath, T. R. Govindachari, P. C. Parthasarathy and N. Viswanathan, *Helv. Chim. Acta*, **44**, 1040 (1961).
155. C. A. Henrick and P. R. Jefferies, *Australian J. Chem.*, **17**, 915 (1964).
156. L. Canonica, B. Rindone, C. Scholastico, G. Ferrari and C. Casagrande, *Tetrahedron Letters*, 2639 (1967).
157. E. Harms, *Arch. Pharm.*, **83**, 144 (1842); H. M. Gordin, *J. Amer. Chem. Soc.*, **20**, 265 (1908).
158. A. Lawson and E. D. Eustice, *J. Chem. Soc.*, 587 (1939); F. Hollis, J. H. Richards and A. Robertson, *Nature*, **143**, 604 (1939).
159. D. G. Hardy, W. Rigby and D. P. Moody, *J. Chem. Soc.*, 2955 (1957).
160. W. Cocker, B. E. Cross, S. R. Duff, J. T. Edwards and T. F. Holley, *J. Chem. Soc.*, 2540 (1953).
161. D. Burn and W. Rigby, *J. Chem. Soc.*, 2964 (1957).
162. W. S. Bennet, G. Eglinton, J. W. B. Fulke, K. Hirao, R. McCrindle, W. Simon and A. Tahara, *J. Chem. Soc. B*, 211 (1967).
163. R. A. Appleton, J. W. B. Fulke, M. S. Henderson and R. McCrindle, *J. Chem. Soc. C*, 1943 (1967).
164. W. H. Castine, D. M. S. Wheeler and M. M. Wheeler, *Chem. Ind.* (*London*), 1832 (1961).
165. L. Mangoni and M. Belardini, *Gazetta*, **93**, 465 (1963); *Tetrahedron Letters*, 4167 (1968).
166. D. M. S. Wheeler, M. M. Wheeler, M. Fetizon and W. H. Castine, *Tetrahedron*, **23**, 3909 (1967); **26**, 1561 (1970).
167. E. Ghigi, *Gazetta*, **78**, 865 (1948); **81**, 336 (1951); **83**, 252 (1953); D. Burn, D. P. Moody and W. Rigby, *Chem. Ind.* (*London*), 928 (1956).
168. M. S. Henderson and R. McCrindle, *J. Chem. Soc. C*, 2014 (1969).
169. J. W. B. Fulke, M. S. Henderson and R. McCrindle, *J. Chem. Soc. C*, 807 (1968).
170. L. Canonica, B. Rindone, C. Scholastico, G. Ferrari and C. Casagrande, *Tetrahedron Letters*, 3149 (1968).
171. D. P. Popa and G. V. Lazur'veski, *Chem. Abstr.*, **66**, 28923 (1967).
172. J. D. White, P. S. Manchard and W. B. Whalley, *Chem. Commun.*, 1315 (1969).
173. E. R. Kaplan, K. Naidu and D. E. A. Rivett, *J. Chem. Soc. C*, 1655 (1970).

174. E. R. Kaplan and D. E. A. Rivett, *J. Chem. Soc. C*, 262 (1968).
175. T. Anthonsen, P. H. McCabe, R. McCrindle and R. D. H. Murray, **25**, 2233 (1969).
176. T. Anthonsen, P. H. McCabe, R. McCrindle, R. D.H. Murray and G. A. R. Young, *Tetrahedron*, **26**, 3091 (1970).
177. D. Chakravarti and R. N. Chakravarti, *J. Chem. Soc.*, 1697 (1952); M. K. Gorter, *Rec. Trav. Chim.*, **30**, 151 (1911); R. Schwyzer, K. Biswas and P. Karrer, *Helv. Chim. Acta*, **34**, 652 (1951); R. J. C. Kleipol and W. Kostermann, *Rec. Trav. Chim.*, **70**, 1085 (1951); *Nature*, **169**, 33 (1952).
178. W. R. Chan, L. J. Haynes and LeRoy Johnson, *Chem. Ind. (London)*, 22 (1960); M. P. Cava and B. Weinstein, *Chem. Ind. (London)*, 851 (1959); M. P. Cava, W. R. Chan, L. J. Haynes, LeRoy Johnson and B. Weinstein, *Tetrahedron*, **18**, 397 (1962); M. P. Cava, W. R. Chan, R. P. Stein and C. R. Willis, *Tetrahedron*, **21**, 2617 (1965).
179. S. W. Pelletier, R. L. Chappell and S. Prabhakar, *Tetrahedron Letters*, 3489 (1966).
180. W. R. Chan, D. R. Taylor, C. R. Willis and H. W. Fehlabar, *Tetrahedron Letters*, 4803 (1968).
181. T. R. Govindachari, *Tetrahedron*, **21**, 3237 (1965).
182. M. Sumimoto, Y. Tanaka and K. Matsufuji, *Chem. Ind. (London)*, 1928 (1963); *Tetrahedron*, **20**, 1427 (1964).
183. C. Kaneko, T. Tsuchiya and M. Ishikawa, *Chem. Pharm. Bull.*, **11**, 1346 (1963).
184. M. Sumimoto, *Tetrahedron*, **19**, 643 (1963).
185. C. Kaneko, T. Tsuchiya and M. Ishikawa, *Chem. Pharm. Bull.*, **11**, 1346 (1963).
186. J. A. Giles and J. N. Schumacher, *Tetrahedron*, **14**, 246 (1961); J. A. Giles, J. N. Schumacher and G. W. Young, *Tetrahedron*, **19**, 107 (1963).
187. T. Kato, M. Tanemura, T. Suzuki and Y. Kitahara, *Chem. Commun.*, 28 (1970); *Tetrahedron Letters*, 1468 (1970).
188. R. Misra, R. C. Pandey and Sukh Dev, *Tetrahedron Letters*, 3751 (1964).
189. W. Cocker, A. L. Moore and A. C. Pratt, *Tetrahedron Letters*, 1983 (1965).
190. R. Misra, R. C. Pandey and Sukh Dev, *Tetrahedron Letters*, 2681, 2685 (1968).
191. T. Anthonsen and R. McCrindle, *Acta Chem. Scand.*, **23**, 1068 (1969).
192. S. Kusumoto, T. Okazaki, A. Ohsuka and M. Kotaki, *Bull. Soc. Chem. (Japan)*, **42**, 812 (1969).
193. D. E. U. Ekong and J. I. Okugun, *J. Chem. Soc. C*, 2153 (1969).
194. P. R. Jefferies and T. G. Payne, *Tetrahedron Letters*, 4777 (1967).
195. M. Kotake and K. Kuwata, *J. Chem. Soc. (Japan)*, **57**, 837 (1936).
196. J. T. Pinkey and R. F. Simpson, *Chem. Commun.*, 9 (1967).
197. G. A. Sim, T. A. Hamor, I. C. Paul and J. M. Robertson, *Proc. Chem. Soc.*, 75 (1961); *J. Chem. Soc.*, 4133 (1962).
198. D. H. R. Barton, H. C. Cheung, A. D. Cross, L. M. Jackmann and M. M. Smith, *Proc. Chem. Soc.*, 76 (1961); *J. Chem. Soc.*, 5061 (1961).
199. J. S. Birtwhistle, D. E. Case, P. C. Dutta, T. G. Halsall, G. Mathews, H. D. Sabel and V. Thaller, *Proc. Chem. Soc.*, 329 (1962).
200. C. E. McEachen, A. T. McPhail and G. A. Sim, *J. Chem. Soc. B*, 633 (1966).
201. W. R. Chan, D. R. Taylor and C. R. Willis, *Chem. Commun.*, 191 (1967).
202. T. J. King and S. Rodrigo, *Chem. Commun.*, 575 (1967).
203. T. J. King, S. Rodrigo and S. C. Wallwork, *Chem. Commun.*, 683 (1969).
204. G. Berti, O. Livi and D. Segnini, *Tetrahedron Letters*, 1401 (1970).
205. K. Feist *et al.*, *Ann. Chem.*, **517**, 119 (1935); **519**, 124 (1935); **522**, 185 (1936); **523**, 289 (1936); F. Wessely and K. Schonol, *Monatsh.*, **68**, 21 (1936); **70**, 30 (1937); **71**, 10 (1938).

206. D. H. R. Barton and D. Elad, *J. Chem. Soc.*, 2085, 2090 (1956).
207. M. P. Cava and E. J. Soboczenski, *J. Am. Chem. Soc.*, 78, 5317 (1956).
208. K. H. Overton, N. G. Weir and A. Wylie, *J. Chem. Soc. C*, 1482 (1966).
209. H. C. Cheung, D. Melville, K. H. Overton, J. M. Robertson and G. A. Sim, *J. Chem. Soc. B*, 853 (1966).
210. D. H. R. Barton, K. H. Overton and A. Wylie, *J. Chem. Soc.*, 4809 (1962); K. Balasubramanian, D. H. R. Barton and L. M. Jackmann, *J. Chem. Soc.*, 4816 (1962).
211. K. Ito and H. Furukawa, *Chem. Commun.*, 653 (1969).
212. T. Hori, A. K. Kiang, K. Nakanishi, S. Sataki and M. C. Woods, *Tetrahedron*, 23, 2649 (1967).
213. A. Sato, M. Kurabayashi, H. Nagahori, A. Ogi and M. Mishima, *Tetrahedron Letters*, 1095 (1970).
214. S. Yamazaki, S. Tamura, F. Marumo and Y. Saito, *Tetrahedron Letters*, 359 (1969).

The Steroidal Sapogenins of the Dioscoreaceae

KEN'ICHI TAKEDA   SHIONOGI RESEARCH LABORATORY, SHIONOGI AND CO., LTD., OSAKA, JAPAN

I.   INTRODUCTION

A.   On the Family Dioscoreaceae

Dioscoreaceae is a family belonging to the order Liliales of the class Monocotyledoneae[1,2]. The family is variously thought to be closely related to Amaryllidaceae[3], Taccaceae[4–6], Liliaceae[7] etc., of the same order. Although Dioscoreaceae is divided into the genera *Dioscorea*, *Epipetrum*, *Rajana*, *Tamus* and others, the boundaries between the genera are not sufficiently clear for unambiguous classification[1,2,5–7]. *Dioscorea* is the largest and most important genus of the family. Burkill has proposed that the immediate ancestors of the Dioscoreaceae are the proto-Dioscoreaceae which, he suggests, emerged from the proto-Liliales by an evolutionary change occurring on the Asian continent[7].

Dioscorea plants are dioecious, but both male and female flowers, which are colored white, pale yellow or pale green, are very tiny and therefore difficult to examine in detail. Also, though the leaves are usually cordate and petiolate, they vary considerably both in size and shape among individuals of the same species or even between different parts of a single plant, making it very difficult to recognize the leaves of corresponding male and female plants. For these reasons, the exact number of the species belonging to the genus is not clear, but it is thought to be about 600.

The majority of the plants in this genus have a tuber or rhizome as a storage organ. Except in a minor number of dwarf species, the aerial parts of the plants are annual vines, climbing up a support by twining, the direction of which, left or right, depends upon the species. Reproduction is by seeding, tuber, or by bulbil. *Dioscorea* plants thrive in tropical and subtropical areas, the centres of the distribution being South East Asia, Africa, and Central and South America, with only a few species indigenous to Europe and North America.

The family Dioscoreaceae contains many genera which are closely related to *Dioscorea*, for example, *Higinbothamia*, *Borderea*, *Epipetrum*, *Rajana*, *Tamus* etc. The unique species of the genus *Borderea* differs from *Dioscorea* only in that plants are of dwarf character and are not vines, while the genus *Higinbothamia* differs only in the number of the seeds in each loculus[5,6]. Some botanists therefore consider that plants belonging to these two genera should be included in the genus *Dioscorea*[1,8–10]. The genus

Tamus, which is limited to a few species found in the Mediterranean and in England, is unique in that its members have berries.

Not only is there uncertainty of classification within the family Dioscoreaceae, there are also diverging opinions concerning the classification of sections within the genus *Dioscorea*. A study of the members of this family which grow in the Amazon area is needed, as, until recently, these were hardly investigated. Furthermore, a detailed study of the constituents of Dioscoreaceae plants should be undertaken.

B. Steroidal Sapogenins (Spirostanes)

1. *General elucidation*[11]

Steroidal sapogenins have been isolated from plants belonging to Scrophulariaceae, Liliaceae, Amaryllidaceae and Dioscoreaceae and also from

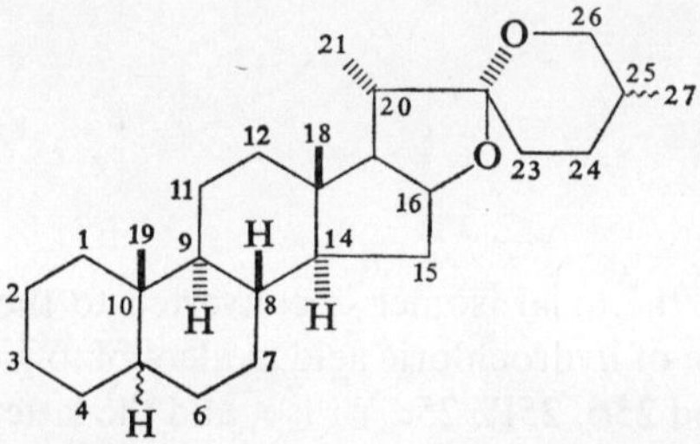

Figure 1. General structure and numbering of the carbon atoms steroidal sapogenins.

ox bile. The first isolation of a steroidal saponin was that of digitonin from *Digitalis purpurea* L. by Schmiedeberg in 1875. Digitogenin, the aglycone of digitonin, was first investigated by Kiliani and later by Windaus in connection with his main work from 1890 to 1918 on gitogenin, a sapogenin closely related to digitogenin. About 25 years later, Marker *et al.* correctly determined the structure of the spiroketal side-chain, i.e. the spiro E and F rings attached to the 16- and 17-positions of the D ring of the steroid nucleus which form the characteristic structural feature of steroidal sapogenins. The general structure and the numbering of the carbon atoms of steroidal sapogenins are shown in Fig. 1. According to the IUPAC nomenclature steroidal sapogenins are also given the general name 'spirostanes'.

Variety of the spirostane skeleton is limited, just as in other natural occurring steroids: i.e. in all cases 18 and 19 angular methyl groups are oriented β and the 21-methyl group in the side-chain has an α-configuration. There are cases of compounds with both *cis* and *trans* A/B ring junctures, but the B/C and C/D ring junctures are always *trans* fused, sometimes with a double bond located between the 5- and 6-positions. With regard to the stereochemistry of the spiroketal side-chain, the configurations of the ether linkage attached at the 16-position and the carbon–carbon bond between the 22- and 23-positions are both designated as β.

There are two isomers depending on the configuration of the 27-methyl group at position 25. In one (**1**) the conformation is equatorial (α) to the chair-like ring F while in the other (**2**) it is axial (β). The former is more stable

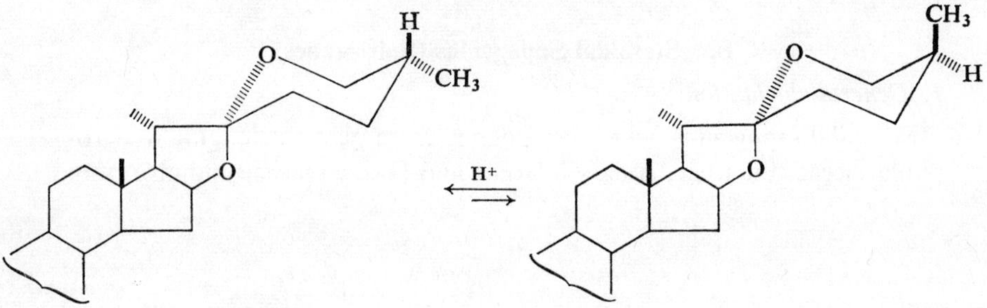

(1) **(2)**

than the latter and the axial isomer is converted to the stable equatorial isomer by the action of hydrochloric acid in alcohol to give an equilibrium. The former is named 25D, 25R, 25α, or iso, and the latter 25L, 25S, 25β, or neo.

Besides the normal steroidal sapogenin types shown in Fig. 1, in some cases a modified form, named furostane and having an opened F ring structure, also exists in the plant body, especially in the form of the 26-glucoside (**3**). The furostane structure is shown in Fig. 2.

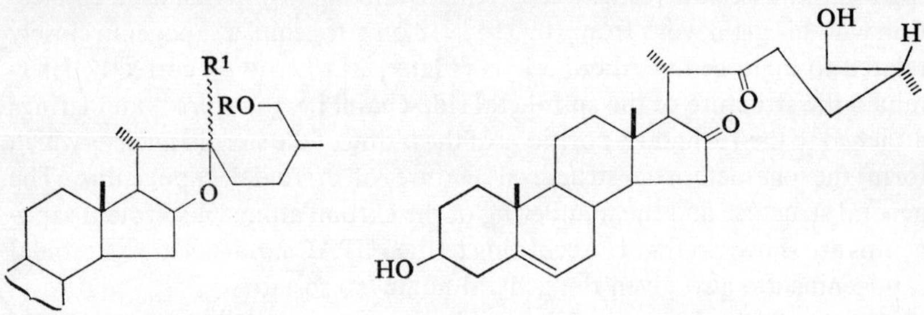

Furostane (3) Kryptogenin (**4**)

R = H or Sugar

Figure 2. Furostane structure and example of modification.

Table 1. Chromatographic characterization of steroidal sapogenins

	Config. at 5	OH or CO	R_f Values in solvent system[b]				
			A	B	C	D	E
1. Neometeogenin[a] (L)	A-aromatic	11α	0.81	0.52	0.63	0.66	0.64
2. Meteogenin[a] (D)	A-aromatic	11α	0.81	0.52	0.63	0.66	0.64
3. Sarsasapogenin (L)	cis	3β	0.65	0.39	0.46	0.51	0.51
4. Diosgenin (D)	Δ^5	3β	0.59	0.35	0.41	0.42	0.46
5. Tigogenin (D)	trans	3β	0.59	0.35	0.39	0.39	0.46
6. Gentrogenin (D)	Δ^5	$3\beta,12(CO)$	0.49	0.24	0.25	0.32	0.26
7. Yonogenin (D)	cis	$2\beta,3\alpha$	0.21	0.11	0.11	0.17	0.13
8. Tokorogenin (D)	cis	$1\beta,2\beta,3\alpha$	0.09	0.02	0.02	0.12	0.03
9. Kogagenin (D)	cis	$1\beta,2\beta,3\alpha,5\beta$	0.03	0.01	0.01	0.08	0.00

[a] Isolated from *Metanarthesium luteo-viridi* (Liliaceae).

[b] Solvent system: A (chloroform: EtOH = 95:5), B (chloroform: Me_2CO = 9:1), C (benzene: Me_2CO = 85:15), D (benzene: MeOH = 92:8), E (n-hexane: EtOAc = 1:1).

One or more of the positions on the steroid nucleus, except positions 7, 8, 9, 14, 18 or 19, may be substituted by functional groups, for example by hydroxyl or ketone. Substitution sometimes also occurs at positions 25 or 27 on the F ring. In some cases a double bond is located between C-25 and C-27 in addition to that at Δ^5 or $\Delta^{9(11)}$. A particularly interesting modified sapogenin is kryptogenin (**4**), which is a 16,22-diketo derivative, i.e. a furostane with an oxidatively opened E ring as shown in Fig. 2.

2. *Characterization of steroidal sapogenins*

a. Chromatographic analysis. It is well known that the steroidal sapogenins can be identified by their characteristic colour reaction with sulphuric acid, perchloric acid or, more commonly, antimony trichloride on silica gel thin-layer chromatograms[12]. The results of silica gel chromatography of some common sapogenins, including 3α-hydroxylated sapogenins, in different solvent systems, are given in Table 1. These colour reactions, in

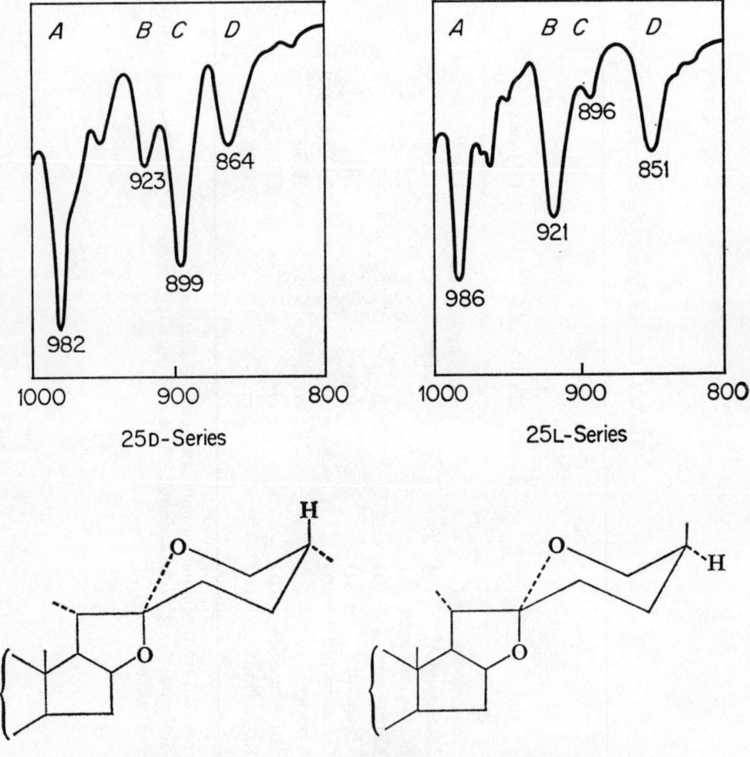

Figure 3(a). Infrared spectra of steroidal sapogenins.

conjunction with photometric measurement, also provide a convenient method for the quantitative determination of steroidal sapogenins, and Akahori *et al.*[13, 14] have reported the quantitative analysis of diosgenin with ferric chloride, phosphoric acid and sulphuric acid, and those of yonogenin and tokorogenin with anisaldehyde and phosphoric acid, on silica gel thin-layer chromatograms.

Gas[15], liquid column[16], or paper chromatographic[17, 18] separation of steroidal sapogenins have been also utilized. Kawasaki *et al.*[19], using a gas–liquid chromatographic method, achieved a satisfactory separation of spirostanol glycosides, though with the exception of some triglycosides.

b. Infrared spectra of steroidal sapogenins. The i.r. spectra of steroidal sapogenins have been investigated by Wall[20] and Jones[21] and four bands, 980 (*A* band), 920 (*B* band), 900 (*C* band) and 860 cm^{-1} (*D* band), among about 20 absorption bands which are due to the spiro-ketal side-chain, have been assigned as characteristic of the E and F rings. With 25L- (S- or neo)

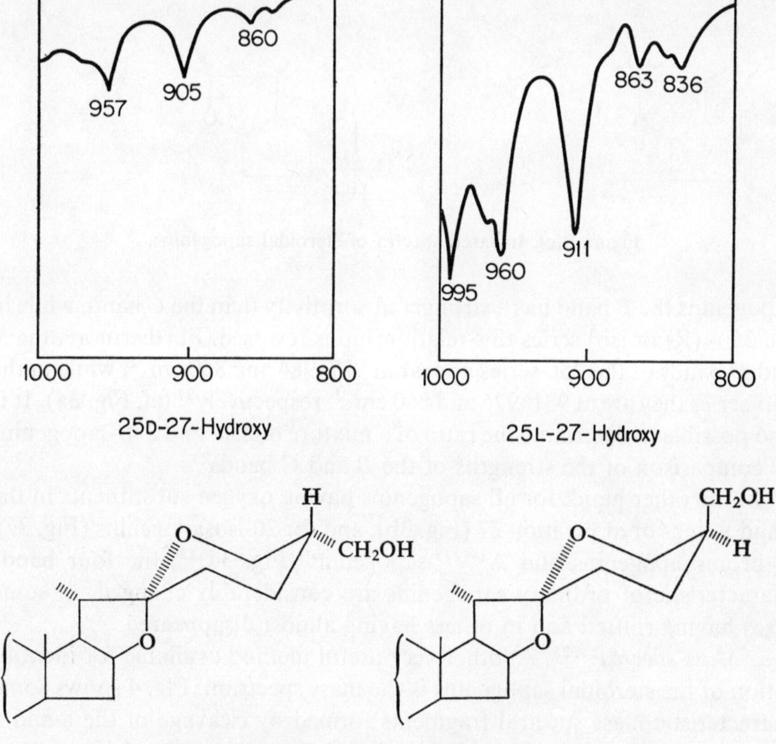

Figure 3(b). Infrared spectra of steroidal sapogenins.

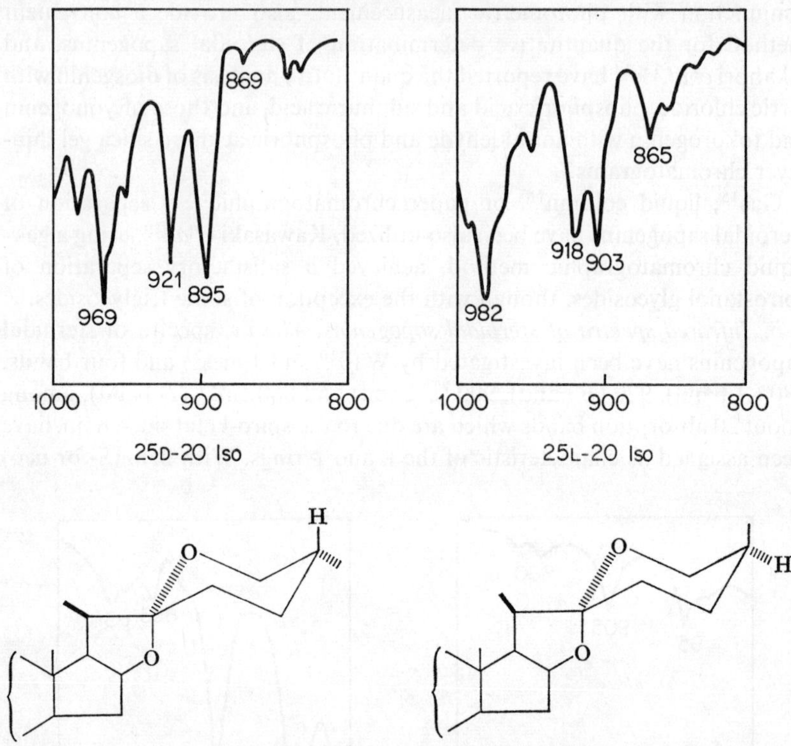

Figure 3(c). Infrared spectra of steroidal sapogenins.

sapogenins the B band has a stronger absorptivity than the C band, while in the 25D- (R- or iso) series this relationship is reversed. Furthermore, the A and D bands of the 25L-series appear at 987–984 and 850 cm^{-1}, while in the 25D-series they are at 981–976 and 860 cm^{-1}, respectively[22] (cf. Fig. 3a). It is also possible to determine the ratio of a mixture of 25L- and 25D-sapogenins by comparison of the strengths of the B and C bands[23].

On the other hand, for all sapogenins having oxygen substituents in the E and F rings or at position 27 (Fig. 3b), and for 20-isosapogenins (Fig. 3c), 23-bromosapogenins and $\Delta^{25(27)}$-sapogenins (Fig. 3d)[22], the four bands characteristic of ordinary sapogenins are considerably changed, in some cases having shifted and in others having almost disappeared.

c. Mass spectra[24,25]. Another very useful method available for identification of the steroidal sapogenins is the mass spectrum. Fig. 4 shows some characteristic mass spectral fragments formed by cleavage of the E and F rings. With 25- or 27-hydroxysapogenins, the two fragments designated as

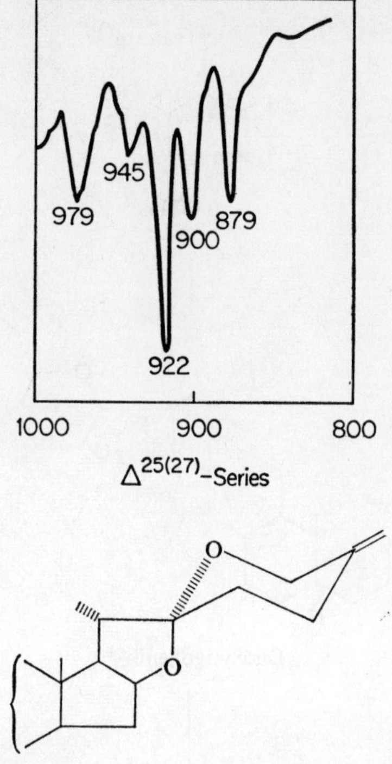

Figure 3(d). Infrared spectra of steroidal sapogenins.

f and *g* in the Figure are all 16 mass units greater than the normal mass number.

d. Nuclear magnetic resonance[26]. Nuclear magnetic resonance spectral data for about 150 steroidal sapogenins and their derivatives have been summarized by Tori and Aono[26]. A particular problem in n.m.r. work on sapogenins lies in their solubility, many of the free sapogenins being sparingly soluble in deuterio-chloroform or deuterio-acetone, though pyridine is found to be a useful solvent in many cases.

II. STEROIDAL SAPOGENINS OF THE DIOSCOREACEAE

A. History and Isolation

1. History

Since the discovery by Marker about 30 years ago of a very simple and economical method for the side-chain degradation of steroidal sapogenins

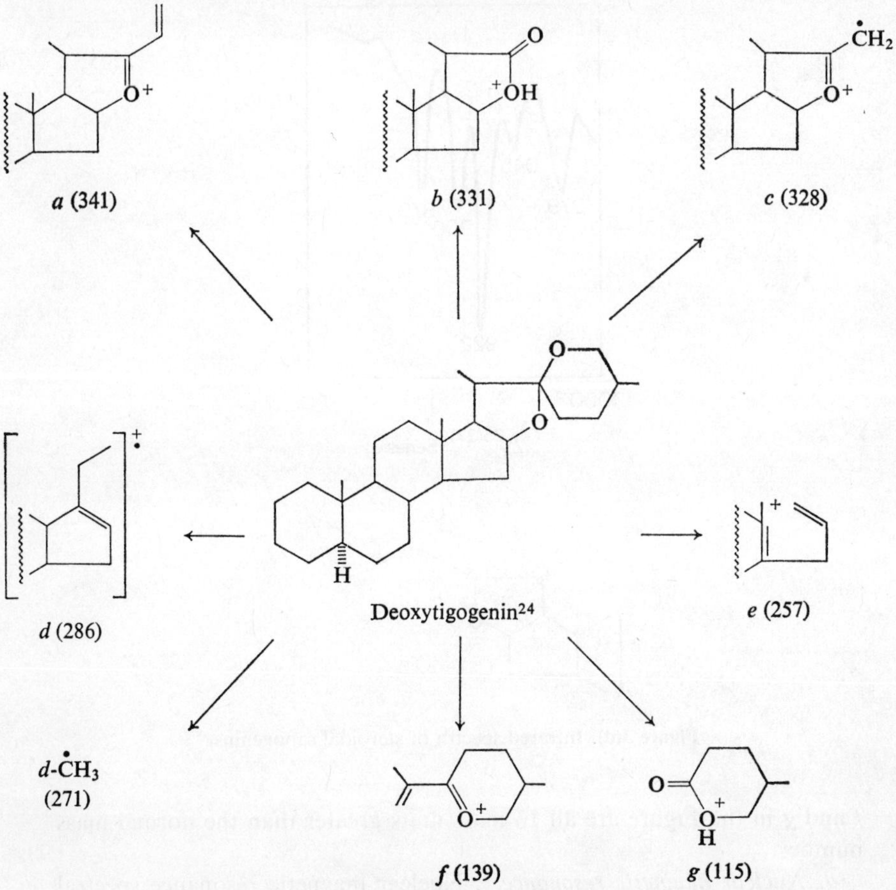

Figure 4. Main mass spectral fragments of steroidal sapogenins.

via pseudo-sapogenin (see below), diosgenin has rapidly become important, both as a source of pregnenolone and for the synthesis of modified steroidal hormones. Diosgenin was first isolated from *Dioscorea tokoro*, one of the *Dioscorea* species indigenous to Japan, by Tsukamoto *et al.* and by Fujii *et al.*[27,28] at about the same time, and was named after the plant.

Following Marker's discovery, chemical interest has been concentrated on modification of the steroidal skeleton in order to obtain synthetic steroidal hormones having more activity than the natural product, and a large number of useful modified steroid hormones have been synthesized, for example prednisolone (**5**), 9α-fluoro-16-methylprednisolone (**6**) (dexamethanone), norethisterone (**7**), metenolone (**8**) etc. (see Fig. 5).

Dioscorea mexicana, the Mexican Yam, is one of the most important sources of diosgenin. Many studies have been made on the steroidal sapogenin components in plants of the Dioscoreaceae, Amaryllidaceae and

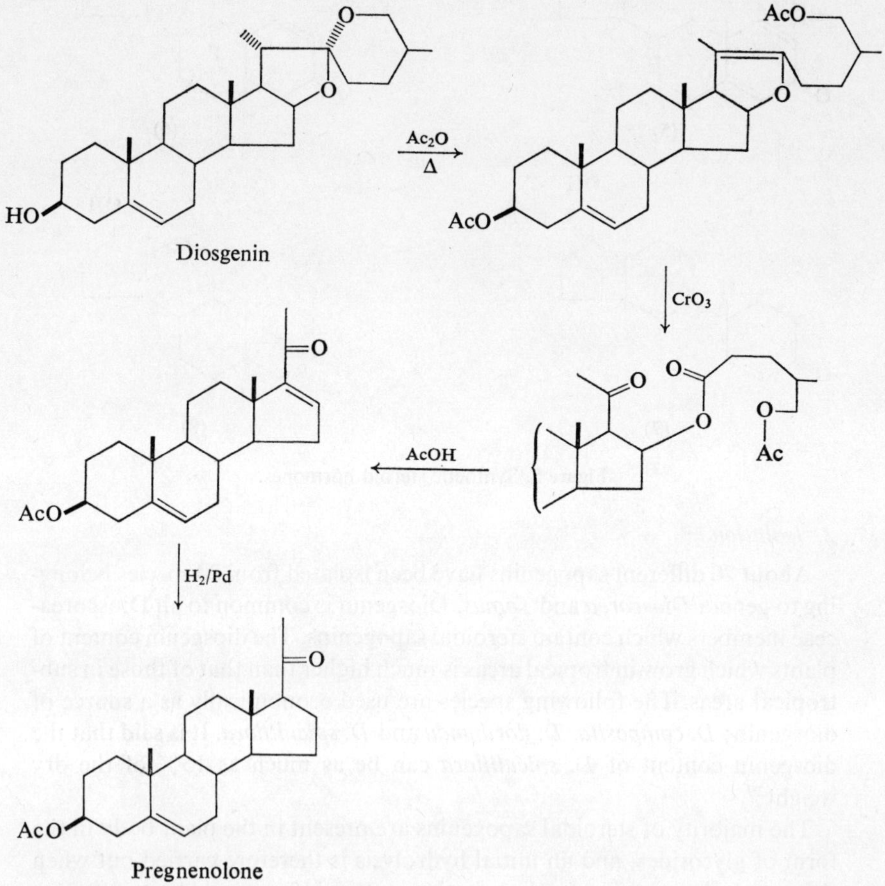

Diosgenin

Ac₂O
Δ

CrO₃

AcOH

Pregnenolone

H₂/Pd

Degradation of diosgenin by method of R. E. Marker.

Liliaceae, and more than 100 species belonging to these families have been examined. However, since plants belonging to the genus *Dioscorea* commonly have a large underground rhizome which provides a very convenient source for the isolation of constituents, chemical studies have been made mainly on this genus.

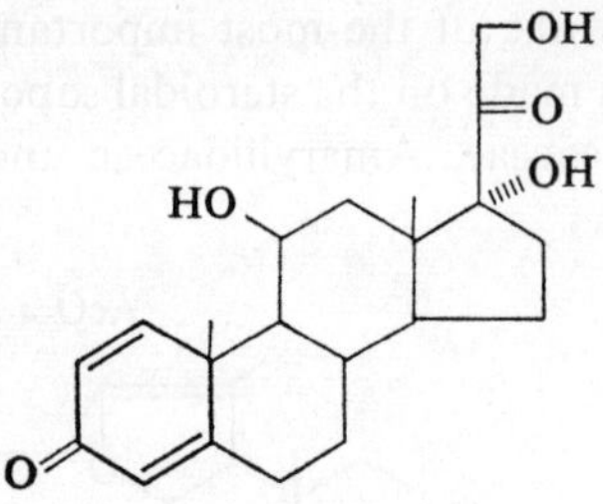

(5)

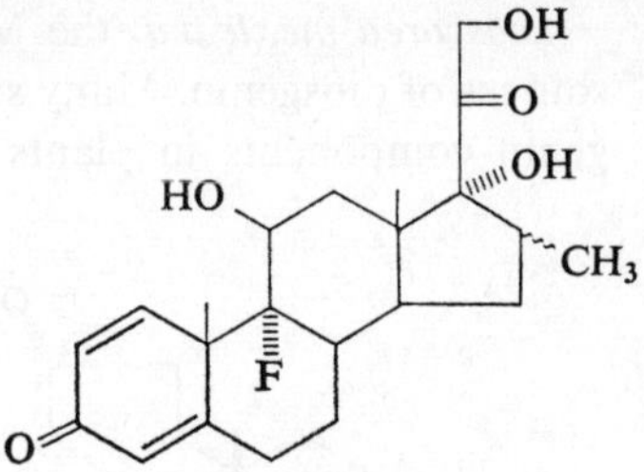

(6)

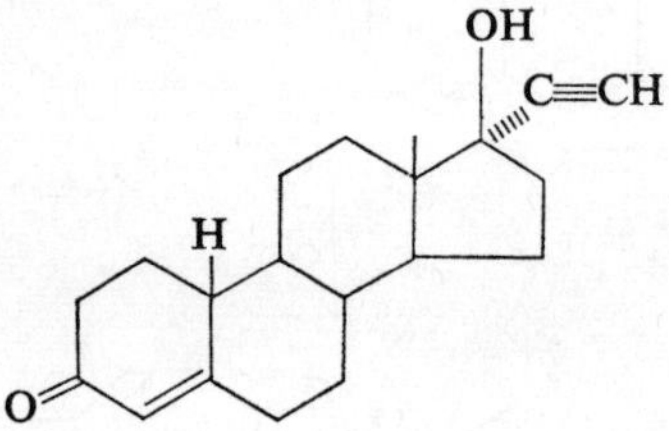

(7)

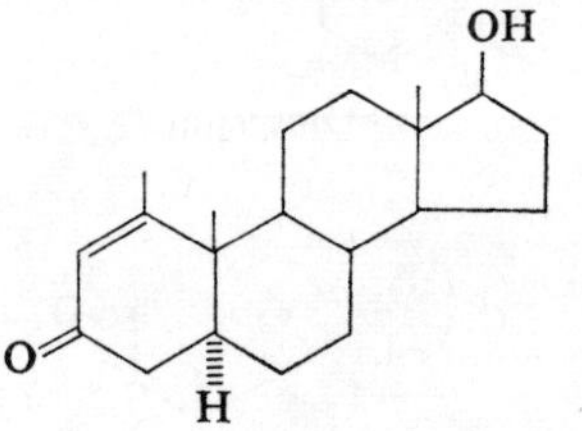

(8)

Figure 5. Synthetic steroid hormones.

2. Isolation

About 20 different sapogenins have been isolated from 70 species belonging to genera *Dioscorea* and *Tamus*. Diosgenin is common to all Dioscoreaceae members which contain steroidal sapogenins. The diosgenin content of plants which grow in tropical areas is much higher than that of those in subtropical areas. The following species are used economically as a source of diosgenin; *D. composita, D. floribunda* and *D. spiculiflora*. It is said that the diosgenin content of *D. spiculiflora* can be as much as 15% of the dry weight[29].

The majority of steroidal sapogenins are present in the plant body in the form of glycosides, and an initial hydrolysis is therefore carried out when the sapogenins are to be used as starting materials for steroid hormones. On an industrial scale hydrolysis and extraction from the plant is usually effected in one of the following two ways. Either the saponin mixture is extracted from dried finely cut pieces of rhizome or tuber with methanol or ethanol, then hydrolysed with acid[30], or acid hydrolysis of the dried plants is followed by extraction with petroleum benzine, benzene, or other non-polar solvent[31,32]. In the latter case, as there is no problem of epimerization at C-25, somewhat drastic hydrolysis is preferable in order to destroy undesirable by-products.

If it is desired to examine the chemical structures of the natural sapogenins, the saponins should be hydrolysed under very mild conditions or by enzymatic hydrolysis with a microorganism[33-35], *Aspergillus* for example, in order to avoid rearrangement or dehydration.

B. Chemistry

1. Sapogenins of dioscoreaceae

The structures of steroidal sapogenins isolated from members of the Dioscoreaceae together with the names of the plants from which they have been isolated are listed in Table 2.

Diosgenin and its 25-isomer, yamogenin, are very common steroidal sapogenins, distributed widely among plants belonging to the families Dioscoreaceae, Liliaceae etc. Gentrogenin, chiapagenin and their corresponding 25-isomers and kryptogenin are also isolated from Liliaceae plants. A unique 27-hydroxylated sapogenin, isonarthogenin (9), which had previously been found in the genus *Metanarthecium*[77] and more recently in *Alteris*[78] of the family Liliaceae, has now also been found in *D. quinqueloba*[70]. The structures of all steroidal sapogenins isolated from Dioscoreaceae plants have been elucidated except for that of tenuipegenin. It is interesting that all of these compounds have structures which belong either to a 5β-steroid series or to a $\Delta^{5(6)}$-unsaturated steroid series. Nothing is known of the structure of tenuipegenin except that it contains four hydroxyl groups.

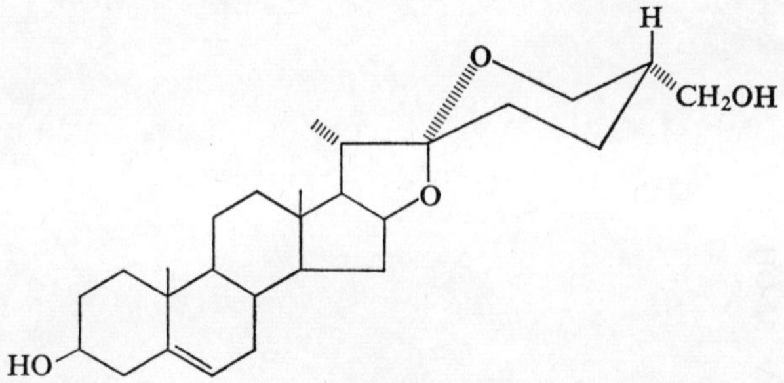

(9)

A characteristic feature of this family is the existence of sapogenins and their 25-isomers in which an hydroxyl group at the 3-position has an α-configuration. The structures of these 3α-hydroxy sapogenins, yonogenin, igagenin, tokorogenin, kogagenin and isodiotigenin and the corresponding 25-isomers, are shown in Fig. 6. In addition to these five 3α-sapogenins, there has been one report of the presence of epismilagenin[37] in *Dioscorea*, though this has not been confirmed.

Table 2. Steroidal sapogenins isolated from members of the Dioscoreaceae

Sapogenin	C-25	C-5	Substituents	M.p. (°C)	$[\alpha]_D$	Source
Epismilagenin	D	β	3α-OH	220–227	−43.8	*D. deltoidea* Wall.[37], *D. prazeri* Prain et Burkill[37]
Smilagenone	D	β	3-keto	188–189	−52	*D. deltoidea* Wall.[37], *D. prazeri* Prain et Burkill[37]
Diosgenin	D	Δ	3β-OH	204–207	−129	*D. althaeoides* R. Knuth[38,39], *D. asclepiadea* Prain et Burkill[40], *D. balcanica* Kosanin[41], *D. bartlettii* Morton[42-44], *D. bernoulliana* Prain et Burkill[45], *D. bulbifera* L.[46-48], *D. caucasica* Lipsky[49], *D. chiapasensis* Matuda[50,51], *D. collettii* Hook, fil.[39], *D. composita* Hemsl.[42,43,52-54], *D. convolvulacea* Cham. et Schlechtd.[46], *D. cyphocarpa* Robinson[46], *D. deltoidea* Wall.[38,39,47,54-56], *D. dugesii* Robinson[46], *D. elephantipes* (*L'Héritier*) Engler[57], *D. escuintlensis* Matuda[45], *D. esculenta* (Lour.) Burkill[56], *D. floribunda* Mart. et Gal.[43,52-54], *D. friedrichsthalii* R. Knuth[58], *D. galeottiana* Kunth[46], *D. glauca* Rusby[43,59], *D. gracillima* Miq.[40,60], *D. hemicrypta* Burkill[57], *D. hirsuticaulis* Robinson[46], *D. hondurensis* R. Knuth[33], *D. izuensis* Akahori[40], *D. jaliscana* Wats.[46], *D. lobata* Uline[46], *D. mexicana* Guillemin[42,43,46,52,53], *D. militaris* Robinson[46], *D. minima* Robins. et Seaton[46], *D. multiflora* Mart.[61], *D. multinervis* Benth.[46,61], *D. nelsonii* Uline[53], *D. nigrescens* R. Knuth[38], *D. nipponica* Makino[38,40,60], *D. panthaica* Prain et Burkill[38,39], *D. platycolpota* Uline[46,53], *D. polygonoides* Humb. et Bonpl.[53], *D. prazeri* Prain et Burkill[47,54,57], *D. pringlei* Robinson[46], *D. quinqueloba* Thunb.[40,60], *D. remotiflora* Kunth[46], *D. septemloba* Thunb.[40,62], *D. sititoana* Honda et Jotani[40], *D. spiculiflora* Hemsl.[43,44,52,53], *D. sylvativa* Ecklon[53,54,57], *D. tenuipes* Franch. et Savat.[40,60], *D. tokoro* Makino[27,63], *D. ulinei* Greenman[46], *D. urceolata* Uline[46], *D. villosa* L.[42,43,52,55,64,65], *D. zingiberensis* C. H. Wright[39], *T. communis* L.[54,66,67], *T. edulis* Lowe[68].

Yamogenin	L	Δ	3β-OH	201	−123	*D. bartlettii* Morton[53], *D. bulbifera* L.[46], *D. chiapasensis* Matuda[57], *D. collettii* Hook fil.[39], *D. cyphocarpa* Robinson[46], *D. deltoidea* Wall.[54], *D. dugesii* Robinson[46], *D. floribunda* Mart. et Gal.[43,52-54], *D. hirsuticaulis* Robinson[46], *D. hondurensis* R. Knuth[54], *D. mexicana* Guillemin[42,52,53,69], *D. militaris* Robinson[46], *D. minima* Robins. et Seaton[46], *D. multinervis* Benth.[46], *D. platycolpota* Uline[46], *D. prazeri* Prain et Burkill[54], *D. remotiflora* Kunth[46], *D. spiculiflora* Hemsl.[52,53,70], *D. sylvativa* Ecklon[54], *D. tenuipes* Franch. et Savat.[40], *D. tokoro* Makino[63], *D. ulinei* Greenman[46], *D. urceolata* Uline[46], *T. communis* L.[54]
Isonarthogenin	D	Δ	3β,27-OH	240–242	−110.2	*D. quinqueloba* Thunb.[70]
Tamusgenin	D	Δ	3β-OH, 11-keto	180–182	−75	*D. edulis* Lowe[36]
Isochiapagenin	D	Δ	3β,12β-OH	257–259	−130	*D. chiapasensis* Matuda[51]
Chiapagenin	L	Δ	3β,12β-OH	249–251	−126	*D. chiapasensis* Matuda[51]
Gentrogenin	D	Δ	3β-OH, 12-keto	215–216	−57	*D. spiculiflora* Hemsl.[44,53,69]
Correllogenin	L	Δ	3β-OH, 12-keto	209–211	−69	*D. spiculiflora* Hemsl.[44,53,69]
Yonogenin	D	β	2β,3α-OH	240–243	−45	*D. tenuipes* Franch. et Savat.[73], *D. tokoro* Makino[63,71,72]
Neoyonogenin	L	β	2β,3α-OH	198–199	−63.7	*D. tenuipes* Franch. et Savat.[73]
Igagenin	D	β	2β,3α,27-OH	253	−43.6	*D. tokoro* Makino[74]
Tokorogenin	D	β	1β,2β,3α-OH	266–268	−49.6	*D. tenuipes* Franch. et Savat.[73], *D. tokoro* Makino[63,75,79]
Neotokorogenin	L	β	1β,2β,3α-OH	244–245	−41.7	*D. tenuipes* Franch. et Savat.[73]
Isodiotigenin	D	β	2β,3α,4β-OH	212.5–213.5	−26.9	*D. tokoro* Makino[76]
Diotigenin	L	β	2β,3α,4β-OH	280–281	−59.8	*D. tenuipes* Franch. et Savat.[40]
Kogagenin	D	β	1β,2β,3α,5β-OH	316	−27	*D. tokoro* Makino[63,71]
Isotenuipegenin	D	β	tetrahydroxy	307–309	−39.5	*D. tokoro* Makino[76]
Tenuipegenin	L		tetrahydroxy	299–300	−51.5	*D. tenuipes* Franch. et Savat.[73]
Kryptogenin		Δ	3β,27-OH	189		*D. bulbifera* L., *D. cyphocarpa* Robinson, *D. dugesii* Robinson, *D. hirsuticaulis* Robinson, *D. militaris* Robinson, *D. minima* Robins. et Seaton, *D. multinervis* Benth., *D. platycolpota* Uline, *D. remotiflora* Kunth, *D. ulinei* Greenman, *D. urceolata* Uline[46]

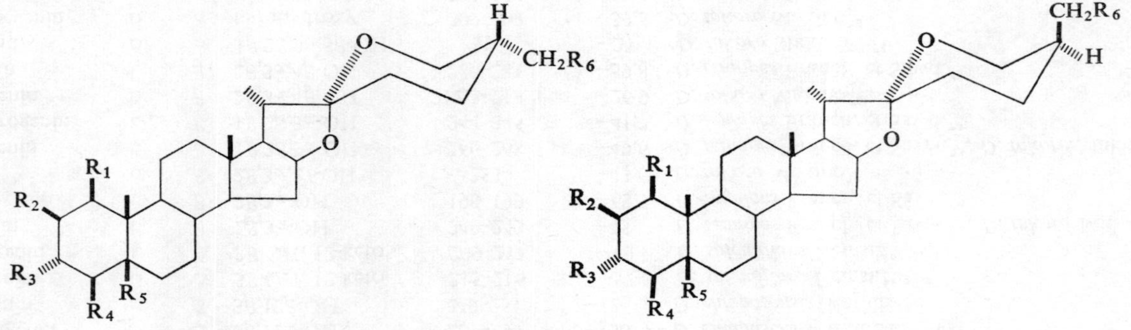

Figure 6. Structures of 3α-hydroxy spirostanes.

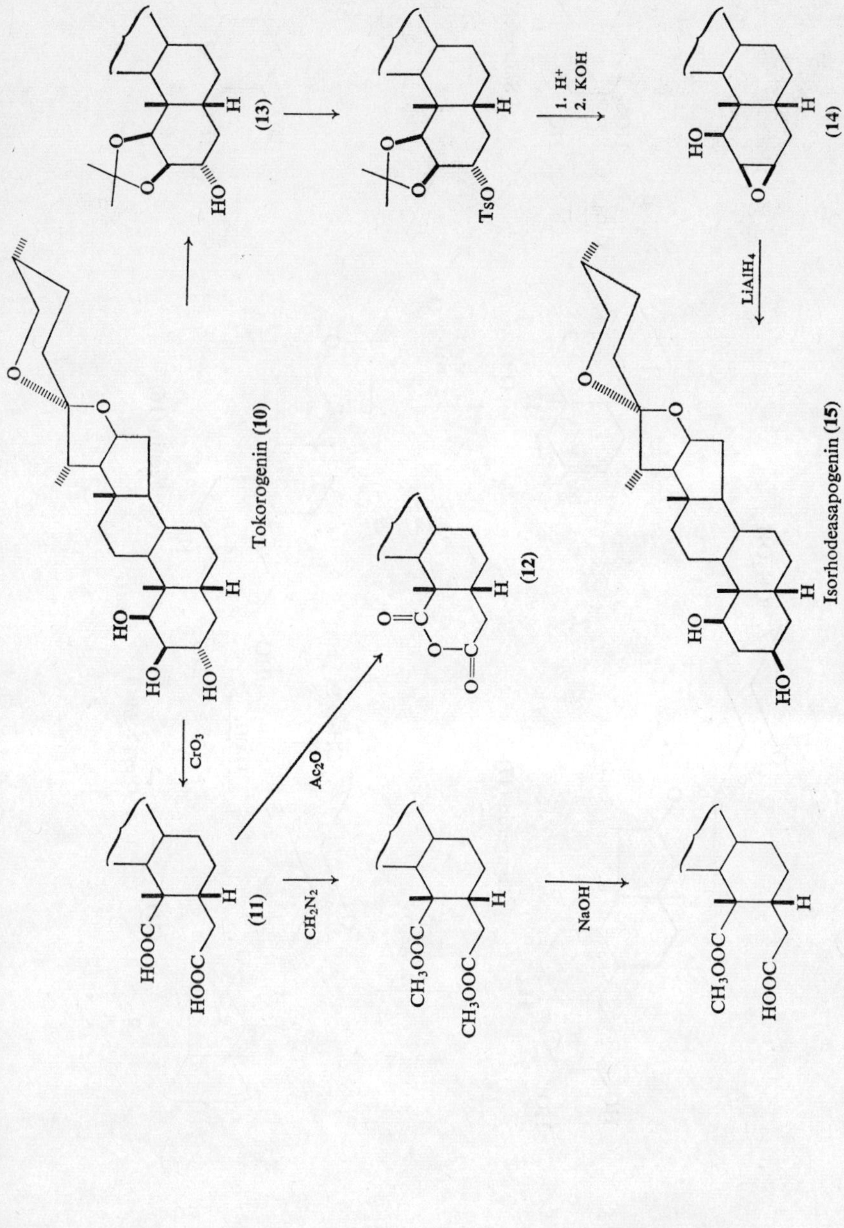

Tokorogenin (10)

(13)

(14)

1. H⁺
2. KOH

LiAlH₄

CrO₃

Ac₂O

(11)

(12)

CH₂N₂

NaOH

Isorhodeasapogenin (15)

Structural elucidation of tokorogenin.

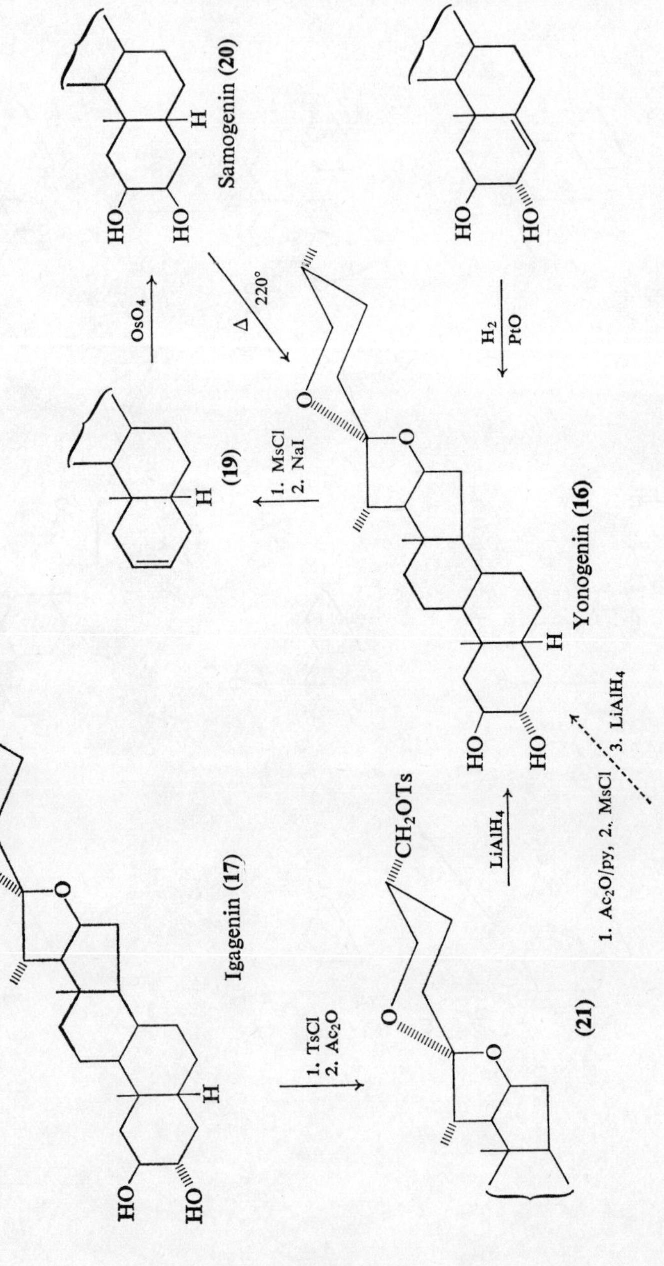

Progress in Phytochemistry

Samogenin (20)

Yonogenin (16)

Igagenin (17)

(19)

(21)

OsO$_4$

Δ 220°

1. MsCl
2. NaI

H$_2$
PtO

LiAlH$_4$

1. Ac$_2$O/py, 2. MsCl
3. LiAlH$_4$

1. TsCl
2. Ac$_2$O

CH$_2$OH

CH$_2$OTs

HO

HO

Structural relationships of 3α-hydroxy steroidal sapogenins.

Steroidal sapogenins with an α-oriented hydroxyl group at position 3 had not hitherto been found in the plant kingdom. Furthermore, they have so far been found only in *D. tokoro* and *D. tenuipes**, and those sapogenins belonging to the 25L- (neo) series are found only in *D. tenuipes*.

Some of the chemical work done on these sapogenins is summarized in the following sections.

2. *Structural elucidation of tokorogenin*[75,79]

Tokorogenin was the first 3α-hydroxylated sapogenin to be isolated from *D. tokoro* and its structure was determined by the following reaction sequences. Chromic acid oxidation of tokorogenin gave tokorogenic acid (**11**), a dicarboxylic acid with one carbon less than tokorogenin itself. This acid was converted to its acid anhydride (**12**) by the action of acetic anhydride. From its chemical nature, one of the carboxylic acid residues of tokorogenic acid was assumed to be tertiary. Next, the remaining hydroxyl group of tokorogenin acetonide (**13**) was tosylated, the acetonide group removed with acid, and the resulting diol converted to a hydroxy epoxide (**14**) by alkali treatment. Lithium aluminium hydride reduction of this epoxide gave the known isorhodeasapogenin (**15**). From these results the structure of tokorogenin was assigned as 25D-1β,2β,3α-trihydroxy-5β-spirostane (**10**).

3. *Chemical relationships among four 3α-hydroxylated sapogenins*

Igagenin (**17**) and kogagenin (**18**) were correlated with yonogenin (**16**), while kogagenin was converted to tokorogenin by the chemical reactions illustrated on pages 304 and 305.

25D-5β-Spirost-2-ene (**19**)[72] was obtained from yonogenin dimesylate by the action of sodium iodide in acetone in a sealed tube at 120°. This spirost-2-ene was converted to samogenin (**20**) by treatment with osmium tetroxide. The samogenin thus obtained reverted to yonogenin (**16**) at 220° by epimerization of the 3β-axial hydroxyl group.

Igagenin 27-monotosylate (**21**)[80] was easily converted to yonogenin by removal of the C-27 hydroxyl group with lithium aluminium hydride.

Anhydro kogagenin (**22**)[81], obtained from kogagenin by the action of thionyl chloride in pyridine, was catalytically hydrogenated to tokorogenin (**10**). On the other hand, diacetyl kogagenin, obtained from kogagenin by the action of acetic anhydride under very mild conditions, was converted to its diacetyl monomesylate (**23**), which on lithium aluminium hydride reduction gave a mixture of 2β,3α-,5β- (**24**) and 2α,3α,-5β-trihydroxy-sapogenins

---

* D. Chakravarti *et al.* reported that epismilagenin was isolated from *D. deltoidea* and *D. prazeri*.

(**25**). Dehydration of the former compound by thionyl chloride, followed by catalytic reduction, gave yonogenin (**16**)[82, 83].

By the above reaction sequence, excluding the dehydration and reduction steps, tokorogenin (**10**) should also be easily converted to yonogenin (**16**), though such a conversion has not yet been confirmed. (See dotted line in scheme.)

4. Synthesis of isodiotigenin[84]

Isodiotigenin (**33**) and its 25L-isomer, diotigenin[85], constituents of *D. tokoro* and *D. tenuipes* respectively, have three hydroxyl groups which, located at C-2β, C-3α and C-4β, are vicinal and have equatorial conformations.

Yonogenin (**16**) was used as starting material for the synthesis of isodiotigenin. Yonogenin was first converted to its 3-monomesylate (**26**), isomeric monomesylate (**27**) and dimesylate also being obtained. The mixture of monomesylates was oxidized to a mixture of ketones which was then separated into its components. 2-Oxo-3-mesylate (**28**) thus obtained was treated with lithium bromide and lithium carbonate to give 25D-5β-2-oxospirost-3-ene (**29**). Osmium tetroxide hydroxylation followed by reduction of the carbonyl group with lithium in liquid ammonia gave a triol (**30**) in which all three hydroxyl groups, at 2, 3 and 4, had a β-configuration. In this reaction, osmium tetroxide should attack from the less hindered β-side of the molecule and a carbonyl group should be reduced to give mainly an equatorial hydroxyl group. As the 3β-hydroxyl group has an axial conformation, the triol could be acetylated to its 2,4-diacetate (**31**) under very mild conditions. The latter was then oxidized by chromium trioxide to the oxo-diacetae (**32**), which in turn was reduced by sodium borohydride to give a mixture of the configurational isomers at the C-3 hydroxyl group. One of these was identical in all respects with natural isodiotigenin (**33**) and the other, the main product, was identical with the above-mentioned 2β,3β,4β-triol (**30**).

C. Steroidal Saponins

1. Diosgenin glycosides

There are several kinds of glycosides of diosgenin, having different kinds and numbers of sugar residues and different positions of linkage between the sugars. Furthermore, certain prosapogenins, in which one or more sugar moieties in the saponin have been removed, probably by enzyme hydrolysis, have also been isolated from the plant tissue. The structures of these diosgenin glycosides together with the glycosides of 3α-hydroxylated sapogenins

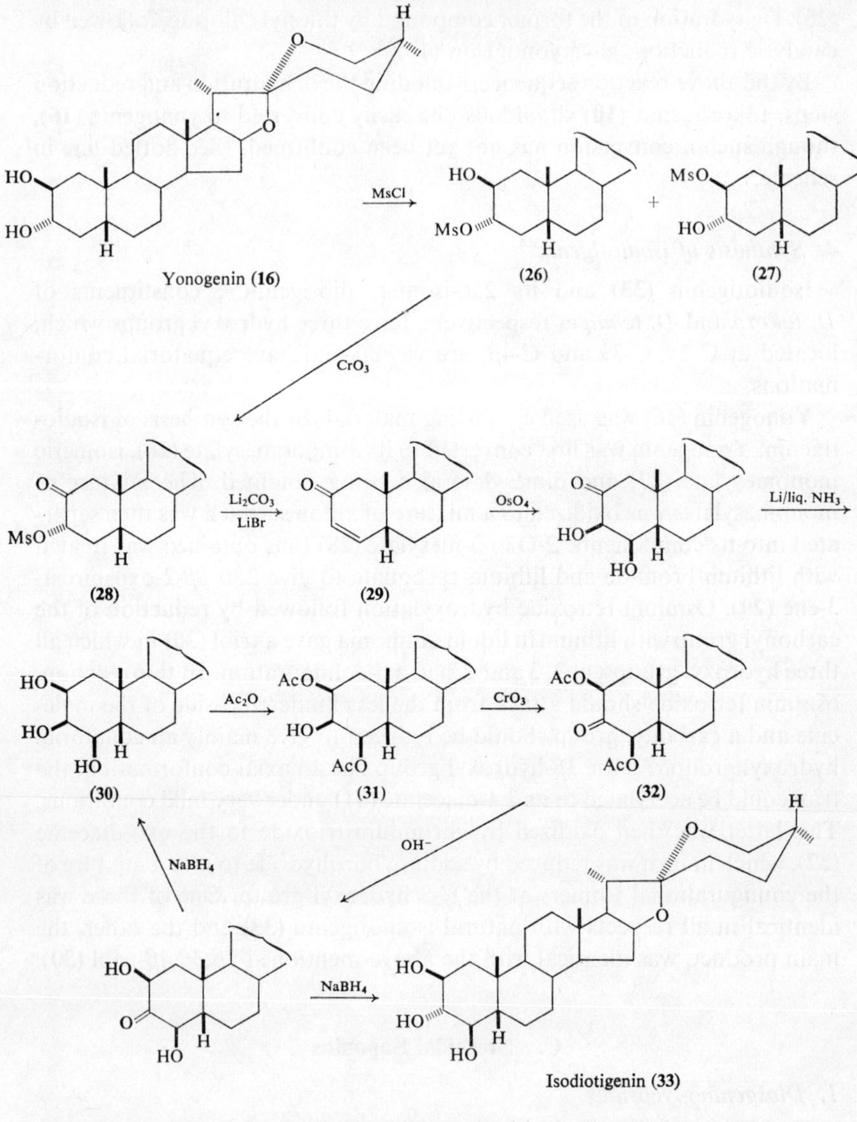

Yonogenin (16)   (26)   (27)

MsCl

CrO$_3$

(28)   (29)   (30)   (31)   (32)

Li$_2$CO$_3$
LiBr

OsO$_4$

Li/liq. NH$_3$

Ac$_2$O

CrO$_3$

NaBH$_4$

OH$^-$

NaBH$_4$

Isodiotigenin (33)

are given in Table 3. The numbers written beside the lines between the sugars
in the Table show the positions of the linkage between the sugars.

An interesting finding was reported recently by Schreiber *et al.*[95] on the
glycoside of a sapogenin alkaloid isolated from *Solanum paniculatum* L.
They assigned to this alkaloid glycoside the structure 25D-3β-amino-5α-
furostane-22α,26-diol 26-*O*-glucoside (34), a structure analogous to that of

Table 3. Glycosides of diosgenin and spirostane-3α-ols

Saponin	M.p. (°C)	$[\alpha]_D$	Aglycone	Sugar	Source
Dioscin	276–280	−115	diosgenin-3——glucose	2-rhamnose[86]; 4-rhamnose	*D. gracillima, D. nipponica, D. polystachya*[87] *D. septemloba*[44], *D. tenuipes*[92], *D. tokoro*
Gracillin	284–289	−85	diosgenin-3——glucose	2-rhamnose[86]; 3-glucose	*D. gracillima*[93], *D. polystachya*[87], *D. septemloba*[44]
Dioscinin	202–203		diosgenin-3——glucose	2-rhamnose; 4-rhamnose ? glucose[87]	*D. polystachya*[87]
Prosapogenin-B	215–220	−96	diosgenin-3——glucose—2-rhamnose[19]		*D. tokoro*[19]
Proto-dioscin	225–230	−69.7	diosgenin	26—glucose[88]; 3—glucose, 2-rhamnose, 4-rhamnose	*D. gracillima, D. septemloba, D. tenuipes*[88]
Proto-gracillin	185–189	−69.2	diosgenin	26—glucose[88]; 3—glucose, 2-rhamnose, 3-glucose	
Yononin	238–240	−14.5	yonogenin-2—arabinose[89]		*D. tokoro*[94]
Tokoronin	270–274	−10.6	tokorogenin-1—arabinose[90]		*D. tokoro*[94]
Compound X	275–284	−43.0	tokorogenin-1—glucose[91]		*D. tokoro*[90]

11

a sarsaparillosid which was isolated from *Smilax aristolochiaefolia* Mill. and reported by Tschesche *et al.*[96] about the same time.

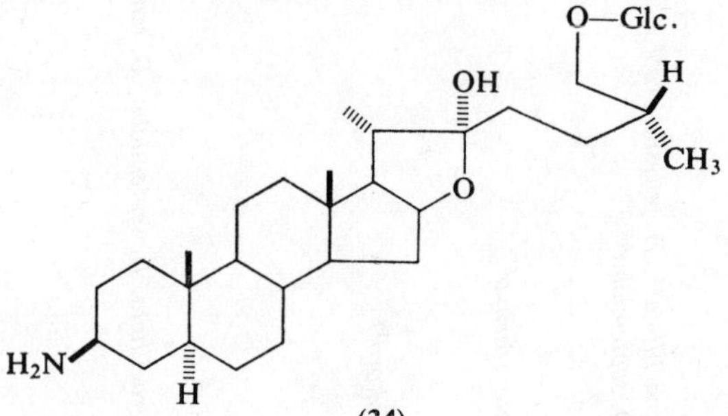

(34)

These 26-*O*-linked saponins also exist in some *Dioscorea* species. For instance, proto-dioscin (**35**) and proto-gracillin (**36**), both furostane-type 22,26-diol 26-glucosides, have recently been reported by Kawasaki *et al.*[88] Besides these, some other furostane-type saponins have also been reported by the same author. Kikubasaponin*, a diosgenin tetraglycoside of the furostane 26-*O*-glucoside type, was isolated from *D. septemloba*, and two more have been isolated from *D. gracillima*. These furostanol 26-*O*-glucosides are easily converted to spirostanols by acid hydrolysis under very mild conditions, accompanied by removal of a sugar moiety from the 26-position.

Furostanol glycosides are also found in certain members of the Liliaceae and Scrophulariaceae. The interesting proto-saponins, proto-pennogenin glycoside (**37**) and $\Delta^{17(20)}$-dehydrokryptogenin glycoside (**38**), have recently been isolated from plants belonging to the genus *Trillium* (Liliaceae), also by Kawasaki *et al.*[97]

These furostane-type 26-*O*-glucosides, including pseudo-sapogenin, give a red colour with Ehrlich reagent but spirostane itself and phytosterols give no coloration.

About 20 years ago, Marker[98] suggested that steroidal sapogenins might be secondarily produced artifacts, derived from glycosides of furostanol-type (opened F ring) sapogenins by acid hydrolysis. The above results support Marker's suggestion. It is also to be expected that these furostanol-type glycosides occur widely in plants as proto-types of spirostanol glycosides.

2. Glycosides of 3α-hydroxylated steroidal sapogenins

Since almost all saponins with a 3β-hydroxyl group have a sugar moiety attached at the 3β-hydroxyl, an interesting question is to which hydroxyl

---

* Recently, kikubasaponin was proved to be identical with proto-type (22-OCH$_3$) gracillan by the Kawasaki group (private communication).

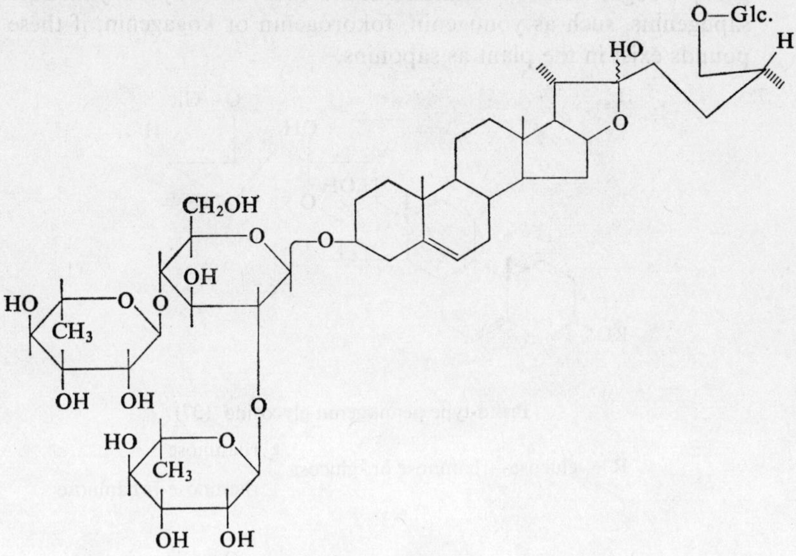

Proto-dioscin (35)

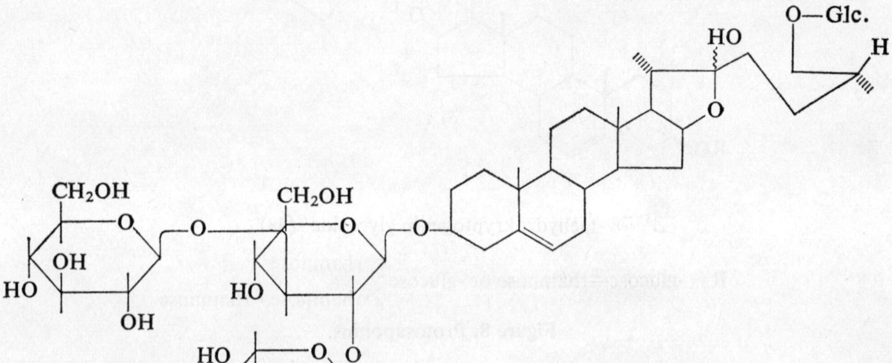

Proto-gracillin (36)

Figure 7. Some 26-*O*-linked saponins.

11*

group a sugar residue is attached in the unusual 3α-hydroxylated steroidal sapogenins, such as yonogenin, tokorogenin or kogagenin, if these compounds exist in the plant as saponins.

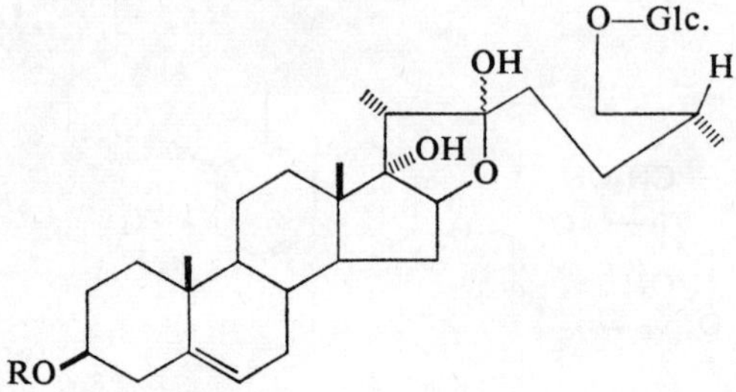

Proto-type pennogenin glycoside (37)

$$R = \text{-glucose}\overset{2}{-}\text{rhamnose} \quad \text{or} \quad \text{-glucose} \begin{cases} \overset{2}{\diagup} \text{rhamnose} \\ \underset{4}{\diagdown} \text{rhamnose}\overset{4}{-}\text{rhamnose} \end{cases}$$

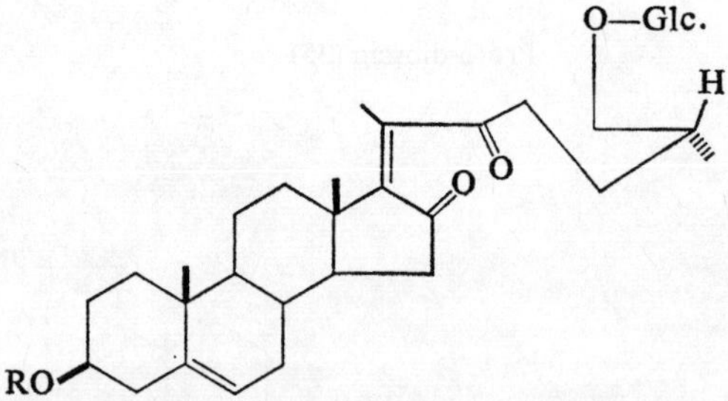

$\Delta^{17(20)}$-Dehydrokryptogenin glycoside (38)

$$R = \text{-glucose}\overset{2}{-}\text{rhamnose} \quad \text{or} \quad \text{-glucose} \begin{cases} \overset{2}{\diagup} \text{rhamnose} \\ \underset{4}{\diagdown} \text{rhamnose}\overset{4}{-}\text{rhamnose} \end{cases}$$

Figure 8. Protosaponins.

These novel types of sapogenin having a 3α-hydroxyl group in the molecule are isolated only from *D. tokoro* and *D. tenuipes*. In the aerial part of both plants they exist as free sapogenins, while in the underground part they exist as saponin, though in very low concentration. In spite of the small amount of these saponins in the rhizome, however, the structures of three of them have been elucidated; these are yononin[89], a saponin of yonogenin, and the tokorogenin derivatives tokoronin[90, 94] and the tentatively named compound X[91].

The structure of yononin (**39**) was determined as yonogenin-2-O-α-L-arabinopyranoside by Kawasaki *et al.*, using chemical and physicochemical methods. Similarly, tokoronin (**40**) and compound X (**41**) were assigned as tokorogenin-1-O-α-L-arabinopyranoside and tokorogenin-1-O-β-D-glucopyranoside, respectively, by the same group.

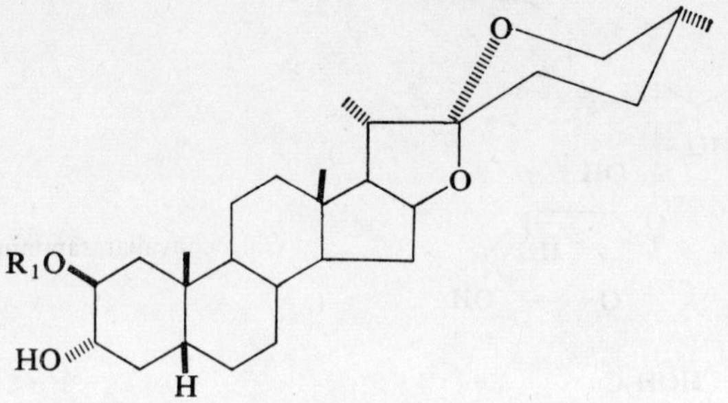

Yononin (**39**)

R₁ = α-L-Arabinose

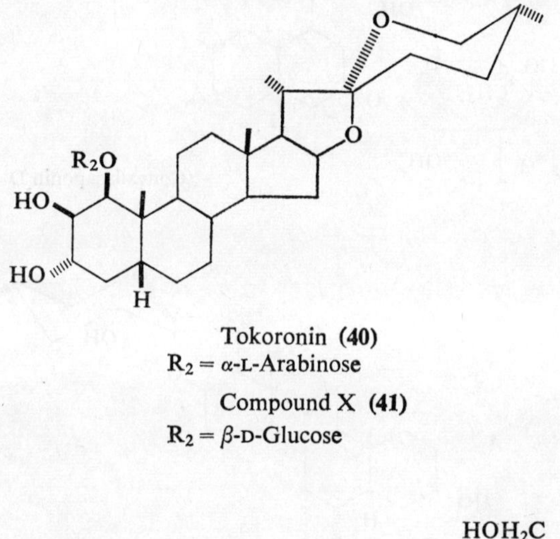

Tokoronin (**40**)

R₂ = α-L-Arabinose

Compound X (**41**)

R₂ = β-D-Glucose

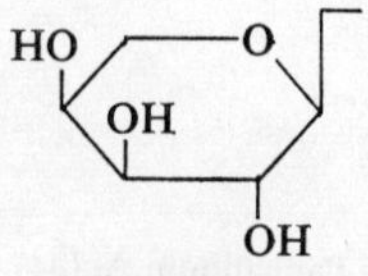

L-Arabinose

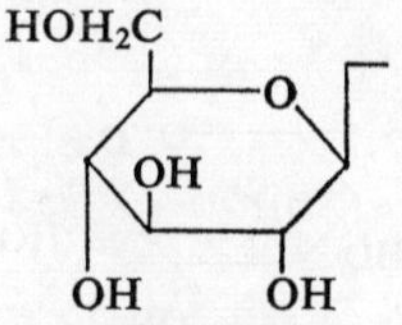

D-Glucose

Figure 9. Some saponins of 3α-hydroxy spirostanes.

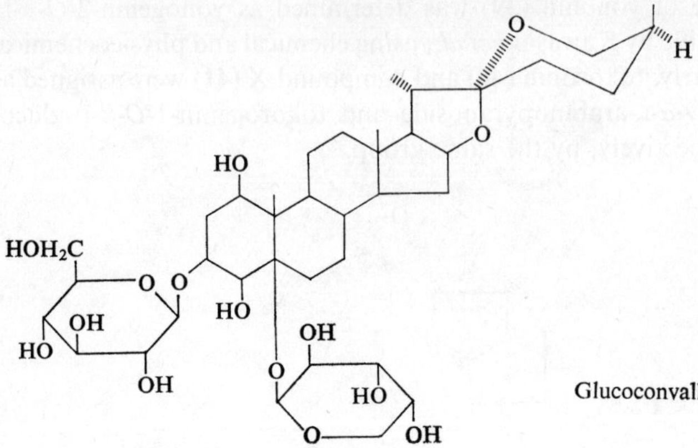

Glucoconvallarasaponin B (42)

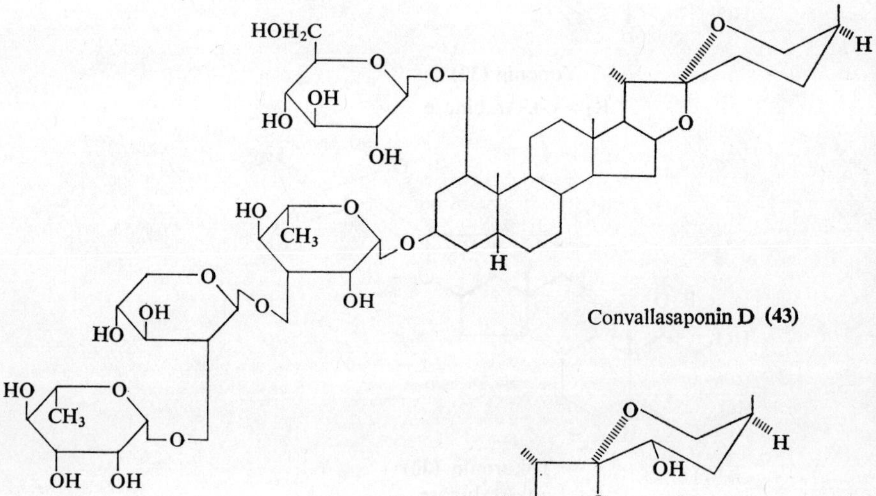

Convallasaponin D (43)

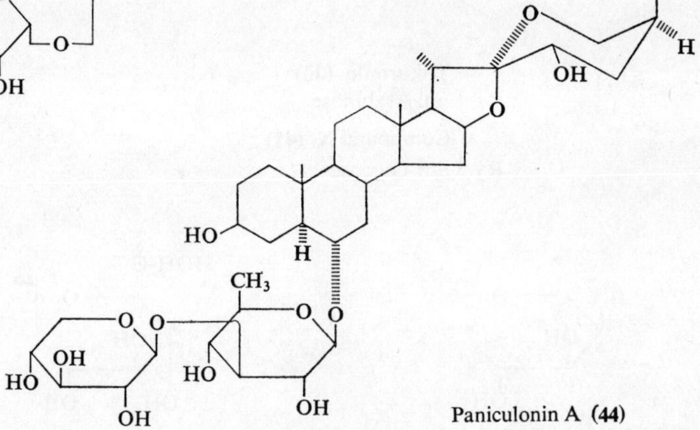

Paniculonin A (44)

Figure 10. Some examples of glycosides in which a sugar is located at other than the 3-position.

In these compounds, the sugar moiety is always attached to a hydroxyl group other than the 3α-hydroxyl group, which remains unsubstituted.

Other examples of saponins in which a sugar linkage is located other than at the 3-hydroxyl group are glucoconvallasaponin B (**42**)[99], convallasaponin D (**43**)[100] and paniculonin A (**44**)[101] and B.

In all these saponins, except for paniculonin, in which the second sugar moiety is linked to a 6α-hydroxyl group, the sugar moiety is attached to a β-oriented hydroxyl group in the steroid molecule.

Finally, if, as reported by Chakravarti, epismilagenin does exist in plants, and if it is present as a saponin, it should be a furostanol glycoside, which is yet another interesting problem to be solved.

D. Biosynthesis

1. Biosynthesis of steroidal sapogenins

Much work has been done and many hypotheses made regarding the biosynthetic pathway from mevalonic acid and from squalene to steroidal sapogenins, but it was Heftmann *et al.*[102] who first reported the role of cholesterol in this pathway by demonstrating the incorporation of labelled cholesterol into diosgenin by *D. spiculiflora*. Subsequently, the conversion of cholesterol to steroidal sapogenins was proved in several species in families other than Dioscoreaceae[103, 104].

On the other hand Ourisson *et al.*[105] isolated cycloartenol together with 24-methylene cycloartenol from tissue culture derived from *D. composita*, and they postulated that cycloartenol might be an alternative to lanosterol as a precursor of cholesterol and steroidal sapogenins.

Recently, Tomita *et al.*[106] in our laboratory found that the callus derived from seedlings of *D. tokoro* retains the ability to synthesize diosgenin, yonogenin, tokorogenin and some phytosterols together with cholesterol. They then took 24-tritiated cycloartenol[107], synthesized by reduction of 24-oxo-acetylcycloartenol with tritiated sodium borohydride followed by dehydration and deacetylation with lithium aluminium hydride, incubated it with the callus derived from *D. tokoro* and confirmed that this tritiated cycloartenol was incorporated into the steroidal sapogenins mentioned above as well as into some phytosterols. These results showed that cycloartenol is not only a precursor of certain phytosterols but also of steroidal sapogenins in *D. tokoro* tissue culture.

As Cornforth *et al.*[108] had confirmed the positions of tritium and radioactive carbons (^{14}C) in squalene (**47–48**), using tritium and ^{14}C-labelled isomeric mevalonic acids (**45–46**) as shown in Fig. 11, the labelled positions of cycloartenol (**49–50**) were also easily deduced. Then in order to

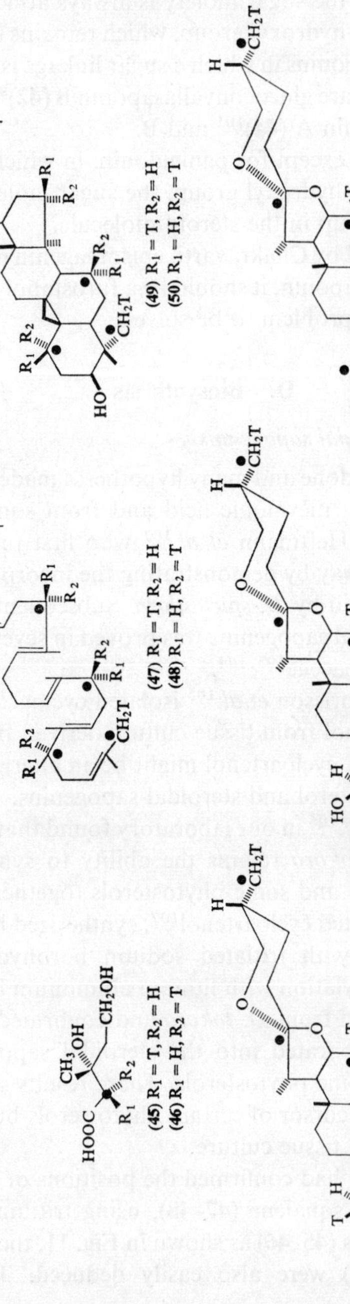

(49) $R_1 = T, R_2 = H$
(50) $R_1 = H, R_2 = T$

(53) $T:{}^{14}C = 3:5$

(47) $R_1 = T, R_2 = H$
(48) $R_1 = H, R_2 = T$

(52) $T:{}^{14}C = 3:5$

(45) $R_1 = T, R_2 = H$
(46) $R_1 = H, R_2 = T$

(51) $T:{}^{14}C = 4:5$

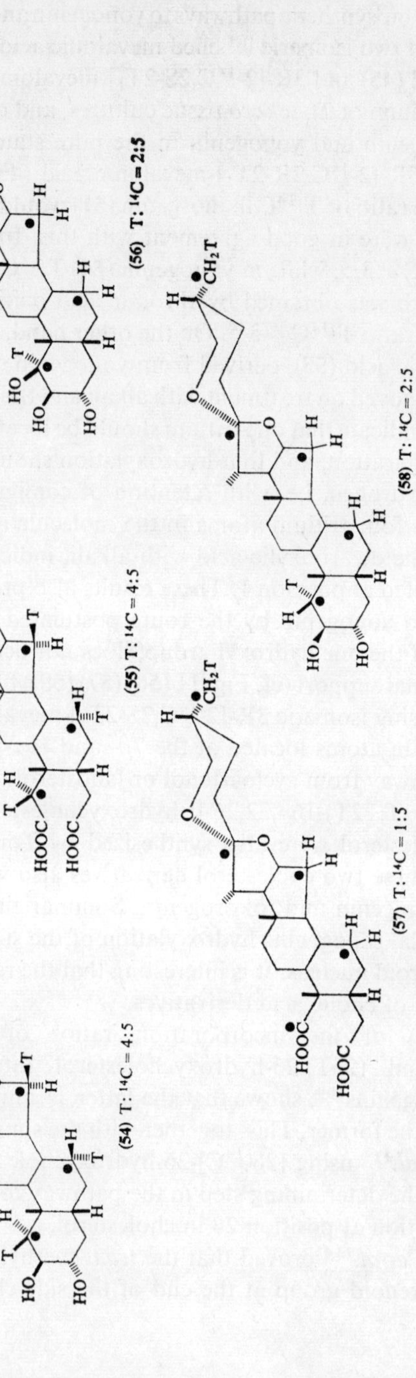

(56) T:^{14}C = 2:5

(58) T:^{14}C = 2:5

(55) T:^{14}C = 4:5

(57) T:^{14}C = 1:5

(54) T:^{14}C = 4:5

Figure 11. Tritiated and ^{14}C-labelled intermediates and sapogenins used in elucidating biosynthetic pathways.

establish[109] the biosynthetic pathways to yonogenin and tokorogenin, Tomita *et al.* incubated two isomeric labelled mevalonic acids, 3R-[2-^{14}C,2R-2T$_1$]-mevalonic acid (45) and 3R-[2-^{14}C,2S-2T$_1$]-mevalonic acid (46), separately in a liquid medium of *D. tokoro* tissue cultures, and obtained labelled diosgenin, tokorogenin and yonogenin in the pure state from the incubated callus. When 3R-[2-^{14}C,2R-2T$_1$]-mevalonic acid (45) was used it was expected that the ratio of T/^{14}C in diosgenin (51) would be 4/5 and the experimental results were in good agreement with this. In tokorogenin (52) the ratio was T/^{14}C = 3/5, while in yonogenin (54) T/^{14}C was 4/5. Tokorogenic acid (53), which was obtained by chromic acid oxidation of tokorogenin, also showed a ratio T/^{14}C = 3/5. On the other hand, although the ratio for the dicarboxylic acid (55), derived from yonogenin (54), was also 4/5, one tritium was removed on treatment with alkali and the ratio became 3/5.

The results indicate that one tritium should be located at the C-1 position, with a β-configuration, and that hydroxylation should occur at the side of the tritiated hydrogen, i.e. with retention of configuration. Furthermore, yonogenin has four tritium atoms in the molecule and one of them is removed from the dicarboxylic acid with alkali, indicating that this tritium should be located at position 1. These results also prove that conversion of tokorogenin to yonogenin by the route postulated by Marker[110], i.e. by the removal of the one hydroxyl group, does not occur under these conditions. Additional support (cf. Fig. 11 (56) (57) (58)) for this was provided by experiments using isomeric 3R-[2-^{14}C,2S-2T$_1$]-mevalonic acid (46). In this case two tritium atoms located at the 7β- and 15α-positions are removed along the pathway from cycloartenol or lanosterol to cholesterol[111, 112].

Further, (16αT, 22T)-16β,22,26-trihydroxycholesterol and (16αT)-16β,26-dihydroxycholesterol were also synthesized by Tomita[109] and incubated with callus. These two cholesterol derivatives also were incorporated into diosgenin, yonogenin and tokorogenin. Summarizing these results, in the case of steroidal sapogenins hydroxylation of the side-chain occurs before that of the steroid nucleus. It is interesting that the reverse is the case in the hydroxylation of cholic acid derivatives.

Comparison of the incorporation ratios of (16αT)-16β-hydroxycholesterol and (26T)-26-hydroxycholesterol into the corresponding steroidal sapogenins[109], shows that the latter is much more highly incorporated than the former. This, together with the similar results obtained by Heftmann *et al.*[113] using [26-^{14}C]-26-hydroxycholesterol in *D. floribunda*, indicate that the determining step in the pathway to steroidal sapogenin is the hydroxylation at position 26 in cholesterol.

Since Birch *et al.*[114] proved that the *trans*-methyl residue of the isopropylidene isoprenoid group at the end of the side-chain of cycloartenol is derived from the carbon atom at position 2 of mevalonic acid, and since Heftmann *et al.*[113] have proved that interconversion of a 25D-sapogenin to a 25L-sapogenin or the reverse reaction cannot occur *in vivo*, the 25D-sapogenin must be derived by hydroxylation of the *cis*-carbon atom of the isopropylidene group of cycloartenol, and the 25L-sapogenin by hydroxylation of the *trans*-carbon atom, as shown below (see page 320).

2. *Biosynthesis of 3α-hydroxylated steroidal sapogenins*[115]

It seemed of interest to us to clarify the biosynthetic pathway to the 3α-hydroxylated sapogenins and to compare this with the biosynthesis of bile acid from cholesterol in the animal kingdom.

We[115] used a homogenate of *D. tokoro* seedlings, taken about 10–15 days after germination, and adjusted to pH 6.5 with phosphate buffer. When labelled diosgenin was incubated with this homogenate it was converted to diosgenone (**59**) and smilagenin (**60**) in the incorporation ratios of 0.84 and 0.11%, respectively, together with a small amount (incorporation ratio = 0.0062%) of yonogenin (**16**). Next, diosgenone (**59**) was incubated and found to be converted mainly to the 3β-hydroxy-Δ^4 derivative (**61**) (0.25%), plus a small amount of smilagenin (0.057%) and traces of epismilagenin (**62**) and yonogenin (**16**). Smilagenin and epismilagenin were converted to smilagenone (**63**) in the incorporation ratios 0.79 and 0.40%, respectively, while smilagenone was converted to smilagenin, epismilagenin and yonogenin, though in very low yield, in the ratios of 0.046, 0.06 and 0.002%, respectively. Although the pathway from the Δ^4-3-one derivative (**61**) to smilagenone (**63**) has not yet been confirmed, it must follow from these results that smilagenone (**63**) is an important intermediate in this pathway.

Smilagenin (**60**) and epismilagenin (**62**) also gave yonogenin (**16**) on incubation, in incorporation ratios of 0.014 and 0.018% respectively. A trace of isodiotigenin (**33**) (0.00083%) was obtained, but only from the incubation of epismilagenin and no tokorogenin was detected.

The results are summarized below (see page 321).

As far as can be gathered from these results, epismilagenin would seem to be a most important precursor in the biosynthesis of 3α-hydroxylated steroidal sapogenins. These results also support the work of Chakravarti *et al.*[37], who isolated epismilagenin and smilagenone from *D. deltoidea* and *D. prazeri*.

III. OTHER PROBLEMS

A. Changes in Sapogenin Content with Age of Plant

Since Marker *et al.*[116] reported that in some plants belonging to genus *Agave* (Amaryllidaceae), the sapogenin content changes according to the

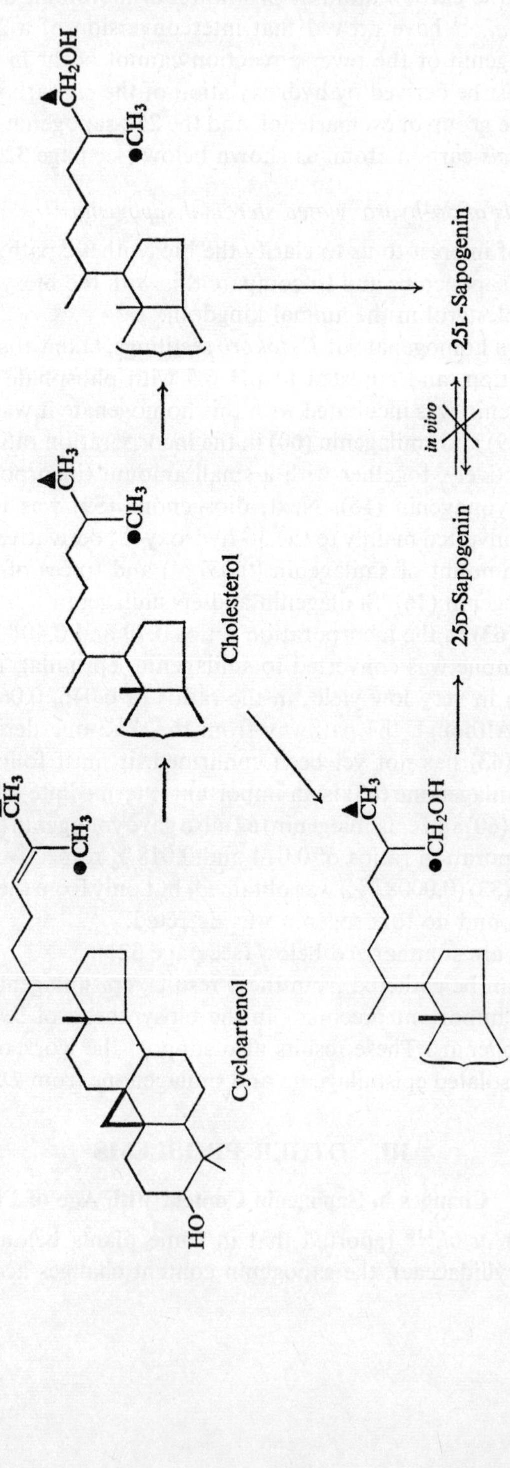

25L-Sapogenin

25D-Sapogenin

in vivo

Cholesterol

Cycloartenol

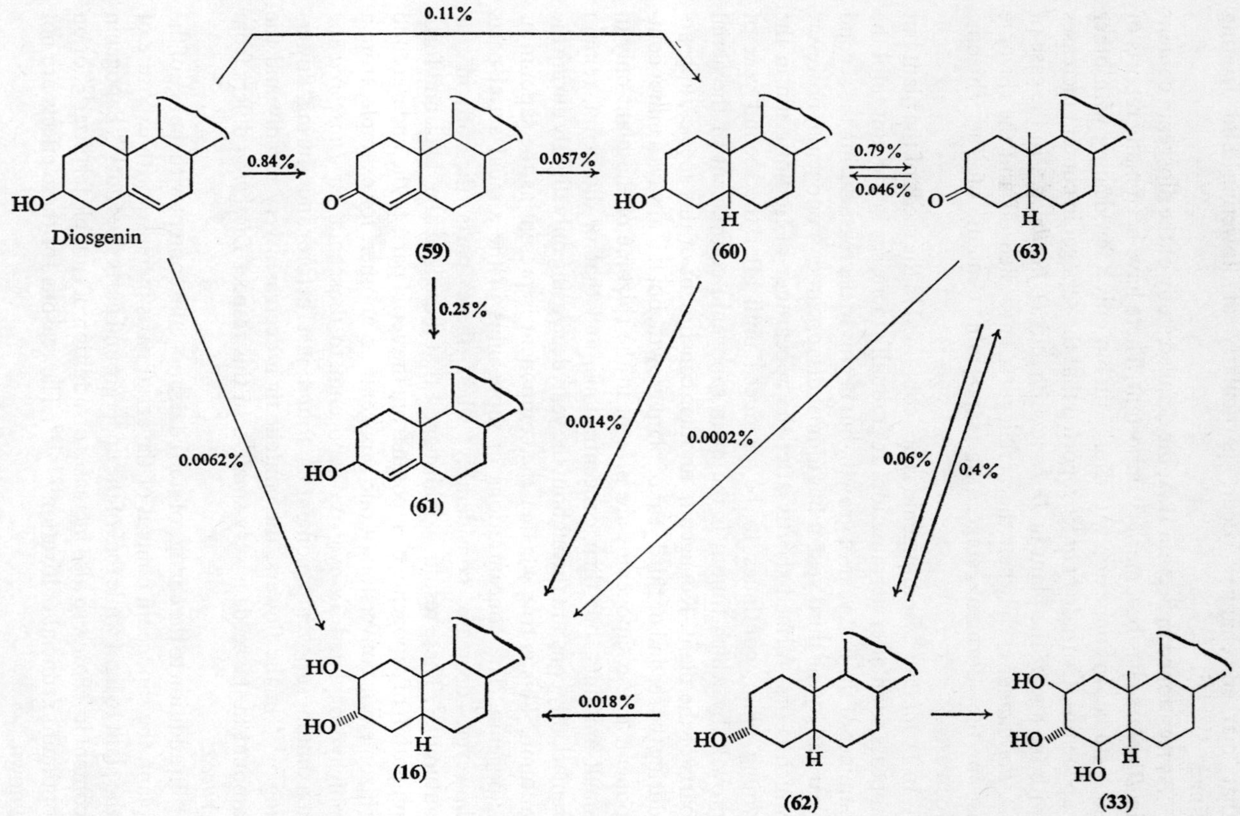
0.11%
Diosgenin
0.84%
(59)
0.057%
(60)
0.79%
0.046%
(63)
0.25%
(61)
0.0062%
0.014%
0.0002%
0.06%
0.4%
HO
HO
(16)
0.018%
HO
H
(62)
HO
HO
HO
H
(33)

age of the plant, many studies have been made[117, 118] and it has been found that marked changes occur, particularly at flowering and fruiting time[54, 119–121].

As mentioned in Section II.A, the detailed study of the diosgenin content in *Dioscorea* has become very important. There have been many reports of such studies in connexion with diosgenin but only a few dealing with other sapogenins. Cruzado *et al.*[122] reported that the sapogenin content increases with the age of the plant in *D. composita* and *D. floribunda*. In the case of *D. floribunda*, even when the aerial parts have died off and the rhizome is in the dormant stage, the sapogenin content of the rhizome increases.

In *D. tokoro*[76], isodiotigenin is not contained in the seeds of the plant but appears soon after germination. It gradually increases in amount until the summer of the first year of growth, but begins to decrease during the second year and by the third year it has almost disappeared. Yonogenin, however, which is first formed shortly after the appearance of isodiotigenin in the young plant, continues to be produced until after the second year of growth, by which time it is the main sapogenin component of the aerial parts of the plant. Kogagenin, another constituent of this species, appears during the second or third year of plant growth, but it is always a minor component. These sapogenins are present in the rhizome of the plant only in small amounts, but their concentrations, and that of diosgenin, remain remarkably constant throughout the year, decreasing only slightly during the autumn. This is true whether the component is present as the saponin or sapogenin. The concentrations of sapogenins in the annual aerial parts however, decrease considerably when these parts die off in the autumn[123]. These results would suggest that there is neither back and forth migration of sapogenin or saponin from the aerial parts to the underground rhizome nor conversion of one sapogenin to another, for example, from a polyhydroxylated 3α-hydroxy sapogenin to diosgenin. In *Dioscorea* the maximum sapogenin content is found just before and during flowering[72, 124] in the flowers, the content then decreases very rapidly and the amount in the seeds is very small[63]. In this respect *Dioscorea* differs from *Yucca*.

In addition to the change of steroidal sapogenin content with the age of the plant, the sapogenin content of the aerial parts increases with exposure of the plant to light while that of the underground parts does not[125]. Sapogenin content in *Dioscorea* also increases on addition of suitable fertilizer[126] or on optimum artificial cultivation[127–129]. The reasons for these effects are not known.

B.   Distribution of Sapogenins in the Plant

The steroidal sapogenin composition of different parts of the same plant differs both in the quantities and in the kinds of sapogenins contained. Studies on plants of the genera *Yucca*[116, 119, 130] and *Digitalis*[131], for example, have shown that the amounts and kinds of sapogenins found in the fruit differ from those in the flowers and leaves.

As *Dioscorea* plants have a large storage organ, the rhizome or tuber, it is the constituents of these parts which have been mainly examined, only minor work having been done on the aerial parts; however, the majority of the steroidal sapogenins contained in the plant is in any case generally found in the underground parts with only minor amounts distributed in the aerial parts of the plant. In studies of the constituents of nine Japanese species of *Dioscorea*[40, 63] which contain steroidal sapogenins, diosgenin and yamogenin have been isolated from the rhizomes of all of the species but these compounds have been scarcely detected in the aerial parts of these plants. In *D. tokoro* and *D. tenuipes* however, although yonogenin (or its 25L-isomer) and diotigenin (or its 25D-isomer), both 3α-hydroxylated sapogenins, have been isolated in considerable amounts from the aerial parts, they are found only in very small amounts in the underground parts. These 3α-sapogenins are usually present in aerial parts as the free sapogenin, while in the rhizome they generally occur as the saponin[132]. A peculiar distribution pattern is found in the case of igagenin or isodiotigenin; the former is isolated only from the female flowers[74] while the latter occurs in the tissue of the buds and young sprouts of the plant[76, 133]. Though diosgenin has generally only been isolated in small amounts from the aerial parts of *D. tokoro*, some specimens exist which contain a considerable amount of this substance[123]. *Dioscorea tenuipes*, a species morphologically very similar to *D. tokoro*, has two types. Only one of these types contains steroidal sapogenins in the aerial parts, but the constituents of the rhizomes are identical, and diosgenin, yamogenin and tokorogenin are isolated as components from each type[40]. Morphological examination shows no distinction between the types but it is clear that *D. tenuipes* is heterogenous as far as its steroidal sapogenin composition is concerned, and Akahori *et al.* have tentatively used the name *D. tenuipes* complex to include both types[73]. Some problems concerning these two species *D. tokoro* and *D. tenuipes* will be discussed in detail in the last section.

The roles of the steroidal sapogenins are not known, but in this connexion it is perhaps relevant that particular sapogenins are found in the young tissues and other specific parts of the plant.

C.   Components Other than Steroidal Sapogenins

The following components other than steroidal sapogenins have been isolated from *Dioscorea*: phytosterols, terpenes, alkaloids, organic and amino
acids, some tannins and antocyanins[134] and carbohydrates[135].

1. Phytosterols and triterpenes

β-Sitosterol and stigmasterol[73,136] have been isolated as the sterol components from *Dioscorea*. The very common triterpene, taraxerol[73], has also
been found in some *Dioscorea* plants. It is interesting that cycloartenol and
24-methylene cycloartenol (**64**) together with β-amyrin have been isolated
only from the cultivated tissue of *D. composita* and not from the fresh
plant[105]. This fact suggests that cycloartenol is an important precursor for
the biosynthesis of the steroidal sapogenins *via* cholesterol, analogous to
lanosterol in the animal body.

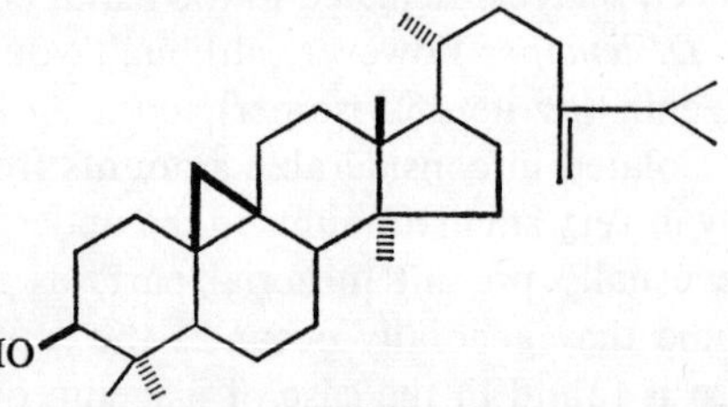

HO

(**64**)

2. Diterpenes

Recently, Kawasaki[137,138] and his coworkers reported the existence of
several derivatives belonging to a new class of furanoid nor-diterpenes having a modified labdane skeleton, named diosbulbin A-G, in the whole plants
of *D. bulbifera* L., forma spontanea Makino et Nemoto (a wild type of
D. bulbifera L.), a plant in which steroidal sapogenins had hitherto scarcely
been detected*. The structures of diosbulbin A, B and C, the main components of this plant, are shown in Fig. 12, (**65**), (**66**) and (**67**). Although the
other four components, diosbulbin D to G, are also assumed to have a
similar skeleton and to be closely related to diosbulbin B, their detailed
structures are not yet clear. Traces of diosbulbins have also been found in
D. bulbifera L., *D. septemloba* and *D. rhipogonoides*[139]. Diosbulbin B is
always the main component among these furanoid nor-diterpene constituents in *Dioscorea*. All these furanoid nor-diterpene derivatives have a bitter
taste.

* A minute amount of diosgenin was detected in the glycoside fraction of this plant[137].

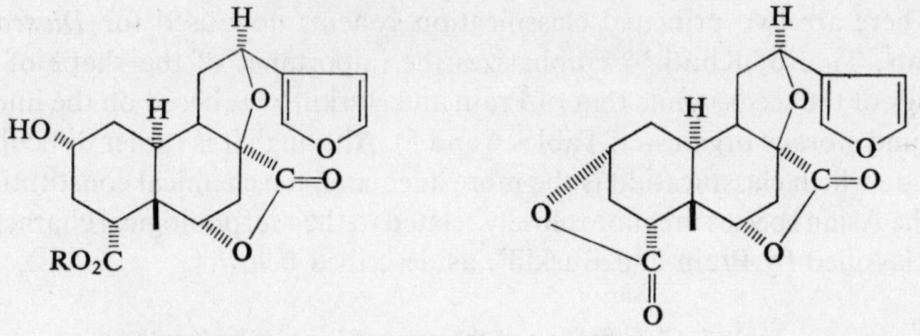

(65) R = CH₃
(67) R = H

(66)

Figure 12. Diosbulbin analogues.

3. Alkaloids

The first paper on the alkaloids of *Dioscorea* appeared as early as 1894. This was by Boorsma[140], who reported the isolation of dioscorine from *D. hirsuta* Bl., and later from *D. hispida* Denstedt[141]. Another alkaloid, dihydrodioscorine was also isolated from *D. sansiberensis* Pax and *D. dumetorum* Pax together with dioscorine. The structure of dioscorine was determined as **68** from the results of the chemical degradation by Pinder *et al.* in 1961[142]. Dihydrodioscorine is a saturated dioscorine but its stereochemistry has not yet been clarified[143].

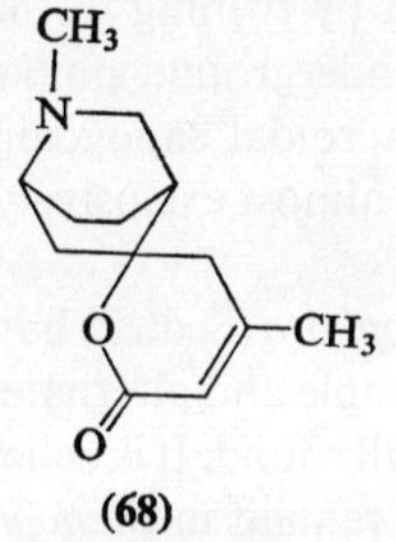

(68)

Dioscorea batatas Decne. cultivar *Tsukune* Makino also contains allantoin[144] and 16 amino acids.

D. Chemotaxonomy of *Dioscorea*

As mentioned in Section I, the constituents of plants belonging to the Dioscoreaceae have not been examined sufficiently, and steroidal sapogenins have so far been isolated only from the genera *Dioscorea* and *Tamus*. For this reason, the discussion here on the relationship between chemical constitution and morphological classification (Sections), will be limited to the genus *Dioscorea*.

There are two principal classification systems now used for *Dioscorea* plants. That of Knuth[5,6] emphasizes the importance of the shape of the wings of the seeds, while that of Prain and Burkill[7,8] is based on the underground storage organs (cf. Tables 4 and 5). Although it is rather difficult to decide which classification is the more adequate, the chemical constitutions of the Asian species are more closely related to the morphological characters as classified by Prain and Burkill[8], as described below.

Table 4. Classification of the genus *Dioscorea* by Knuth

Seeds winged towards the base of the capsule
 Subgen. I. *Helmia* (Kunth) Benth.   (Sect. 1–Sect. 19)

Seeds winged all round
 Subgen. II. *Eudioscorea* Pax   (Sect. 20–Sect. 58)

Seeds winged towards the apex of the capsule
 Tuber underground
  Subgen. III. *Stenophora* (Uline) R. Knuth   (Sect. 59, 60)

 Tuber above ground and large with a very thick coat of cork
  Subgen. IV. *Testudinaria* (Salisb.) Uline

Plants belonging to section *Enantiophyllum* (Sect. 9) have the following characteristics: (1) climbing by twining to the right, (2) leaves simple, (3) seeds winged all round, (4) underground annual tubers. Plants of this section however, contain neither steroidal sapogenins nor alkaloids. The plants cultivated as food crops are almost exclusively limited to the species of this section[145].

Plants of the section *Stenophora* (Sect. 2) have the characteristics of climbing by twining to the left, simple and alternate leaves, rhizomes, and except for *D. tokoro*, seeds winged all round. It is believed that more of the ancestral characteristics of the genus remain in *Stenophora* plants than in the other sections. Steroidal sapogenins have been isolated from all species belonging to this section so far examined, about twenty out of approximately thirty known *Stenophora* species. Steroidal sapogenins have been isolated from only two species belonging to other sections; *D. esculenta* of the section *Combilium* (Sect. 4) and *D. bulbifera* of *Opsophyton* (Sect. 7).

Alkaloids have been isolated from plants belonging to the section *Lasiophyton* (Sect. 8), the members of which show the following characteristics: male flower without a disc, leaves usually compound, and seeds winged towards the base of the capsule. This is also the only Asian section in the genus whose plants have compound leaves.

Thus, the chemical constitution and morphological character of the plants

Table 5. Classification of the genus *Dioscorea* in the East by Prain and Burkill

Dwarf and not climbing	Sect. 1. *Borderea* (Miég.) Benth. et Hook.
Climbing by twining to the left	
Underground parts a rhizome; leaves simple; seeds winged all round except *D. tokoro*	Sect. 2. *Stenophora* Uline
Underground parts an annual tuber	
Male flowers with a disc; leaves simple; seeds winged all round or towards the apex of the capsule	Sect. 3. *Stenocorea* Prain et Burkill Sect. 4. *Combilium* Prain et Burkill Sect. 5. *Paramecocarpa* Prain et Burkill Sect. 6. *Shannicorea* Prain et Burkill
Male flowers without a disc; leaves simple or compound; seeds winged towards the base of the capsule	
Leaves simple	Sect. 7. *Opsophyton* Uline
Leaves usually compound	Sect. 8. *Lasiophyton* Uline
Climbing by twining to the right; underground parts an annual tuber; leaves simple; seeds winged all round	Sect. 9. *Enantiophyllum* Uline

are closely related to the Prain and Burkill classification of the eastern *Dioscorea*. Furthermore, as it is thought by many botanists that the plants growing in the Japanese area are the oldest in the genus, and as steroidal sapogenins, especially 3α-hydroxy spirostanes, are contained for the most part in the *Stenophora* section of the eastern *Dioscorea*, it would seem to be important and interesting to study the evolutionary progress of the plants belonging to the order Liliales from this point of view.

Though morphologically *D. tokoro* and *D. tenuipes* differ from each other only slightly in their flowers and seeds[8, 146] (cf. Fig. 13), there is a marked

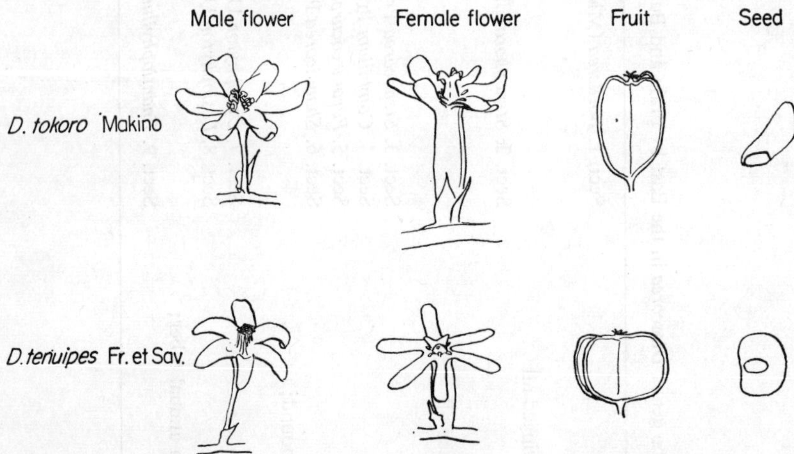

Figure 13. Flowers, fruit and seeds of *Dioscorea tokoro* and *Dioscorea tenuipes*.

difference in their chemical constitution. As shown in Table 6, the major constituents of the steroidal sapogenins isolated from *D. tokoro* all belong to the 25D-(iso) series, while, except for yonogenin and tokorogenin, the constituents of *D. tenuipes* are all in the 25L-(neo) series. Thus it would seem that the chemical composition of the plant is a very important factor in classification of the species. *Dioscorea tenuipes* particularly is an interesting plant, having two morphologically identical but constitutionally distinct types. One type contains steroidal sapogenins (mainly diosgenin, yamogenin and tokorogenin) only in the rhizome, whereas the other contains them in both the aerial and underground parts (the 3α-hydroxy sapogenins neoyonogenin and diotigenin are contained in the aerial parts in this species[73]). The former type grows in the eastern part of Japan and the latter in the western part, though some specimens of the latter growing in Kyushu show a somewhat different pattern in the spirostanes isolated from their aerial parts[123].

Table 6. Constituents of *D. tokoro* and *D. tenuipes*

Sapogenins	*D. tokoro*	*D. tenuipes*
Yonogenin	+	+
Neoyonogenin		+
Igagenin	+	
Tokorogenin	+	+
Neotokorogenin		+
Isodiotigenin	+	
Diotigenin		+
Kogagenin	+	
Isotenuipegenin	+	
Tenuipegenin		+

Furthermore the chromosome number of the former is $2_n = 20$ while that for the latter is $2_n = 40$[147]. From these facts it may be assumed that this plant is now in the process of evolution.

IV. CONCLUSION

Sufficiently detailed morphological and chemotaxonomical studies on Dioscoreaceae have not yet been made.

The following points especially would seem to be worth investigating further.

1. Certain 3α-hydroxy spirostanes have been isolated from *Dioscorea* species, this being the first time these compounds have been found in the plant kingdom. They have been found in two species only, *D. tokoro*, and *D. tenuipes*. both species indigenous to Japan and closely related morphologically.

2. *Dioscorea tenuipes* exhibits two types which, though morphologically identical, are heterogenous in their sapogenin content and distribution, one type containing steroidal sapogenins in its aerial parts while the other does not.

3. The relationships between botanical evolution, morphological change and chemical constitution of *D. tokoro* and *D. tenuipes* thus merit attention.

V. ACKNOWLEDGEMENTS

The author would like to thank Professor T. Kawasaki of Kyushu University for his useful discussions and for providing unpublished data. Thanks are also due to Dr. H. Minato, Dr. A. Akahori and Dr. Y. Tomita for their kind help in preparing this manuscript and also to Mr. R. Meyer for his checking of the English translation.

VI. REFERENCES

1. J. Hutchinson, *The Families of Flowering Plants*, Vol. II, 2nd ed., Oxford University Press, Oxford, 1959, p. 652.
2. H. Melchior, *Syllabus der Pflanzenfamilien*, Vol. II, 2nd ed. (Ed. A. Engler), Gebrüder Borutraeger, Berlin, 1964, p. 533.
3. E. B. Uline, *Bot. Jahr.*, **25**, 125 (1898).
4. C. Queva, *Mem. Soc. Sci. Agr.*, Lille, Ser., 4, **20**, 1 (1894).
5. R. Knuth, 'Dioscoreaceae,' in *Das Pflanzenreich* (Ed. A. Engler), IV-43, 1924, p. 1.
6. R. Knuth, 'Dioscoreaceae,' in *Die naturliche Pflanzenfamilien*, 2nd ed. (Ed. A. Engler), 1936, p. 15a, 438.
7. I. H. Burkill, *J. Linn. Soc. (Bot.)*, **56**, 319 (1960).
8. D. Prain and I. H. Burkill, *Ann. Roy. Bot. Garden Calcutta*, **14**, 1 (1936).
9. E. Matuda, *An. Inst. Biol. (Mex.)*, **24**, 279 (1953).
10. B. G. Schubert, *J. Arnold Arboretum*, **47**, 147 (1966).
11. L. Fieser and M. Fieser, *Steroids*, 3rd ed., Reinhold Publishing Corp., New York, 1959, p. 810.
12. K. Takeda, S. Hara, A. Wada and N. Matsumoto, *J. Chromatog.*, **11**, 562 (1963).
13. A. Akahori, K. Murata, F. Yasuda, S. Nagase, M. Togami and T. Okanishi, *Ann. Rept. Shionogi Res. Lab.*, **16**, 74 (1966).
14. T. Okanishi and M. Togami, *Chem. Pharm. Bull.*, **17**, 315 (1969).
15. W. J. A. Vandenheuvel and E. C. Horning, *J. Org. Chem.*, **26**, 634 (1961).
16. H. A. Walens, A. Turner, Jr. and M. E. Wall, *Anal. Chem.*, **26**, 325 (1954).
17. C. Sannie and H. Lapin, *Bull. Soc. Chim. France*, 1080 (1952).
18. E. Heftmann and L. Hayden, *J. Biol. Chem.*, **197**, 47 (1952).
19. K. Kawasaki and T. Yamauchi, *Chem. Pharm. Bull.*, **16**, 1070 (1968).
20. M. E. Wall, C. R. Eddy, M. L. McClennan and M. E. Klumpp, *Anal. Chem.*, **24**, 1337 (1952).
21. R. N. Jones, E. Katzenellenbogen and K. Dobriner, *J. Am. Chem. Soc.*, **75**, 158 (1953).
22. K. Takeda, H. Minato, A. Shimaoka and Y. Matsui, *J. Chem. Soc.*, 4815 (1963).
23. K. R. Brain, F. R. Y. Falzi, R. Hardman and A. B. Wood, *Phytochemistry*, **7**, 1815 (1968).
24. H. Budzikiewicz, J. M. Wilson and C. Djerassi, *Monatsh*, **93**, 1033 (1962).
25. H. Budzikiewicz, K. Takeda and K. Schreiber, *Monatsh.*, **101**, 1003 (1970).
26. K. Tori and K. Aono, *Ann. Rept. Shionogi Res. Lab.*, **14**, 136 (1964).
27. T. Tsukamoto and Y. Ueno, *Yakugaku Zasshi, Japan*, **56**, 802 (1936).
28. K. Fujii and T. Matsukawa, *Yakugaku Zasshi, Japan*, **56**, 408 (1936).
29. F. W. Martin and H. Delpin, *Turrialba*, **15**, 296 (1965).
30. N. Applezweig, *Steroid Drugs*, Vol. 1, McGraw-Hill, London, 1962.
31. J. W. Rothrock, P. A. Hammes and W. J. McAleer, *Ind. Eng. Chem.*, **49**, 186 (1957).
32. E. B. Hershberg and D. H. Gould, *U.S. Pat.*, 2,774,714 (1956).
33. I. Yoshioka, M. Fujio, M. Osamura and I. Kitagawa, *Tetrahedron Letters*, 6303 (1966).

34. M. M. Krider, T. C. Cordon and M. E. Wall, *J. Am. Chem. Soc.*, **76**, 3515 (1954).
35. Y. Nagai, M. Sawai and Y. Kurosawa, *J. Agr. Chem. Soc., Japan*, **44**, 15 (1970).
36. R. Freire, A. G. González and E. Suarez, *An. Quim.*, **64**, 745 (1968).
37. D. Chakravarti, R. N. Chakravarti and M. N. Mitra, *Nature*, **186**, 236 (1960).
38. Pu-Chou Wang and T. H. Chou, *Acta Pharm. Sinica*, **11**, 235 (1964).
39. Jun Chow, Da-Gang Wu and Wei-Guang Huang, *Acta Pharm. Sinica*, **12**, 392 (1965).
40. A. Akahori, *Phytochemistry*, **4**, 97 (1965).
41. V. I. Kichenko, *Med. Prom. SSSR*, **15**, 17 (1961).
42. M. E. Wall, M. M. Krider, C. R. Eddy, J. J. Willaman, D. S. Correll and H. S. Gentry, U.S. Dept. Agr., Agr. Research Service Circ. AIC-363 (1954).
43. M. E. Wall, C. R. Eddy, J. J. Willaman, D. S. Correll, B. G. Schubert and H. S. Gentry, U.S. Dept. Agr., Agr. Research Service Circ., AIC-367 (1954).
44. H. A. Walens, S. S. Serota and M. E. Wall, *J. Am. Chem. Soc.*, **77**, 5196 (1955).
45. W. Koch, *Ceiba*, **12**, 58 (1966).
46. R. E. Marker, R. B. Wagner, P. R. Ulshafer, E. L. Wittbecker, D. P. J. Goldsmith and C. H. Ruof, *J. Am. Chem. Soc.*, **65**, 1199 (1943).
47. R. N. Chakravarti, D. Chakravarti and A. K. Barua, *Indian Med. Gaz.*, **88** (8), 1 (1953).
48. J. J. H. Simes, J. G. Tracey, L. J. Webb and W. J. Dunstan, *CSIRO, Bulletin*, No. 281 (1959).
49. F. W. Martin, *Econ. Bot.*, **23**, 373 (1969).
50. S. Serota, R. T. Koob and M. E. Wall, *J. Org. Chem.*, **25**, 1768 (1960).
51. I. T. Harrison, M. Velasco and C. Djerassi, *J. Org. Chem.*, **26**, 155 (1961).
52. M. E. Wall, C. S. Fenske, J. J. Willaman, D. S. Correll, B. G. Schubert and H. S. Gentry, U.S. Dept. Agr., Agr. Research Service Circ. ARS-73 (1955).
53. M. E. Wall, C. S. Fenske, H. E. Kenney, J. J. Willaman, D. S. Correll, B. G. Schubert and H. S. Gentry, *J. Am. Pharm. Assoc., Sci. Ed.*, **46**, 653 (1957).
54. G. Blunden, C. L. Briggs and R. Hardman, *Phytochemistry*, **7**, 453 (1968).
55. M. E. Wall, C. S. Fenske, J. W. Garvin, J. J. Willaman, Q. Jones, B. G. Schubert and H. S. Gentry, *J. Am. Pharm. Assoc., Sci. Ed.*, **48**, 695 (1959).
56. R. N. Chakravarti, M. N. Mitra and D. Chakravarti, *Bull. Cal. S.T.M.*, **8**, 58 (1960).
57. C. R. Podmore, *African Wild Life*, **14**, 209 (1960); *B.A.*, **36**, 2071 (1961).
58. F. W. Martin, N. E. Delfel and H. J. Cruzado, *Turrialba*, **13**, 159 (1963).
59. R. N. Chakravarti and S. N. Dash, *Bull. Cal. S.T.M.*, **8**, 59 (1960).
60. T. Tsukamoto and T. Kawasaki, *Yakugaku Zasshi, Japan*, **74**, 72 (1954).
61. J. Labat, *Anales Assoc. Quim. Arg.*, **47**, 5 (1959); cf. *C.A.*, **54**, 14301d (1960).
62. T. Tsukamoto, T. Kawasaki and Y. Shimauchi, *Yakugaku Zasshi, Japan*, **77**, 1221 (1957).
63. A. Akahori, *Ann. Rept. Shionogi Res. Lab.*, **11**, 93 (1961).
64. R. E. Marker, D. L. Turner and P. R. Ulshafer, *J. Am. Chem. Soc.*, **62**, 2542 (1940).
65. R. E. Marker, R. B. Wagner and P. R. Ulshafer, *J. Am. Chem. Soc.*, **64**, 1283 (1942).
66. R. Laorga and M. Pinar, *An. Real Soc. Españ. Fís. Quim.*, **56B**, 797 (1960).
67. G. Held and D. Vagujfalvi, *Bot. Közlem*, **52**, 201 (1963).
68. A. G. González, R. Freire and E. S. López, *An. Real Soc. Españ. Fís. Quim.*, **63B**, 966 (1967).
69. M. E. Wall, J. W. Garvin, J. J. Willaman, Q. Jones and B. G. Schubert, *J. Pharm. Sci.*, **50**, 100 (1961).
70. A. Akahori and F. Yasuda, *Chem. Pharm. Bull.*, **19**, 846 (1971).
71. T. Okanishi and A. Shimaoka, *Ann. Rept. Shionogi Res. Lab.*, **6**, 78 (1956).

72. K. Takeda, T. Okanishi and A. Shimaoka, *Chem. Pharm. Bull.*, **6**, 532 (1958).

73. A. Akahori, F. Yasuda and T. Okanishi, *Chem. Pharm. Bull.*, **16**, 498 (1968).

74. A. Akahori, I. Okuno, T. Okanishi and T. Iwao, *Chem. Pharm. Bull.*, **16**, 1994 (1968).

75. M. Nishikawa, K. Morita, H. Hagiwara and M. Inoue, *Yakugaku Zasshi, Japan*, **74**, 1165 (1954).

76. A. Akahori, F. Yasuda, I. Okuno, M. Togami, T. Okanishi and T. Iwao, *Phytochemistry*, **8**, 45 (1969).

77. H. Minato and A. Shimaoka, *Chem. Pharm. Bull.*, **11**, 896 (1963).

78. A. Akahori, private communication.

79. K. Morita, *Bull. Chem. Soc. Japan*, **32**, 791 (1959).

80. F. Yasuda, Y. Nakagawa, A. Akahori and T. Okanishi, *Tetrahedron*, **24**, 6535 (1968).

81. K. Takeda, T. Kubota and A. Shimaoka, *Tetrahedron*, **7**, 62 (1959).

82. T. Kubota, *Chem. Pharm. Bull.*, **7**, 898 (1959).

83. T. Kubota and K. Takeda, *Tetrahedron*, **10**, 1 (1960).

84. K. Takeda, G. Lukacs and F. Yasuda, *J. Chem. Soc.*, 1041 (1968).

85. M. Ogata, F. Yasuda and K. Takeda, *J. Chem. Soc.*, 2397 (1967).

86. T. Kawasaki and T. Yamauchi, *Chem. Pharm. Bull.*, **10**, 703 (1962).

87. O. S. Madaeva, V. K. Pyzhkova and V. V. Panina, *Khim. Prir. Soedin*, **3**, 155 (1967).

88. S. Kiyosawa, M. Hutoh, T. Komori, T. Nohara, I. Hosokawa and T. Kawasaki, *Chem. Pharm. Bull.*, **16**, 1162 (1968).

89. T. Kawasaki and K. Miyahara, *Tetrahedron*, **21**, 3633 (1965).

90. K. Miyahara and T. Kawasaki, *Chem. Pharm. Bull.*, **17**, 1369 (1969).

91. K. Miyahara, F. Isozaki and T. Kawasaki, *Chem. Pharm. Bull.*, **17**, 1735 (1969).

92. T. Tsukamoto, T. Kawasaki, A. Naraki and T. Yamauchi, *Yakugaku Zasshi, Japan*, **74**, 984 (1954).

93. T. Tsukamoto and T. Kawasaki, *Yakugaku Zasshi, Japan*, **74**, 1127 (1954).

94. T. Kawasaki and T. Yamauchi, *Yakugaku Zasshi, Japan*, **83**, 757 (1963).

95. H. Rippenger, H. Budzikiewicz and K. Schreiber, *Chem. Ber.*, **100**, 1125 (1967).

96. R. Tschesche, G. Lüdke and G. Wulff, *Tetrahedron Letters*, 2785 (1967).

97. T. Nohara, Y. Ogata, K. Miyahara and T. Kawasaki, 14th Symposium on 'The Chemistry of Natural Products,' Fukuoka, Japan, 1970.

98. R. E. Marker and J. Lopez, *J. Am. Chem. Soc.*, **69**, 2389 (1947).

99. M. Kimura, M. Tohma, I. Yoshizawa and H. Akiyama, *Chem. Pharm. Bull.*, **16**, 25 (1968).

100. M. Kimura, M. Tohma and I. Yoshizawa, *Chem. Pharm. Bull.*, **16**, 1228 (1968).

101. H. Ripperger and K. Schreiber, *Chem. Ber.*, **101**, 2450 (1968).

102. R. D. Bennett and E. Heftmann, *Phytochemistry*, **4**, 577 (1965).

103. R. D. Bennett, E. R. Lieber and E. Heftmann, *Phytochemistry*, **6**, 837 (1967).

104. R. Tschesche and H. Hulpke, *Z. Naturforsch.*, **21b**, 494 (1966).

105. J. D. Ehrhardt, L. Hirth and G. Ourisson, *Phytochemistry*, **6**, 815 (1967).

106. Y. Tomita, A. Uomori and H. Minato, *Phytochemistry*, **9**, 111 (1970).

107. Y. Tomita and A. Uomori, *Chem. Commun.*, 284 (1971).

108. J. W. Cornforth, G. Popják and L. Yengoyan, *J. Biol. Chem.*, **241**, 3970 (1966); G. Popják and J. W. Cornforth, *Biochem. J.*, **101**, 553 (1966).

109. Y. Tomita, unpublished work.

110. E. Heftmann, *Lloydia*, **31**, 293 (1968).

111. G. F. Gibbons, L. J. Good and T. W. Goodwin, *Chem. Commun.*, 1212 (1968).

112. G. F. Gibbons, L. J. Good and T. W. Goodwin, *Chem. Commun.*, 1458 (1968).

113. R. D. Bennett, E. Heftmann and R. A. Joly, *Phytochemistry*, **9**, 349 (1970).

114. A. J. Birch, M. Kocor, N. Sheppard and J. Winter, *J. Chem. Soc.*, 1502 (1962).
115. K. Takeda, H. Minato and A. Shimaoka, *J. Chem. Soc.*, Parkin Trans I, 957 (1972).
116. R. E. Marker, R. B. Wagner, P. R. Ulshafer, E. L. Wittbecker, D. P. J. Goldsmith and C. H. Ruof, *J. Am. Chem. Soc.*, **69**, 2167 (1947).
117. A. A. Dawidar and M. B. E. Fayes, *Arch. Biochem. Biophys.*, **92**, 420 (1961).
118. C. R. Karnick, *Planta Med.*, **16**, 269 (1968).
119. A. M. Woodbury, M. E. Wall and J. J. Willaman, *Econ. Bot.*, **15**, 79 (1961).
120. M. E. Wall, B. H. Warnock and J. J. Willaman, *Econ. Bot.*, **16**, 266 (1962).
121. G. Blunden, R. Hardwan and F. J. Hind, *Planta Med.*, **19**, 19 (1970).
122. H. J. Cruzado, H. Delpin and B. A. Roark, *Turrialba*, **15**, 25 (1965).
123. A. Akahori, unpublished work.
124. A. Akahori, *Ann. Rept. Shionogi Res. Lab.*, **13**, 68 (1963).
125. A. Akahori, M. Togami and T. Iwao, *Chem. Pharm. Bull.*, **18**, 436 (1970).
126. H. J. Cruzado, H. Delpin and F. W. Martin, *J. Agr. Univ. Puerto Rico*, **49**, 254 (1965).
127. W. C. Kennard and M. P. Morris, *Agronomy J.*, **48**, 485 (1957).
128. W. H. Preston, Jr., J. R. Haun, J. W. Garvin and R. J. Daum, *Econ. Bot.*, **18**, 323 (1964).
129. H. J. Cruzado, H. Delpin and B. Roark, *Trop. Agr.*, **41**, 345 (1964).
130. C. A. Davila and F. M. Panizo, *Anal. Fis. Quim.*, **54B**, 697 (1958).
131. A. Akahori and F. Yasuda, *J. Pharmacogn. Soc. Japan*, **15**, 149 (1961).
132. A. Akahori, *Ann. Rept. Shionogi Res. Lab.*, **11**, 97 (1961).
133. A. Akahori, F. Yasuda, M. Togami, K. Kagawa, T. Okanishi and T. Iwao, *Phytochemistry*, **8**, 2213 (1969).
134. V. Rasper and D. G. Coursey, *Experientia*, **23**, 611 (1957).
135. R. Hegnauer, *Chemotaxonomie der Pflanzen*, Vol. II, Birkhauser Verlag., Basel u. Stuttgart, 1963, p. 133.
136. R. D. Bennett and E. Heftmann, *Phytochemistry*, **4**, 577 (1965).
137. T. Kawasaki, T. Komori and S. Setoguchi, *Chem. Pharm. Bull.*, **16**, 2430 (1968).
138. T. Komori, S. Setoguchi and T. Kawasaki, *Chem. Ber.*, **101**, 3096 (1968).
139. T. Kawasaki, private communication.
140. W. G. Boorsma, *Meded. uit's Lands Plant.*, **13** (1894).
141. A. R. Pinder, *Nature*, **168**, 1090 (1951).
142. W. A. M. Davis, L. G. Morris and A. R. Pinder, *Chem. Ind. (London)*, 1410 (1961).
143. A. C. Alves and L. N. Prista, *Chem. Abstr.*, **59**, 6452 (1963).
144. H. Ueda and T. Sasaki, *Yakugaku Zasshi, Japan*, **76**, 745 (1956).
145. A. W. Waitt, *Field Crop Abstr.*, **16**, 145 (1963).
146. A. Akahori, *Acta Phytotax. Geobot.*, **21**, 149 (1965).
147. Y. Takeuchi, T. Iwao and A. Akahori, *Acta Phytotax. Geobot.*, **24**, 168 (1970).

Author Index

This author index is designed to enable the reader to locate an author's name and work with the aid of the reference numbers appearing in the text. The page numbers are printed in normal type in ascending numerical order, followed by the reference numbers in brackets. The numbers in *italics* refer to the pages on which the references are actually listed.

If reference is made to the work of the same author in different chapters the above arrangement is repeated separately for each chapter.

12

Subject Index

Note: The letter 't' following a page number indicates that the entry will be found in a table on that page.